Handbook of Cryobiology

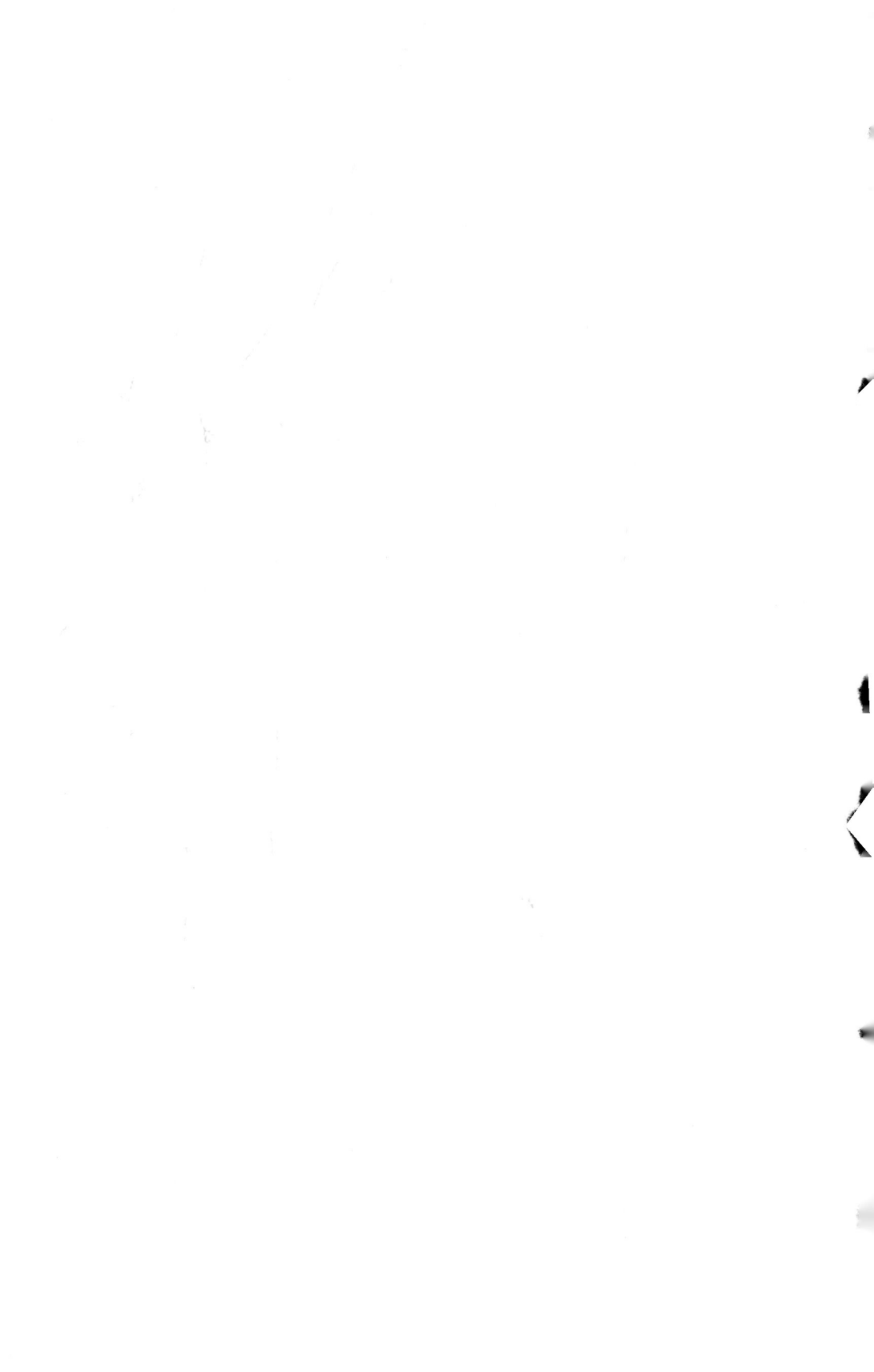

Contents

Preface

Every book is a source of knowledge and this one is no exception. The idea that led to the conceptualization of this book was the fact that the world is advancing rapidly; which makes it crucial to document the progress in every field. I am aware that a lot of data is already available, yet, there is a lot more to learn. Hence, I accepted the responsibility of editing this book and contributing my knowledge to the community.

Significant researches on cryobiology have been included in this book. Lately, there have been certain severe tectonic shifts in cryobiology though not visible on the surface but will have significant impact on both the advancement of novel cryopreservation techniques and the future of cryobiology. This comprehensive book discusses the existing applications and emerging practical protocols for the purpose of cryopreservation along with the description of the novel cryobiological ideas. The topics discussed in this book are as follows: Basic Cryobiology and Kinetic Vitrification, Stem Cells and Cryopreservation in Regenerative Medicine, Human Assisted Reproduction Techniques (ART), and Farm/Pet/Laboratory Animal ART. It consists of contributions made by researchers as well as scientists from across the planet and will serve as an important source of reference for those interested in cryobiology.

While editing this book, I had multiple visions for it. Then I finally narrowed down to make every chapter a sole standing text explaining a particular topic, so that they can be used independently. However, the umbrella subject sinews them into a common theme. This makes the book a unique platform of knowledge.

I would like to give the major credit of this book to the experts from every corner of the world, who took the time to share their expertise with us. Also, I owe the completion of this book to the never-ending support of my family, who supported me throughout the project.

Editor

Part 1

Basic Cryobiology and Kinetic Vitrification

Vitrification Technique – New Possibilities for Male Gamete Low-Temperature Storage

E. Isachenko, P. Mallmann, G. Rahimi, J. Risopatròn,
M. Schulz, V. Isachenko and R. Sànchez
Woman Hospital, University of Cologne
Germany

1. Introduction

According to the recent scientific achievements in cryobiology at present the vitrification belong to perspective technologies.

What is the reason?

Because:

- The method is cheap and quick.
- Due to lowering the temperature of solution's glass transition the permeable cryoprotectants prevents the actual freezing of solution and allows to maintain its some flexibility in a glassy phase.
- Vitrification without permeable cryoprotectants allows to avoid the cryoprotectants toxicity and osmotic stress; the damage of plasmatic and mitochondrial membrane during equilibration with cryoprotectants; protects of plasmatic and mitochondrial membrane against lipid peroxidation and formation of reactive oxygen species and DNA damage.

Cryobiology is a rapidly evolving field which only relatively recently has found broad applications in reproductive medicine. However, as any emerging technology, it has both a great potential and a need for further developments (Petrunkina, 2007).

According to worldwide experience, successful cryopreservation of spermatozoa from different kind of animal including human for long-term storage (cryobanking of genome) or relatively short-term storage (artificial insemination) in conjunction with assisted reproductive technologies allows to promote long-term cryopreservation programmes. The use of programmable or non-programmable "slow" (conventional) freezing (McLaughlin et al., 1990; Yin and Seibel, 1999; Stanic et al., 2000) allows to preserve relatively large volume of diluted ejaculate or prepared spermatozoa from 0.25 to 1.0 mL with good rates of motility after thawing (Sawetawan et al., 1993; larson et al, 1997) and acceptable levels of integrity of acrosomal and cytoplasmic membranes, in other words with sufficiently high quality post-thaw characteristics (Hammadeh et al., 1999; Duru et al., 2001; Meseguer et al., 2004; Isachenko et al., 2003, 2008). It does also provide acceptable protection against membrane changes and destabilization induced by cryopreservation (Glander and Schaller, 2000; Schuffner et al, 2001).

To avoid the lethal intracellular ice formation, the cryoprotective solution as ruler contains buffers, carbohydrates (glucose, lactose, raffinose, sucrose and trehalose), salts (sodium citrate, citric acid), egg yolk and antibiotics including permeable cryoprotectant glycerol or other cryoprotectives (Mazur, 1963; Barbas & Mascarenhas, 2009) in combination with comparably slow rates of freezing (Gao et al., 1997) are widely used for these purpose. The aim of slow cooling rates is to maintain a very delicate balance between ice crystal formation and growing concentration of dissolved substances. Conventional freezing procedure for mammalian spermatozoa traditionally includes the following stages of manipulations:

1. slow, step-wise adding of freezing solution to the ejaculate
2. cooling during 20 to 45 min
3. warming in water bath and
4. treatment of spermatozoa by the density gradient or the swim-up procedure
    (McLaughlin et al., 1990; Yin and Seibel, 1999; Stanic et al., 2000).
5. The ultimate target of the last manipulation is the removal of permeable
    cryoprotectants.

As a rule, a pre-requisite for that is a dilution of semen suspension with culture medium in order to reduce the toxicity of permeable cryoprotectant (according to manufacturer's instructions for cryoprotectant of choice). This is associated with additional costs and, not at least, with environmental and adaptation challenges for spermatozoa. In fact, sensitivity of spermatozoa to additional mechanical manipulation is increased after freezing-thawing, and the negative effects of cryopreservation on cell viability and functional competence can be aggravated by additional procedures. The problem is that the addition and, in particular, the removal of permeable osmotically-active cryoprotective agents (permeable cryoprotectants) before cooling and after warming can induce lethal stress due to intracellular ice formation, intracellular eutectic formation or so-called 'dilution (toxic) effects' (Fraga et al., 1991, Petrunkina, 2007) including chilling injury, cytoplasm fracture or even effects on the cytoskeleton (Critser et al., 1988; Fraga et al., 1991; Pérez-Sánchez et al., 1994). The further problems include the chemical toxicity of cryoprotectants and their possible repercussions on the genome or genome- related structures of mammalian spermatozoa (Hammadeh et al., 1999; Gilmore *et al.*, 1997). Moreover, spermatozoa which survive the cryopreservation stress are likely to have undergone subtle functional changes associated with biophysical and biochemical factors influenced by cryoprotectants, which will affect their fertilizing ability (Petrunkina, 2007).

Actually, the problem of the cooling and warming processes is the lethality which closely associated with the intermediate zone of temperature (-10 to -60 °C) that cells must traverse twice during cooling and once during warming (Mazur, 1963).

One of relatively recent and much discussed cryobiological emerged technologies within the field of the reproductive cryobiology is the spermatozoa vitrification (cryopreservation by direct plunging into liquid nitrogen). Vitrification is an alternative method that can also be applied to achieve the same purpose and does not use the special extenders. This method is based on the rapid cooling of the cells by immersion into liquid nitrogen, and, thereby, is the key factor reducing the chance of the formation of big ice crystals. In contrast to the programmable ("slow") conventional freezing, vitrification has series of technological advantages useful for the practice: it renders the use of permeable cryoprotectants superfluous and, in addition, is much faster, simpler in application and more cost-effective than conventional freezing. In spite of that this method has been investigated extensively and successfully applied to female gametes and embryos of different mammalian species including humans (Rall & Fahy, 1985; Chen et al., 2001; Reed et al., 2002; Cervera & Garcia-Ximénez, 2003; Isachenko et al., 2005b, 2007; Silva & Berland, 2004), however, it cannot be directly extrapolated to male gametes, due to deleterious osmotic effect of high concentrations of permeable cryoprotectants.

To date, publications dedicated to this topic are rare (Nawroth, et al., 2002; Koshimoto & Mazur, 2002; Isachenko et al., 2004a, b, 2008). Recent work has reversed this situation in that favorable results have been obtained in human spermatozoa after excluding permeable cryoprotectants from cryopreservation solutions, increasing the cooling rate and using carbohydrates, proteins and other extracellular agents, to increase the viscosity of the surrounding medium of cells and prevent the formation of any intra- and big extracellular crystals (Isachenko et al., 2004a, b). It was shown that permeable-cryoprotectants-free vitrification only with protein (Nawroth et al, 2002; Isachenko et al, 2003, 2004a,b, 2005a) or in combination with sucrose (Isachenko et al., 2008 2011a,b,c,d; Sanchez et al., 2011a, b) as a non-permeable cryoprotectant provides a high recovery rate of motile cells and effectively protects the mitochondrial membrane and the DNA integrity of spermatozoa after warming (Isachenko et al., 2004a, b; 2008). And it is not surprising. According to common point of view the non-permeable cryoprotectants plays the supporting role at permeable cryoprotectants. They binds of extracellular water and at the same time plays anti-toxic role (Kuleshova et al., 1999) decreasing of harmful properties of permeable cryoprotectants. In general, the inclusion of osmotically active, non-permeating compounds into the vitrification solution leads to additional rehydration of cells and, as a result, to decreasing toxic effects of the permeable cryoprotectants on intracellular structures. The non-permeable cryoprotectant sugars possess a unique property: stabilization of a cell membrane (Nakagata & Takeshima, 1992, 1993; Koshimoto et al., 2000; Koshimoto & Mazur, 2002).

Also, the application of this modified cryopreservation technique to human spermatozoa allowed to avoid the toxic effect caused by adding and removing of permeable cryoprotectants including the negative effects on the cells' genetic material (Pérez-Sánchez et al., 1994). In our earlier works we have shown that cryopreserved without permeable cryoprotectants human spermatozoa preserved their relatively high motility rate with ability to fertilize oocytes in vitro (Nawroth et al., 2002; Isachenko *et al.*, 2003; 2004a, b). No statistical differences in parameters such as viability, recovery rate or percentage of morphologically normal spermatozoa with undamaged DNA were noted between vitrified and conventionally frozen cells (Nawroth et al., 2002). However, it was observed that the number of cryopreserved spermatozoa displaying features of acrosome reaction was statistically different from that in freshly prepared swim-up spermatozoa (Isachenko *et al.*, 2004a,b, 2005a, 2008, 2011a,b).

In contrast to the programmable (slow) conventional freezing the vitrification renders redundant the need for special cooling programs addition of permeable cryoprotectants. It is much faster, simpler and more cost-efficient while still effectively protecting spermatozoa from cryo-injuries (Nawroth et al, 2002; Isachenko et al, 2003, 2004a, b, 2005, 2008) and does not require expensive equipment or special cooling procedures. Spermatozoa, vitrified by such technology, would be ready for further use without any additional treatment (centrifugation, separation in the gradient, removal of cryoprotectant and others) immediately after thawing.

Successful pregnancies and births have been reported when using vitrified oocytes and embryos, and vitrification protocols have started to form an important part as well of human as of animal reproductive medicine. Although sperm vitrification techniques have been studied in vitro, first successful pregnancies and live birth after fertilization with vitrified spermatozoa have been reported.

This chapter we would like to present contents the interesting results which we have had in the first time achieved using developed us vitrification technique based on using only of protein and carbohydrates as non-permeable cryoprotectants and applying to some mammalian and fish species.

In our presentation we will not touch the historicalquestions of vitrification of spermatozoa as well, but will concentrate us only on own experience according to vitrification of spermatozoa with using only non-permeable cryoprotectants. This theme was well covered in our previous publications (E. Isachenko et al., 2003, 2008, 2011a; Katkov et al., 2006).

In this chapter we will in detail discuss our new data which we have got in our investigation after vitrification of human, dog and fish spermatozoa.

2. New capillary technology (Isachenko et al., 2011c) for vitrification of small volume of human spermatozoa and practical application

Varied methods to vitrify spermatozoa have been described previously: cryo-loops, droplets- and open pulled straw method (Nawroth et al, 2002; Isachenko et al., 2004a,b, 2005, 2008). According to these results it is possible to achieve up to 60%- and 20% -motility levels after thawing in normospermic and oligo-astheno-terato-zoospermic patients, respectively, depending on vitrification method selected and the quality of the original ejaculate. Independent from the vitrification technique the vitrified spermatozoa can be processed for further use immediately after warming without additional treatment such as centrifugation, gradient separation, removal of cryoprotectants etc is required. This simplicity for practical purposes represents one of the most attractive advantages of our technology. It is worth to mention that the protocol for vitrification does include swim up treatment, therefore, after swim up, vitrification and warming spermatozoa are also free from seminal plasma with potential pathogens. "Slow" freezing of human spermatozoa traditionally proposes the removing of permeable cryoprotectant after thawing.

This «removal of permeable cryoprotectants» is the ultimate target of the last manipulation. As a rule, a pre-requisite for that is a dilution of semen suspension with culture medium in order to reduce the toxicity of permeable cryoprotectant (according to manufacturer's instructions for cryoprotectant of choice). This is associated with additional costs and, not at least, with environmental and adaptation challenges for spermatozoa. In fact, sensitivity of spermatozoa to additional mechanical manipulation is increased after freezing-thawing, and the negative effects of cryopreservation on cell viability and functional competence can be aggravated by additional procedures (Petrunkina, 2007). Cryopreservation induces extensive damage to cells during both freezing and thawing. According to present knowledge, the effective induction of anabiosis in cells at very low temperatures (in liquid nitrogen at -196°C, for example) can be achieved by optimizing the multi-factorial freezing process (Lozina-Lozinski, 1982), commonly with the use of permeable cryoprotectants (Levin, 1982). Acting by depressing the freezing point and by binding intracellular water, the permeable and non-permeable cryoprotectants help to prevent ice formation, and thereby to reduce the cryo-damage (Andrews, 1976; Franks, 1977).

Several protocols of spermatozoa separation are available (e.g. swim-up from the ejaculate, single wash of ejaculate and swim-up from pellet, double wash of ejaculate and swim-up from pellet). However, any methodology needs the use of previous centrifugation. Most of the current technologies for sperm vitrification have an obvious shortcoming in terms of standardization of the portion volume. In particular, as the diameter of the pulled part of straw is not uniform, the volumes of the portions packaged in that way can not be standardized. Here we have reported the vitrification methodology using standard capillaries which can be supplied by industrial manufacturers. The technique was performed as follow (Isachenko V et al., 2011c). All specimens used for this study had fulfilled following quality criteria for spermatozoa concentration, motility and morphology: less than 20 millions spermatozoa/mL, 35 % progressive motile and minimum 3 % morphologically normal spermatozoa. Semen analysis was performed according to published guidelines of the World Health Organization (WHO, 1999). Prior to vitrification, the sedimented spermatozoa were diluted with 0.25 M sucrose (end concentration) in sperm preparation medium at room temperature (Isachenko et al., 2008). The final concentration of spermatozoa was approximately of 0.5×10^6 spermatozoa/mL. Diluted suspensions were maintained at room temperature for 5 min before the cooling procedure. Spermatozoa were prepared and portioned for aseptic vitrification in the following way. Specially for our purposes, 50 µL-plastic capillaries (Fig.1) were manufactured from hydrophobic material as vehicles for cooling sperm cell suspensions (Gynemed GmbH & Co. KG, Lensahn, Germany). The end of the straw was labeled on the top to mark the cutting-off position (Fig.1, arrows). The capillary was filled with 10µL of spermatozoa suspension by aspiration (Fig. 1a). It was absolutely crucial to avoid that the inner surface of the capillary become moist during packaging procedure. Aspirating the volume of sperm cell suspension above the mark and correcting it by lowering the fluid level inside the capillary after aspiration is technologically wrong and would result in excess of portion's volume after thawing. After the aspiration was completed, the capillary was inserted into 0.25 ml straw (Medical Technology GmbH, Bruckberg, Germany). One end of this straw was sealed in advance using heat-sealer (Cryo Bio System, Paris, France). After sealing the second end of the straw (Fig. 1b), the straw was plunged into liquid nitrogen and cooled at a cooling speed of 600°C/min. The speed of cooling was determined using a Testo 950 electrical thermometer (Testo AG, Lenzkirch, Germany) using 0.2 mm electrode located inside of the capillary. Hermetical heat-sealing of 0.25 mL straw can be achieved using flame of alcohol burner and forceps or any commercial equipment (including ultrasound equipment because of large distance between spermatozoa suspension and focus of sonographic appliance). Spermatozoa were stored in liquid nitrogen at least for 24 h before warming. For warming, capillary was removed from isolating 0.25 mL straw. The straw was disinfected with ethanol in the area where the marked end of capillary was (Fig. 1, arrows). The second end of capillary is fixed tightly on the inner surface of the straw, and the part of straw containing spermatozoa is still half submerged in liquid nitrogen (Figs. 1c,d). The upper part of the straw was cut off with sterile scissors as close as possible to the marked end of the capillary, just above the mark without touching the marked end of capillary. The capillary was expelled with a conical bolt (Fig. 1d). For this purpose, conical bolt (instead the conical bolt forceps can be used in place) is inserted into inner part of the capillary and pulled off the straw. The final warming up of spermatozoa is achieved by immersing of capillary without conical apex (capillary must be open from both sides) with vitrified spermatozoa into 1.8 mL centrifuged tube with 0.7 mL pre-warmed to 37°C vitrification medium for approximately 20 sec. (Fig. 1e). It is important to note that the volume of vitrified suspension after warming is not decreased (Fig 1e). Finally, the suspension of spermatozoa was expelled from the capillary for immediate evaluation of spermatozoa quality. Using this technique the exactly quantifiable volumes of spermatozoa samples were obtained: 10 µL suspension of spermatozoa were vitrified, 10 µL were thawed and the same 10 µL added to the respective volume of medium for ICSI or IVF. Thus, one of the most important features of this novel method of vitrification in capillaries is its potential for standardization which can be used for the routine clinical practice. The results of the present study let suggest that cryopreservation by vitrification helps to preserve essential determinants of spermatozoa function, such as motility and plasma membrane integrity. It is well known that spermatozoa cryopreservation is associated with a large decline in spermatozoa viability and other sperm functional parameters (Petrunkina, 2007). In the present study we have compared spermatozoa quality after vitrification by our method with spermatozoa quality after conventional freezing with addition of permeable cryoprotectant. The outcomes indicated that vitrification in capillaries compare to conventional freezing preserved better the motility of spermatozoa (after warming/thawing: 28.0 +6.0 % vs 18.0 + 9.2 %, respectively, P<0.05 and in fresh control 35.0 + 9.5%; after 24 h $in\ vitro$ culture: 12.0 +2.8 % vs 5.0 + 3.1 %, respectively, P<0.05 and in fresh control 20.0 + 3.9%; after 48 h in vitro culture: 6.0 +1.0 % vs 0.5 + 0.02 %, respectively P>0.1 and in fresh control 10.0 + 1.9% [Fig. 2]) and their plasma membrane integrity (56.0 ± 5.1 % vs 22.0 ± 3.5 %, respectively, P<0.05 and in fresh control 96.0 ± 0.6 %, P<0.05 [Figure 3]) which was assessed with LIVE / DEAD sperm viability kit (LIVE/DEAD Sperm Viability Kit, Molecular Probes cat no. L-7011, Eugene, OR, USA). Pilot results have been obtained with respect to evaluating capacitation-like changes associated with cryopreservation, so called "cryo-capacitation" (Cormier and Bailey, 2003). A body of evidence suggests that some spermatozoa' intracellular signaling pathways can be affected during cryopreservation, and after warming spermatozoa display features commonly observed in capacitating or capacitated spermatozoa (Green and Watson, 2001; Petrunkina et al., 2005; Vadnais and Roberts, 2010). It is important, however, to emphasize that the changes induced by cryo-preservation are similar to those of capacitation only at the functional level, and they seem to differ at the molecular level, and with respect to pathways and signaling mechanisms involved (Cormier and Bailey, 2003). Our observations imply that permeable cryoprotectant-free aseptic vitrification is associated with lesser damage to acrosomes compare to conventional freezing (55.0 ± 5.8 % vs 21.0 ± 3.8 %, respectively, P<0.05 and in fresh control 84.0 ± 3.1%, P<0.05 [Fig. 4]). However, the levels of membrane changes related to "cryo-capacitation" assessed by CTC in vitrified spermatozoa were comparable with those after conventional freezing (8.0 ± 1.1% vs 9.0 ± 2.2%, respectively, P <0.01 and in fresh control 2.0 ± 0.3%, P<0.05, [Figure 5]). Changes in the acrosomal membrane status and permeability associated with the capacitation we have evaluated by using the double fluorescence chlortetracycline (CTC)-Hoechst 33258 staining technique (Kay et al, 1994). Nevertheless, the exposure to low temperatures can affect those crucial signaling mechanisms which can not be monitored by CTC. Thus, further studies with additional, advanced techniques are needed to investigate the changes induced by vitrification in its complexity (e.g. targeting specific pathways and membrane processes such as changes in lipid architecture and/or protein kinases/phosphatases regulated pathways). Given the fact that the outcome of basic spermatozoa quality was comparable (or even better) that after conventional freezing, other advantages of the vitrification process must be taken into account. During conventional procedure, the success of applying permeable cryoprotectants for cryopreservation of varied cells and tissues is inseparably linked to such cryoprotectant properties as their ability to permeate rapidly through cellular membrane and their toxicity (Gilmore et al, 1997). These properties are directly connected to osmotic damages of cells during saturation with permeable cryoprotectants before freezing and then at time of cryoprotectants removing after thawing (Gao et al, 1995, Petrunkina, 2007). It is known that human spermatozoa contain large quantities of proteins, sugars, and other components that may act as natural cryoprotectants. Our technology does not presuppose the use of permeable cryoprotectant. In practical terms, permeable cryoprotectant-free vitrification technology for the cryopreservation of spermatozoa (in straws) instead traditional slow freezing with permeable cryoprotectants is already used in following centers: our university's maternity hospital (www.uniklinik-ulm.de): IVF Centers in Temuco, Chile (about 200 IUI cycles/year) and in Ulm, Germany (www.kinderwunsch-ulm.de) (about 1,000 IVF cycles/year). First successful pregnancies and birth of healthy babies has been recently achieved with spermatozoa vitrified without permeable cryoprotectants (Isachenko et al., 2011b). In summary, the newly developed technology of aseptic vitrification of human spermatozoa in capillaries can effectively preserve these cells from cryo-injures. Spermatozoa, vitrified by this technology, are free from seminal plasma owing to swim up procedure preceding vitrification and are free from permeable cryoprotectants. They are ready for further use immediately after warming without any additional treatment. Therefore, the reported technology has a great potential for use in ICSI / IVF.

As successful application of this vitrification technology for routine practice is born of two healthy babies (Isachenko et al., 2011b). We would like shortly present here the history of this case. A couple, both 39 years old, underwent assisted reproduction due to severe endometriosis and oligo-astheno-terato-zoospermia (13 x 10^6 motile spermatozoa/ml; with 42% of progressive motility and 8% morphologically normal spermatozoa [WHO, 1999]). Cryopreserved spermatozoa were used because the partner was absent during the oocyte retrieval procedure. The swim up-processed spermatozoa were diluted and proceeded with vitrification solution according our technique, described above, to achieve a final concentration of 2,5 x 10^6 spermatozoa/ml and 10μl aliquots were vitrified with using of Cut-Standard-Straws (CSS, Isachenko *et al.*, 2007) which was chosen as the prototype of our capillary technology. The spermatozoa were kept frozen in liquid nitrogen (at -196°C) for 7 months. Only the warming technique was different from newly developed one and supposed the concentrations of spermatozoa by centrifugation after warming. The changes in following physiological and morphological parameters of thirty minutes after warming of vitrified spermatozoa and the freshly prepared swim-up were investigated for progressive motility, capacitation-like membrane changes due to determining of phosphatidylserine translocation (PST). The capacitation-like membrane changes was investigated due to determining of phosphatidylserine translocation (PST) in the sperm with appling the anexin V-FITC staining technique (APOPTESTTM-FITC, Nexins Research, the Netherlands).

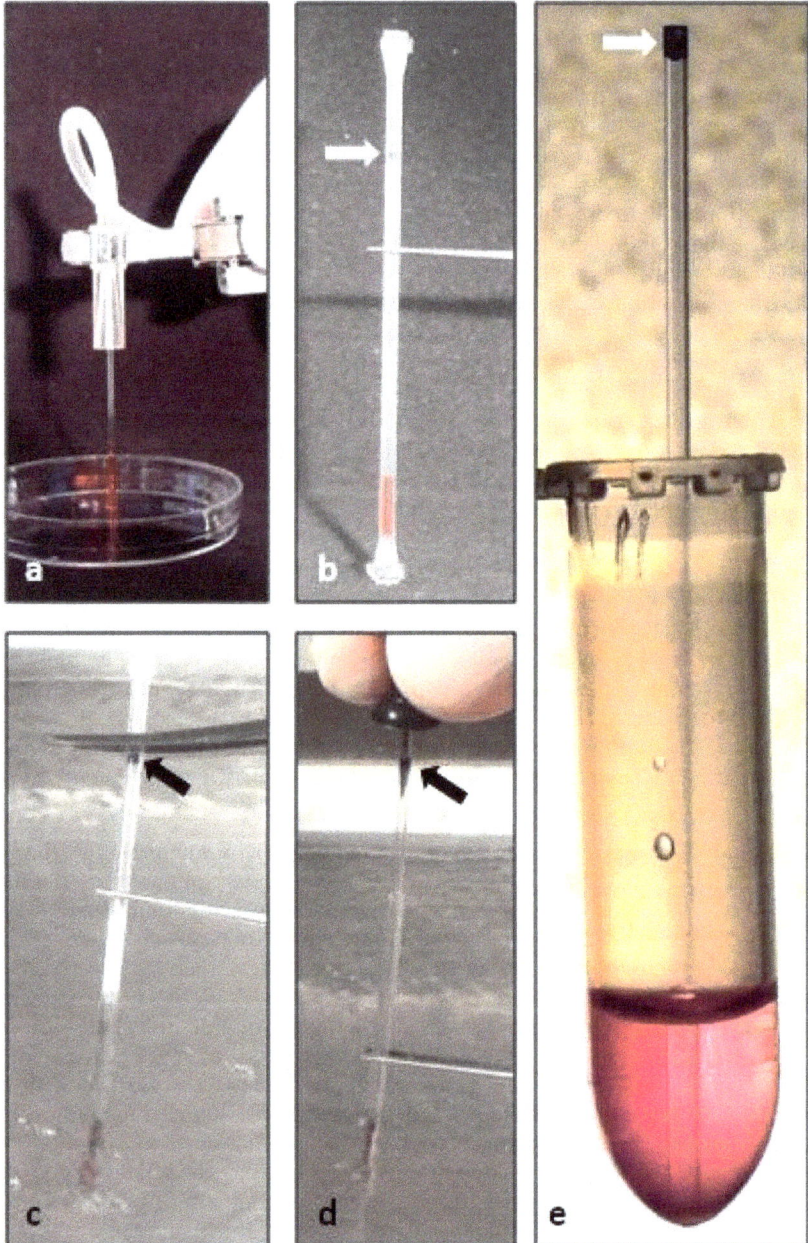

(Arrows) marked end of 50 µL capillary, (a) aspiration of spermatozoa suspension in straw, (b) 50 µL capillary sealed in 0.25 mL straw, (c) cutting of 0.25 mL straw, (d) expelling of 50 µL capillary from 0.25 mL straw, (e) warming of spermatozoa.

Fig. 1. Schematic illustration of human spermatozoa vitrification with 50 µL capillary.

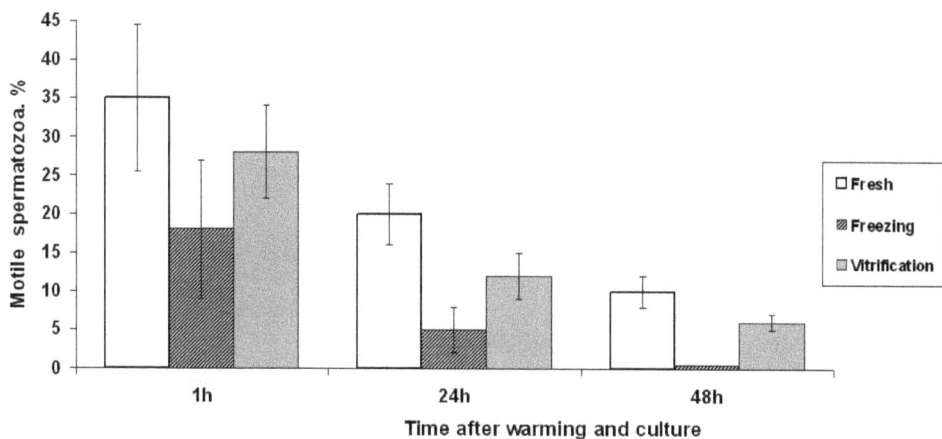

All rates in respective groups are significantly different (P<0.05).

Fig. 2. Motility of human spermatozoa after conventional freezing and vitrification.

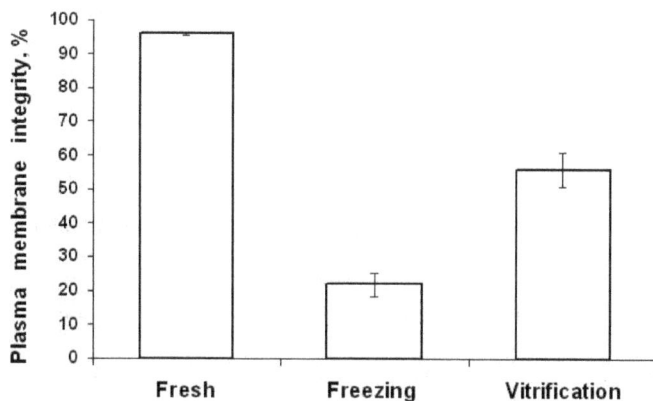

All rates in respective groups are significantly different (P<0.05).

Fig. 3. Plasma membrane integrity of human spermatozoa after conventional freezing and vitrification.

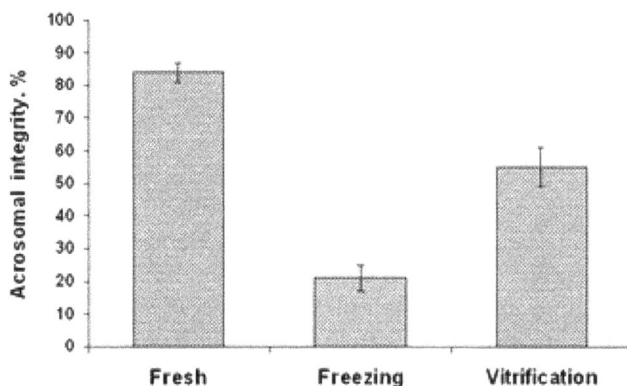

All rates in respective groups are significantly different (P<0.05).

Fig. 4. Acrosomal integrity of human spermatozoa after conventional freezing and vitrification.

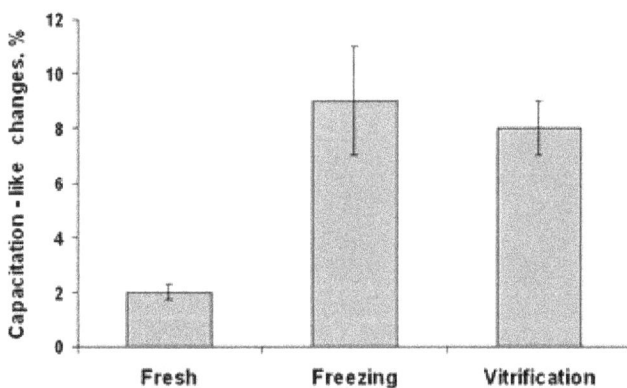

Rates in groups after freezing and vitrification are similar (P>0.5).

Fig. 5. Capacitation-like changes of human spermatozoa after conventional freezing and vitrification.

The mitochondrial membrane potential integrity was evaluates due to measurement of the changes in the (M $\Delta\Psi$) using a unique fluorescent cationic dye, 5,5', 6,6'-tetachloro-1-1', 3,3'-tetraethyl-benzamidazolocarbocyanin iodide. The results were as following: progressive motility 60% vs 90%, correspondingly, 10% were identified as displaying a 'capacitation' CTC pattern and 5% as displaying an 'acrosome reaction' pattern, as compared to 8% and 5% in freshly prepared swim-up sperm respectively; 63% of spermatozoa were classified as having high mitochondrial membrane potential (vs 96% in freshly prepared spermatozoa).

From ten ICSI-ed with vitrified spermatozoa oocytes 6 oocytes showed signs of normal fertilization and two PN-oocytes were culture subsequent 24 hours. At day of embryo transfer two 4-blastomere embryos of Grades "a" (4a) and "b" (4b) (Steer et al., 1992) were transferred to the uterus cavity under ultrasonographic guidance. Fifteen days after embryo transfer, the maternal ß-hCG level was 360 IU/L and two healthy boys were born at term.

These data supports the notion that: i) cells can be frozen effectively without toxic permeable cryoprotectants, and ii) such frozen material could in principle be lyophilized. It is, however, critical to ensure that freeze-drying is not associated with the genetic and developmental abnormalities that have been observed after fertilization with mouse freeze-dried sperm (Ward *et al.*, 2003).

Cryopreservation is normally achieved through a tertiary combination of cells, permeable cryoprotectants and low temperature environment. In contrast, our cryopreservation protocol can be considered as a simplified binary combination of cells (in a simplified medium containing sucrose as a natural cryoprotectant) and a cold environment. The birth of two healthy babies using this *in vitro* fertilization technique is not only the first report on successful fertilization using vitrified spermatozoa (which has obvious practical advantages for assisted reproduction techniques). The above protocol also demonstrates that highly organized cells (human spermatozoa) may be effectively frozen-dried (lyophilized) with the recovery of their most important physiological function after thawing – propagation of genetic hereditary information and subsequent birth of new individuals. Of course, it would need to be proved on a large number of ejaculates that the damage produced by vitrification does not exceed the damage produced by conventional freezing and that there are no deleterious effects on the genetic integrity of sperm after vitrification (Ward *et al.*, 2003). These aspects, however important, are outside the scope of this case report.

3. New technology for vitrification of spermatozoa in big volume (Isachenko et al., 2011d)

Actually, the technique which is not acceptable for different volumes of the same object is incomplete and needs subsequent investigations and development. In this case the next aim of our research was development the acceptable vitrification methodology for big volume of spermatozoa with possibility to use cryopreserved ejaculate for intrauterine insemination. At the beginning of 2011 we have published (E. Isachenko et al., 2011a) the prototype of our big-volume vitrification technology the success of which a healthy baby was born after intrauterine insemination with vitrified spermatozoa (Sánchez et al., 2011a). We would like shortly present the history of this case. A 39-year-old patient and her 35-year-old husband, with a 3-year history of primary infertility, were referred to our center for infertility treatment. Laparoscopy revealed patency of the Fallopian tubes and no evidence of endometriosis or pelvic adhesions. Semen analysis of the husband showed oligo-astheno-terato-zoospermia (WHO, 1999). Despite the poor quality of ejaculate parameters, for financial reasons the patients decided to try intra-uterine insemination (IUI). For IUI the spermatozoa from two ejaculates obtained 3 days apart were vitrified. The volume of the first ejaculate was 1.9 ml, concentration: 37.8 x 10^6 spermatozoa/ml, 8% of progressive "a" and "b" motility, 10% of morphologically normal spermatozoa, and 0.2x10^6 round cells/ml. The volume of the second ejaculate was 3.9 ml, concentration 11.2 x 10^6 spermatozoa/ml, 27% progressive motility, 10% of morphologically normal spermatozoa and 1.2x10^6 round cells/ml. The swim up-processed spermatozoa were diluted and proceeded with vitrification solution according our technique, described above, to achieve a final concentration of 1 x 10⁶ spermatozoa/ml. All subsequent manipulations were performed at room temperature strictly in a horizontal position to prevent a loss of suspension (E. Isachenko et al., 2011a). Aliquots (100 µl) of the diluted sperm suspension were aspirated into one half of 0.25 ml plastic straws (MTG, Bruckberg, Germany); these were then placed in 0.5 ml plastic straws (MTG) and hermetically sealed from both sides to protect the suspension from direct contact with liquid nitrogen. The closed straw-systems, strictly maintained horizontal, were then immersed into liquid nitrogen and stored until use. From two ejaculates three straws were cryopreserved, each with 100 µl of spermatozoa suspension in concentration of 1x10⁶ spermatozoa/ml. Special for this case we have decided to investigate the presence of reactive oxygen species (ROS) in ejaculated and prepared spermatozoa before and after vitrification. The reason was the following. It is known that poor ejaculate quality is closely associated with elevated concentrations of leucocytes (normal values <1 million/ml). The presence of leucocytes can lead to oxidative stress (Henkel and Schill, 2003; Henkel et al., 2005, 2010). Therefore, we determined the concentration of leucocytes in ejaculates due to leucocytes quantifying by an indirect immunofluorescence (IIF) method (Villlegas et al., 2002) and presence of the following antibodies were checked: anti CD45 for all leukocytes (M 855-DAKO, Hamburg, Germany, in concentration of 1/50 in PBS with 5% BSA), anti CD15 for granulocytes (M 733-DAKO, Hamburg, Germany, in concentration of 1/100 in PBS with 5% BSA) or anti CD68 for macrophages (M 718-DAKO, Hamburg, Germany , in concentration of 1/600 in PBS with 5% BSA). However, in spite of the presence of a large numbers of round cells, the IIF was negative for all tested monoclonal antibodies, indicating high levels of spermatogenic cells. The presence of ROS in ejaculates was tested using a chemiluminescence assay (Aitken and Clarkson, 1987). Only the mild increasing of ROS to 76.960 RLU x 10⁷/live sperm was noted (normal value: 35.000 RLU x 10⁷/live sperm (Henkel et al, 1997). However, it is known that ROS in semen samples of oligozoospermic patients usually is slightly increased (Kumar et al., 2009). On the day of ovulation all three cryopreserved samples of spermatozoa suspension were thawed as described in E. Isachenko et al. (E. Isachenko et al., 2011a), the sperm pellet was resuspended in 500 µl of sperm preparation medium pre-warmed to 37°C and used immediately for intrauterine insemination. The suspension of spermatozoa before insemination (30 min post-warming) had a concentration of 2.7x10⁶ spermatozoa/ml with 60% of progressive motility. Fifteen days after IUI, biochemical pregnancy was confirmed by ß-hCG level of 125 IU/L and on 29 December 2010 a healthy male baby was born.

Our finding has confirmed that the aseptic vitrification technique (without use of permeable cryoprotectants) is not only instrumental in effectively preserving spermatozoal function (Isachenko et al., 2011a, b, c, d, Sánchez et al., 2011), but could also have a massive potential for storage of motile spermatozoa for intrauterine insemination, for example, in cases of oligo-astheno-zoospemic patients.

However, the described methodology for vitrification of big-volume spermatozoa suspension is complicated, because exist often dangerous that sperm suspension will flows out the specimen straw and stick together to the inner wall of packaging straw during vitrification procedure. In this case it will be difficult to remove the specimen straw from the packaging one before warming. According to our opinion the technique must be as simple as possible and at the same time with absolute repeatability and the results have to be compatible with slow conventional freezing.

In our lectures we have often mentioned that there is a simplified point of view that vitrification is the solidification without formation of crystals. Extending this description, one could say that vitrification is solidification of vitrifying solution without formation of hexagonal (big, lethal) intracellular structures by extreme elevation in viscosity during cooling. Obviously, thereby vitrification appears beneficial in terms of avoiding cryo-injuries traditionally associated with the formation of intracellular ice. Therefore we developed and for the first time reported (V. Isachenko et al., 2011d) the vitrification methodology where a relatively large volume of spermatozoa suspension can be frozen in one cooling pocket (straw). Vitrification medium described here does include sucrose (Isachenko et al., 2008). As a rule, in routine practice the carbohydrates are the standard part of any cryoprotective solution. They are used for spermatozoa cryopreservation to compensate osmotic effects caused by the permeable cryoprotectants and do play an important role as an additional dissolving, membrane stabilizing and dehydrating agents (Wakayama et al., 1998). Therefore, sucrose can be considered as a natural cryoprotectant, lacking most of toxic properties of permeable cryoprotectants. Human spermatozoa can be successfully frozen in the absence of permeable cryoprotectants, using protein- and sugar-rich extracellular non-permeable cryoprotectants (Koshimoto et al., 2000; Karlsson and Cravalho, 1994). The ability of sucrose to prevent the artificial induction of membrane damages and acrosome reaction during vitrification/warming (Isachenko et al., 2008) corroborated our previous conclusions that the inclusion of sucrose in combination with human serum albumin in the vitrification medium has a visible cryoprotective effect.

In our study (Isachenko et al., 2011d) we reported for the first time a novel technology of aseptic 'cryoprotectant-free' vitrification of human spermatozoa in large volumes. It allows:

1. to obtain 0.5 mL of spermatozoa suspension, free both from seminal plasma (because of swim up procedure preceding vitrification) and free from additives which are part of conventional freezing procedures;
2. to cryopreserve spermatozoa, which are ready for further use immediately after thawing without any additional treatment (centrifugation, separation in the gradient, removal of cryoprotectant and others).

The technology includes:

- cryoprotective medium with only non-permeable cryoprotective agents (0,25 M sucrose in end concentration and 1% human serum albumin). As basal medium is the Human Tubal Fluid (Quinn et al., 1985).
- the end-concentration of prepared for vitrification spermatozoa is 5×10^6 spermatozoa / mL. It is possible to vitrify the different concentrations of prepared spermatozoa without influence on warming resalts (non-published data).
- using of 0.5 mL plastic straws with subsequent sealing from both side before cooling in liquid nitrogen.
- The warming up of spermatozoa is achieved by immersing straw with vitrified spermatozoa into warmed water bath at 42°C.

Technological procedure, shortly

Vitrification. Prior to vitrification, spermatozoa were processed by swim-up technique with subsequent dilution with cryoprotectant medium according to Isachenko (Isachenko et al.,

2008). Diluted suspensions were maintained at room temperature for 5 minutes before the cooling procedure. The packaging of spermatozoa for aseptic vitrification was performed in the following way. Spermatozoa suspensions were cooled in 0.5 mL plastic CBS straws (CryoBio System, Paris, France) (Figure 6). The straw was labeled with asterisk (1 cm from the inner end of cotton-polyvinil plunge, arrows on Figures 6a-g). The straw was filled up to asterisk with 0.5 mL of spermatozoa suspension by aspiration (Figure 6a). Then the filled straw was expelled from the tube while aspiration of air continued. Subsequently, when the suspension reached the polyvinyl plunge, the polymerization of polyvinyl was initiated due to humidification. After aspiration was completed, and the top end of straw was sealed by polymerized polyvinyl, straw was hermetically heat-sealed at both sides using flame of alcohol burner and forceps (Figures 6b,e). The hermetically sealed straw with spermatozoa was allowed to cool briefly (~ 2 seconds). This procedure ensured that spermatozoa at any time were not in contact with the heat-sealing area. Alternatively, any commercial equipment (with exception of ultrasound equipment) could be used for thermo-hermetic sealing. The straws were immersed into liquid nitrogen in horizontal position (approximately for 8 seconds) (Figure 6c) and stored there at least for 24 hours before use.

Warming. The warming up of spermatozoa is achieved by immersing straw with vitrified spermatozoa into water bath at 42°C and dangling it gently in water for 20 seconds (Figure 6 d). After warming, the residual fluid was removed from the straw with paper towel, and straw disinfected with 70% ethanol. The heat-sealed part of straw (opposite to the cotton-polyvinyl plunge) was cut off with sterile scissors, and the aspirator was connected with the straw (Figure 6f). A low differential negative pressure was applied by aspiration. That ensured that after subsequent cutting of the cotton-polyvinyl plunge fluid was not leaking out (Figure 6f). Finally, the suspension was expelled from the straw (Figure 6g) for immediate evaluation of sperm quality, loading into catheter and intrauterine insemination.

The results were compared to slow frozen spermatozoa. For this purpose the Freezing Medium TYB, IrvineScientific, with 12 % (v/v) glycerol and 20 % (v/v) egg yolk were used. The suspension of swim up-prepared spermatozoa was 1:2 diluted with freezing medium (to achieve the concentration of 0.5×10^6 spermatozoa / mL and equilibrated at room temperature for 10 minutes then the 500 µL of spermatozoa suspension was packaged into 0.5 mL plastic straws (Cryo Bio System, Paris, France), the straws were sealed from both sides, kept in horizontal position at 4 °C for 30 minutes and put in the horizontal position into liquid nitrogen vapor (-80 °C, 10 cm over liquid nitrogen surface), kept for 30 minutes and finally placed into liquid nitrogen where they were stored minimum 24 hours until evaluation. For thawing of samples, the straws were taken from liquid nitrogen, hold in air for 30 seconds, immersed into 37°C water bath in horizontal position and hold in this bath for 20 seconds until ice melted. After thawing, 10 mL of basic (HTF-HSA) medium was added to thawed sample and centrifuged for 5 minutes at 340g. The supernatant was removed and pellet resuspended with the same basic medium in order to obtain a final concentration of 0.5×10^6 spermatozoa/mL. The changes in following physiological and morphological parameters of thirty minutes after warming of fresh, vitrified and conventional frozen spermatozoa were investigated for progressive motility (WHO, 1999); cytoplasmic membrane integrity (CMI) with applying of a LIVE/DEAD sperm viability kit, which is used to stain nucleic acid probe molecular (SYBR-14 dye) and propidium iodide (IP) and Acrosomal membrane integrity (AMI). The acrosome-reacted, and capacitated spermatozoa were detected using the double fluorescence chlortetracycline (CTC)-Hoechst

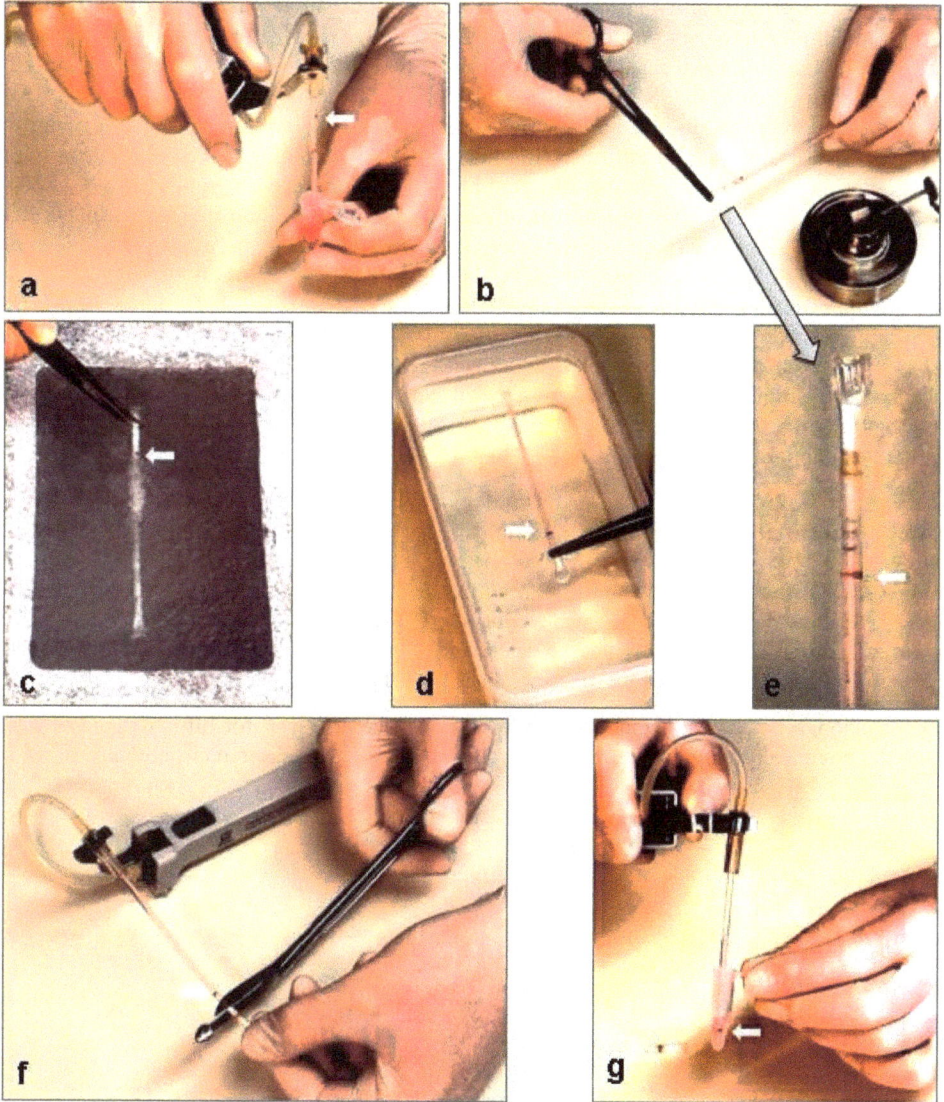

(Arrows) line, (a) aspiration of spermatozoa suspension in straw, (b, e) flame-sealing of straw, (c) cooling of straw, (d) warming of straw, (f) cutting of straw, (g) expelling of spermatozoa suspension from straw.

Fig. 6. Schema of human spermatozoa vitrification using 0.5 mL straws.

33258 staining technique (Kay et al. 1994). The results of that comparative investigation have shown that motility of spermatozoa vitrified in large volume (500 μL) in absence of permeable cryo-protectants displayed statistically higher levels of motility as compared to slow conventional freezing (76.0 +4.7 % vs 52.0 + 3.9 %, respectively, P<0.05; in fresh 85.0 +

5.1%) as well as after 24 and 48 hours in vitro culture (Figure 7). It was observed, that higher rates of membrane integrity (Figure 8) were achieved in vitrified sperm as compare to slow conventional freezing (54.0 ± 5.0 % vs 28.3 ± 3.5 %, respectively, P<0.05), but lower then in non-treated fresh control (98.2 ± 0.5 %, P<0.05). The effect of two procedures used for cryopreservation on sperm functional state as assessed by CTC staining is shown on Figure 8. There was a statistically significant difference between percentages of spermatozoa with intact acrosome after vitrification as compared to conventional freezing (44.4 ± 4.5 % *vs* 30.0 ± 3.9 %, respectively, P<0.05), but statistically lower then in fresh non-treated samples (95.4 ± 5.0 %; P<0.05). There were no statistically significant difference between percentages of sperm identified as 'capacitated' in CTC staining after vitrification as compared to conventional freezing (10.0 ± 1.8 % vs 11.0 ± 1.1 %, respectively, P <0.01), but significantly higher in fresh non-treated control (4.0 ± 0.2%, P<0.05). Described technology has a massive potential for applications in reproductive assisted procedures (ICSI, IVF and IUI) not only because of its simplicity but also because this procedure can effectively protect these cells from cryo-injures, at a level at least comparable to conventional freezing as judged by basic parameters of spermatozoa quality.

This cryoprotectant-free vitrification technology for the cryopreservation of spermatozoa instead traditional slow freezing with permeable cryoprotectants is already used in following centers: our university maternal hospital (www.uniklinik-ulm.de): IVF Centers in Temuco, Chile (about 200 IUI cycles/year) and in Ulm, Germany (www.kinderwunsch-ulm.de) (about 1,000 IVF cycles/year). First successful pregnancies and birth of healthy babies after insemination of vitrified spermatozoa has been recently achieved with vitrified spermatozoa (Isachenko et al., 2011d; Sanchez et al., 2011a).

In conclusion, a basic protection from cryo-injury can be achieved for human spermatozoa using the novel technology of aseptic cryoprotectant-free vitrification in large volumes.

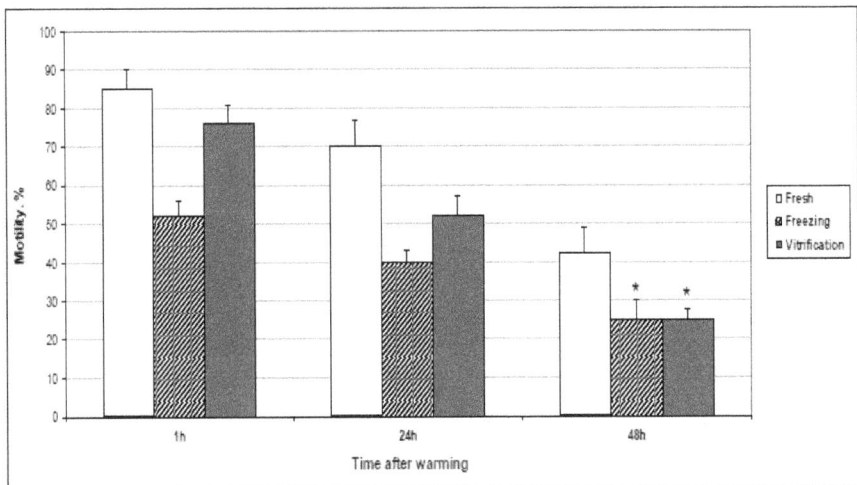

All rates in respective groups are significantly different (P<0.05) instead columns marked with asterisks (P>0.1).

Fig. 7. Motility of human spermatozoa after conventional freezing and vitrification.

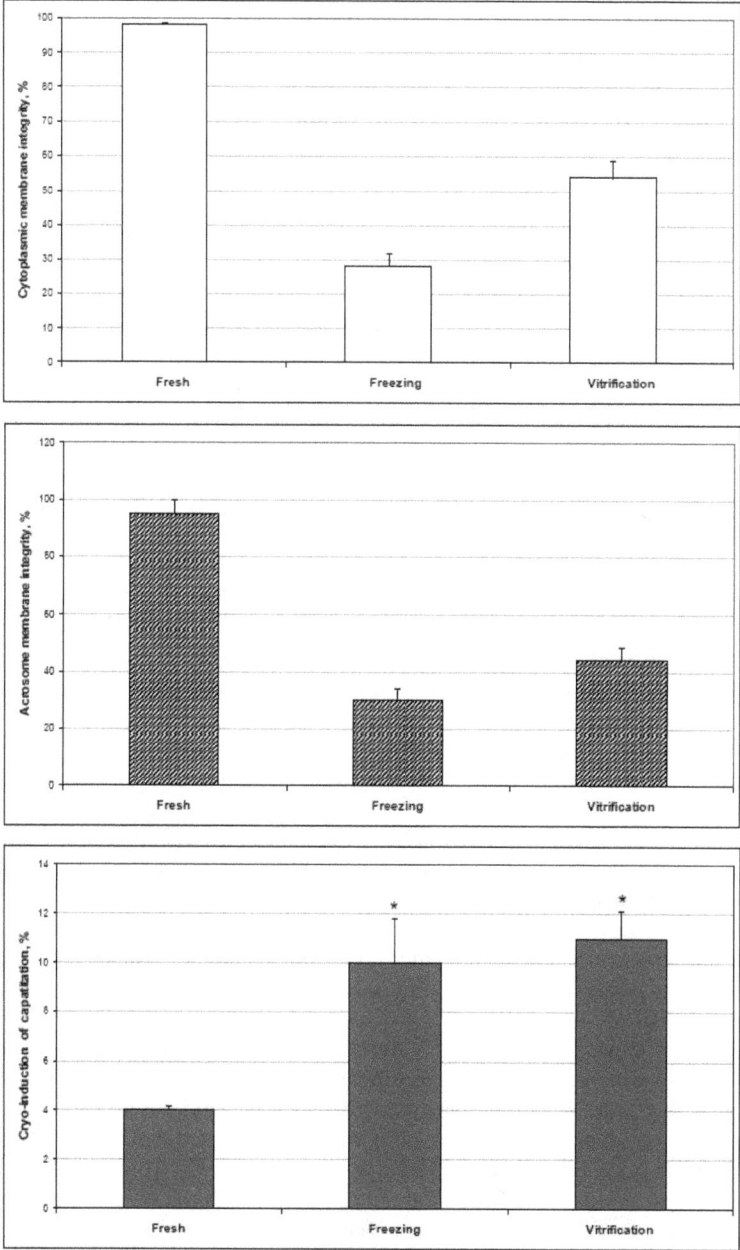

All rates in respective groups are significantly different (P<0.05) instead columns marked with asterisks (P>0.1).

Fig. 8. Cytoplasmic and acrosomal membranes integrity as well as cryo-induction of capacitation of human spermatozoa after conventional freezing and vitrification.

4. Technology for vitrification of dog spermatozoa (Sánchez et al, 2011)

If we will investigate the history of reproductive cryobiology we will see that all routine-used technique, excluding intracytoplasmic sperm injection, was firstly approved on the animal model. The same happened and with vitrification technique. After first promising investigation with frog (Luyet and Hodapp, 1938), human (Jahnel, 1938; Parkes, 1945) fowl (Schaffner, 1942) and human, and rabbit (Hoagland and Pincus1942) spermatozoa the vitrification technique was successful re-discovered in 2002 (Nawroth et al., 2002) on human spermatozoa. Recently we have decided to extrapolate the results of our investigation on animal model, thus we have with high attention examined the work of Watson and Plummer (Watson and Plummer, 1985) about responses of spermatozoa from different kind of animals to cold shock. According to this work most sensitive to cold injury are spermatozoa of animals, which produce gametes with big blade-shaped flat head. The spermatozoa of human, stallion, dog and cat have the highest stability to cold shock due to smallest blade-shaped flat head compare to the rabbit spermatozoa (have middle stability), ram, bull and boar (have the lowest stability to cold shock). Took into account these data we have decided, that the spermatozoa from human, stallion, dog and cat could be similar well preserved using vitrification technique. In this case we have decided to investigate the ability of dog spermatozoa, which stay on third place after human one according to head's size, to maintain their physiological function after vitrification without use of permeable cryoprotectants. In cryobiological routine practice, carbohydrates were already used for sperm cryopreservation (Nakagata, Takeshima, 1992, 1993; Wakayama et al., 1998). It has been suggested that raffinose plays the role of a membrane stabilizing and dehydrating agent. Comparative investigation of three different sugars, monosaccharide glucose, disaccharide sucrose and trisaccharide raffinose, showed that protection against freezing/thawing injuries is independent of the kind of sugar itself, but depends more on the sugar's concentration (Koshimoto & Mazur, 2002). Based on this evidence, we have decided to investigate the different concentrations of sucrose on the viability of cryopreserved spermatozoa. However, the problem was that the dog spermatozoa have the special physiological property which connect to capacitation process. It is well known that spermatozoa of different kind mammalian species are very sensitive to the negative effect of cryoprotectants dependent on temperature (Sánchez & Schill, 1991; Deppe et al., 2004). It is also proved (Pérez et al., 1996) that low survival and fertilizing capacity of cryopreserved mammalian spermatozoa has been attributed to an early state of capacitation resulting from the procedures by which spermatozoa are preserved. These kind of changes have been called as 'early state capacitation' or 'cryocapacitation'. These data of Pérez and colleagues (Pérez et al., 1996) later were supported (Samper, 1997; Maxwell et al., 1997). These authors showed the negative effect of cryocapacitation on the fertilizing capacity and viability of spermatozoa. The described study showed that the canine spermatozoa starts rapidly with capacitation process as soon as have been separated from seminal plasma with subsequent 20–40% of capacitated and spontaneous acrosome reacted spermatozoa in culture media. This rate is higher than in other mammals (Risopatrón et al., 2002; Santiani et al., 2004) and the spermatozoa are therefore more affected by the cryocapacitation process than those of other species. Probably the sperm membrane in this species is especially sensitive to cooling in the range of temperatures between 20°C and 5°C and to heating to 30°C at thawing (Holt & North, 1991; Sánchez & Schill, 1991). In this case for dog spermatozoa which are high sensitive to capacitation the cryopreservation protocol with a very increased cooling speed should be used, because these temperature ranges by ultra-rapid freezing (vitrification) will be just eliminated.

Took into account all mentioned above in our investigation (Sánchez et al., 2011b) to decrease the sensitivity of dog spermatozoa to different manipulations before cryopreservation we have chosen the Human tubal fluid (HTF, Quinn et al., 1985) as basic medium, which was served as control. The centrifugation for removing seminal plasma before dilution with cryoprotective media and subsequent cryopreservation at 700 g for 6 min was performed. This allowed us to achieve very high (~ 80%) amount of spermatozoa with intact acrosome in control.

(Figure 9). Integral membrane proteins are associated with the lipid bilayer and their function may be expected to be altered, especially those that perform the function of transport channels for calcium absorption. The permeability of these channels is increased on cooling, affecting calcium regulation (Robertson & Watson, 1986; Robertson et al., 1988). These facts have serious consequences for cell function (Bailey & Bhur, 1994) and many changes may be incompatible with sperm viability. In this case we have decided to apply to dog spermatozoa the early developed us vitrification protocol (Isachenko et al., 2008) for human sperm cells. The following tested groups were compared: HTF (Control); HTF–bovine serum albumin (BSA, 1% end-concentration); HTF–BSA + 0.1 M sucrose; HTF–BSA + 0.25 M sucrose and HTF–BSA + 0.4 M sucrose.

The vitrification procedure was done as follow. Briefly, aliquots of 30 µl of sperm suspension (different vitrification media) were dropped directly into LN_2. After solidification, the spheres were packaged in cryotubes and stored for at least 24 h in liquid nitrogen before use. The warming was performed by quickly submerging spheres one by one (not more than five spheres) in 5 ml of HTF–BSA 1% pre-warmed to 37°C accompanied by gentle agitation for 5–10 sec. The post-thaw sperm suspension was maintained at 37°C and 5% CO_2 for 10 min and then centrifuged at 300 g for 5 min. The cell pellet was finally re-suspended in 50 µl of HTF only for sperm evaluation.

The influence of tested media on the following physiological parameters of dog spermatozoa we have checked with such screening methods: viability and condition of acrosome with double stain technique (Trypan blue–Giemsa) with subsequent evaluation of acrosome pattern according to Didion (Didion et al., 1989); DNA fragmentation was detected with using of TUNEL technique (Gorczyca et al., 1993); detection of the change in mitochondrial permeability was done according to Smiley (Smiley et al., 1991); the motility of spermatozoa was checked as well.

According to our investigtion the percentage of spermatozoa with acrosome-intact membrane was high in all treatment groups (Figure 9) independent from concentration of sucrose in vitrification solution, but lower then in control (P<0.05).

The best progressive motility after warming (Figure 10) was significantly increased in the sperm vitrified with 0.25 M sucrose and 1% BSA (42.5 ± 2.3%), compared to other treatment groups (P < 0.01). However, lower or higher concentration of sucrose did not significantly improve the progressive motility post-vitrification. Comparable results (60.7% of motility) was reported (Tsutsui et al., 2003) when the dog semen was chilled in egg yolk-**Tris at 4°C for over 4 days, but the spermatozoa lost their fertilizing capacity.

The presence of sucrose in vitrification solution independent from the concentration has strong positive influence on viability of spermatozoa (Figure 10) and was ~70% (P<0.001) for all sucrose-treatment groups.

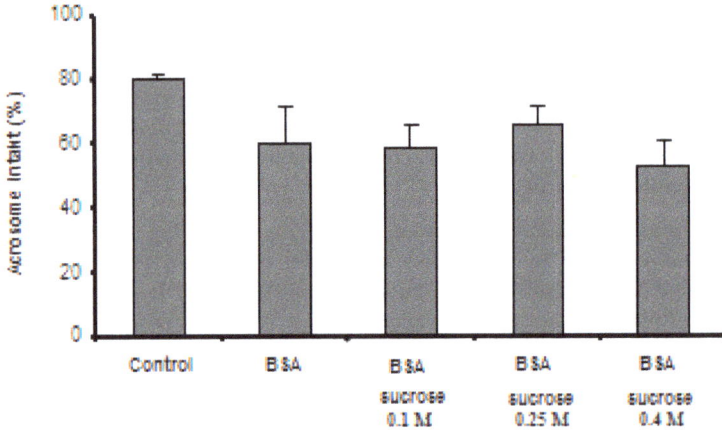

Fig. 9. Acrosome intact in canine spermatozoa after vitrification with 1% BSA and different concentrations of sucrose. Percentage of acrosome intact spermatozoa was determined by dual stain (Trypan blue– Giemsa). Data are expressed as mean ± SD from six experiments. Control = Sperm vitrified with medium HTF only. BSA, bovine serum albumin.

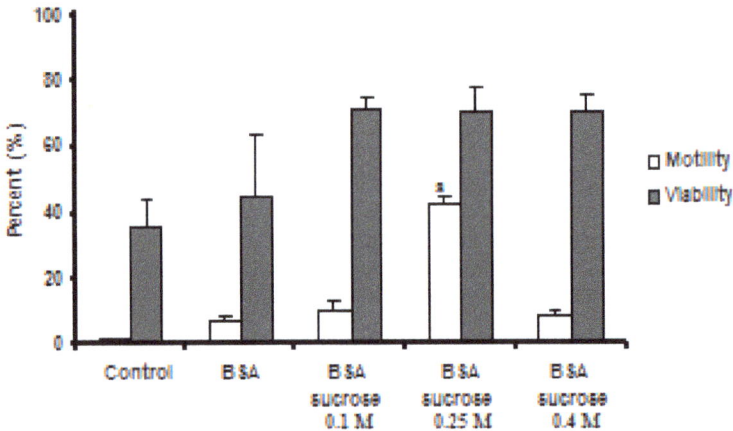

Fig. 10. Progressive motility and viability of canine spermatozoa after vitrification with 1% BSA and different concentrations of sucrose. Motility was determined by microscopic examination using a phase contrast microscope and viability by dual stain (Trypan blue–Giemsa). Data are expressed as mean ± SD from six experiments. A significant difference with respect to the control is indicated by a ($P < 0.01$). Control: Sperm vitrified with medium HTF only. BSA, bovine serum albumin.

Our data have shown that the vitrification significantly protect the sperm DNA (Figure 11) against fragmentation when used 0.25 M sucrose in combination with 1% BSA compare to control (97,2 ± 0.5% vs 94,4 ± 0.6%, respectively, $P < 0.05$). However, the lower (0.1 M) as well as higher concentrations (0.4 M) of sucrose had not significantly protective effect against DNA fragmentation. These data support our previous results (Isachenko et al., 2004a, b, the vitrification medium included only 1% HSA) and we can assume that vitrification itself due to very fast speed of cooling can provide protective effect on DNA and protect against fragmentation. It is very important results, because damage of DNA in sperm is strongly correlated with mutagenic events (Moreno et al., 2004) and how have showed Paasch with colleagues (Paasch et al., 2004) cryopreservation and thawing can be associated with varying extent of activation of apoptotic machinery in human spermatozoa. The danger is that such spermatozoa are still able to fertilize the oocyte, however, the mutations and defects did not possible to discover until the embryo has divided and the fetus has developed (Twigg et al., 1998). At present exist the opinion that DNA decondensation or fragmentation may occur in different magnitudes, which will depend on the process or the kind of cryoprotectant used (Schuffner et al., 2001; Chohan et al., 2004; Ngamwuttiwong & Kunathikom, 2007; Yildiz et al., 2007). Unfortunately, until now this question is still open, because it is not entirely clear what the effect of cryopreservation on DNA integrity is, and what would be the ideal conditions of slow freezing to reduce this effect. In this case the method of vitrification is more successful, because allows to obtain low levels of DNA fragmentation by protection due to applying of very high speed of cooling and exclusion of permeable cryoprotectants from vitrification solution.

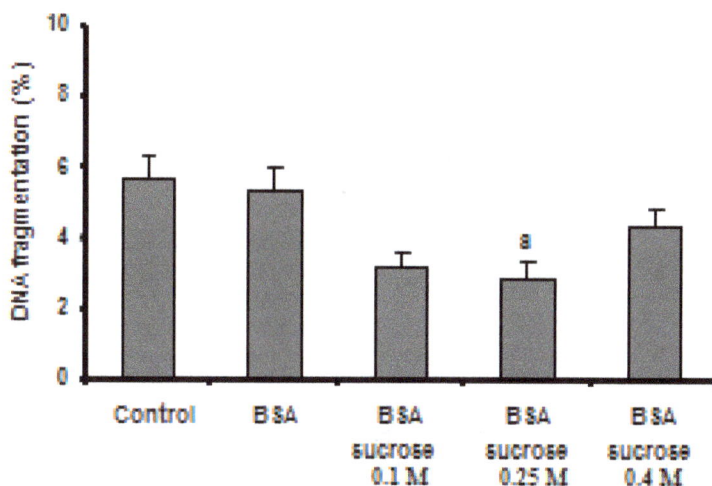

Fig. 11. DNA fragmentation in canine sperm after vitrification with 1%BSA and different concentrations of sucrose. DNA fragmentation was determined by the TUNEL assay. Data are expressed as mean ± SD from six experiments. A significant difference with respect to the control is indicated by a (P < 0.05). Control: Sperm vitrified with medium HTF only. BSA, bovine serum albumin.

Fragmentation DNA has been interpreted at present as apoptosis or apoptosis-like events (Paasch et al., 2004). This has been verified in cryopreserved / thawed sperm, with presence of increased caspase's activity induced by cryopreservation (caspase-3, -8, -9), decreased M $\Delta\Psi$ due to release of regulating proteins associated with mitochondria, evidence of DNA fragmentation, externalization of phosphatidylserine in the plasma membrane (Paasch et al., 2004) and production of reactive oxygen substances (Roca et al., 2005). Among the first events that occur in early apoptosis are changes in mitochondrial permeability which alter the transmembrane potential (M $\Delta\Psi$). Changes in the M $\Delta\Psi$ are caused by the insertion of proapoptotic proteins within the membrane, and oligomerisation may create pores, dissipating the transmembrane potential and thus releasing cytochrome c into the cytoplasm (Zamzami et al., 1995). In our work (Figure 12) we achieved the reduction of apoptotic-like process in canine spermatozoa and have got after warming more then 40% of spermatozoa with intact M $\Delta\Psi$ using of the vitrification solution with 0.25 M sucrose and 1% BSA compare to other treatment groups (P<0.001).

These results have demonstrated that vitrification without the use of permeable cryoprotectants allows to avoid the cryoprotectants toxicity caused by their addition and removal with subsequent negative effects on the spermatozoa genome. The use of sucrose in concentration of 0.25 M in combination with 1% BSA and ultrarapid speed of cooling can effectively preserve important physiological parameters of canine spermatozoa.

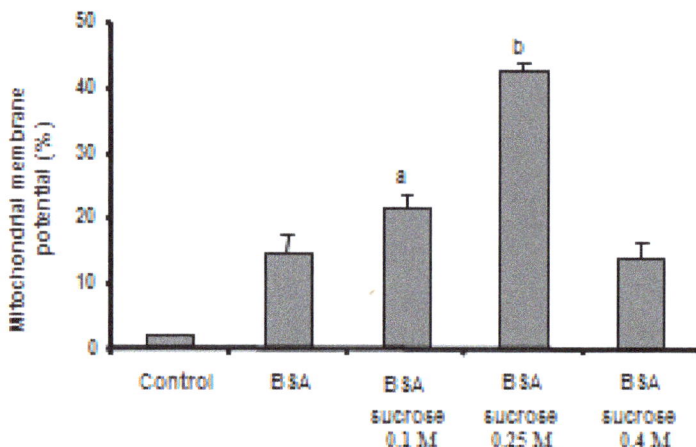

Fig. 12. Integrity of mitochondrial membrane potential in canine spermatozoa after vitrification with 1% BSA and different concentrations of sucrose. Mitochondrial membrane potential was determined by staining with the cationic fluorescent JC-1. Data are expressed as mean ± SD from six experiments. A significant difference with respect to the control is indicated by a (P < 0.05) and b (P < 0.01). Control: Sperm vitrified with medium HTF only. BSA, bovine serum albumin.

5. New technology for vitrification of fish (Oncorhynchus mykiss) spermatozoa Merino et al., 2011a, b.

At the beginning of this sub-chapter we would like to mention that the fish spermatozoa of both sea and river fish species have a very special peculiarities compare to all mammalian species. The fish sperm cells are homogenous; all spermatozoa can be activated at the same time and then swim with very similar characteristics at a certain time point post-activation. In many fish species, the flagellum is 50–60 mm long with a ribbon shape (presence of fins) instead of cylindrical; thus, the flagellum appears brighter by dark-field microscopy, allowing clear visualization of wave shapes (Cosson et al. 2008). Just for knowledge, the head of investigated us rainbow trout spermatozoa is ovoid-shaped, measuring about 3 x 1.3 μm in diameter and possess any acrosome. In middle piece present only one mitochondrial body (several mitochondria are sometimes identified in the middle piece, but later during evolution they are fused together) is shaped like an incompletely closed ring. The middle piece is completely separated from the flagellum by an invagination of the cell membrane, which reaches from the head to the base (Billard, 1983; Tuset et al., 2008). During spermatogenesis, sperm cells are prepared for accomplishing their fertilizing task for which they need to fully exploit their swimming ability immediately and as fast as possible in order to encounter the egg. The initial velocity is very high at activation, but motility duration lasts for periods ranging only 40 s to 20 min as an energetic consequence of the high velocity (Cosson et al. 2008). As possible to see the fish spermatozoa are much different from mammalian one.

Since the first successful cryopreservation of herring sperm 50 years ago (Blaxter, 1953) considerable improvement has been achieved in sperm cryopreservation and developed technology of conventional freezing of fish spermatozoa has been used in agricultural practice very broadly (Scott and Baynes, 1980; Stoss and Holtz, 1981; Dreanno et al., 1997; Wheeler and Thorgaard, 1991; Conget et al., 1996; Lahnsteiner et at., 2000; Fabbrocini et al., 2000; Zhang et al., 2003; Chen et al., 2004; Viveiros and Godinho, 2009). Usually, for protection of spermatozoa from the negative effects of low temperatures caused by conventional freezing ('slow', with controlled rate of cooling), permeable cryoprotectants are used. At present, applied cryobiology practically always uses only four permeable cryoprotectants: three spirits (ethylene glycol, propylene glycol and glycerol) and the highly polarized organic solvent dimethyl sulfoxide. However, as reported for mammalian spermatozoa, these cryoprotectants can produce osmotic and cytotoxic effects, including parthenogenesis (Gilmore et al., 1997). And for fish spermatozoa these problems are still very actual, because post-thaw viability and fertility of the cryopreserved sperm are reduced dramatically as a result of accumulated cellular damage that arise throughout the freezing-thawing process. The same like for other species the cryopreservation results in considerable damage to cellular structures such as plasma membrane, nucleus, mitochondria, and flagellum (Lahnsteiner et al., 1992; 1996; Drokin et al., 1998; Conget et al., 1996; Zhang et al., 2003). So, according to Ogier de Baulny (Ogier de Baulny, 1997) the percentage of spermatozoa with an intact membrane and a functional mitochondrion after cryopreservation varied below 18% only. According to our results which we have achieved on human spermatozoa (Isachenko et al. 2003, 2004a,b, 2005, 2008, 2011a, b, c, d) and dog (Sánchez et al., 2011b) with applying of cryoprotectant-free vitrification protocol we have decided to investigate the method on fish spermatozoa (Oncorhynchus mykiss) (Merino et al., 2011a, b). This decision we have got because the authors of these studies were able to establish statistically higher motility and in vitro fertilization ability of vitrified spermatozoa compared with spermatozoa cryopreserved using conventional slow freezing.

The standard Cortland® culture medium (Trus-Cott et al., 1968) for fish spermatozoa (per liter: 1.88 g NaCl, 0.23 g CaCl2, 7.2 g KCl, 0.41 g NaH2PO4, 1 g NaHCO3, 0.23 g MgSO4 ·7 H2O, 1.0 g Glucose, 10% Glycol and 10% Tris Base and prepared to pH 8 at 268mOsm) was used for all manipulation and served as control. Fresh-retrieved semen was diluted 1:3 in the non-activating Cortland® medium with subsequent determination of the motility and concentration by phase-contrast microscopy. Subjective evaluations of motility were performed by placing 2μl of this sperm suspension on a glass slide and immediately adding 10μl of the activator Powermilt® (Católica of Temuco University, Chile) at 10°C. The motility of the spermatozoa was observed in 12μl sperm activated by subjective microscopic examination under phase contrast optics at 400x magnification. Motility assessments were made in triplicate for each sample at 5 s following activation with Powermilt®. Sperm concentrations were determined with a Neubauer hemocytometer after dilution of 1μl of sperm suspension in 1200μl of standard culture medium. Only samples with high motility (>80%) and concentration 12×10^9 spermatozoa/mL (Drokin et al., 1998) were used in this study.

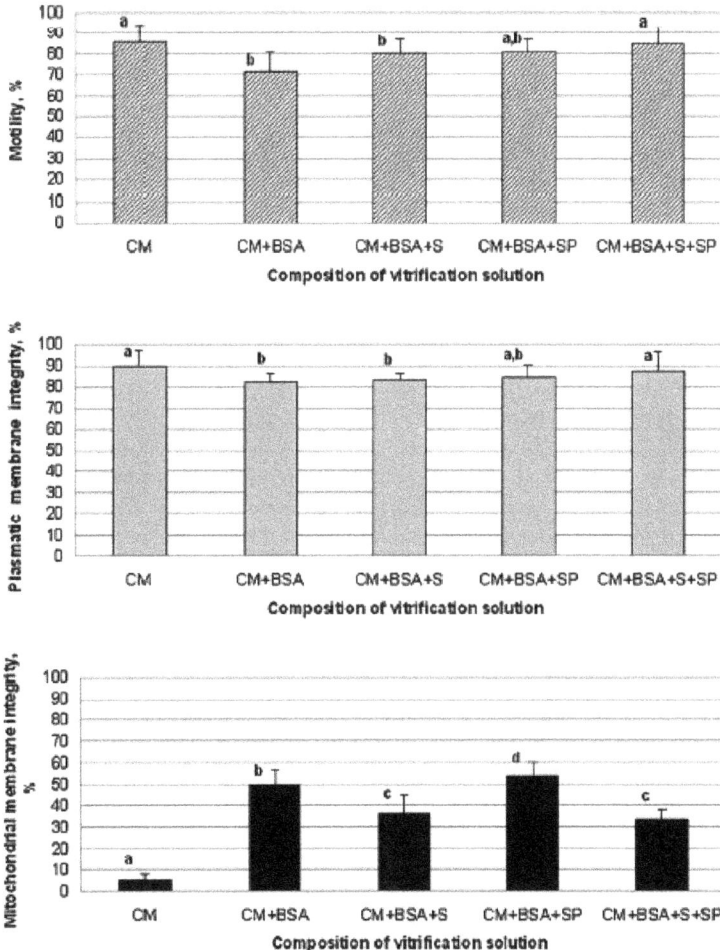

Fig. 13. Motility, cytoplasmic membrane integrity and mitochondrial membrane integrity of vitrified rainbow trout spermatozoa. (CM) Cortland®, (BSA) 1% bovine serum albumin, (SP) 40% of seminal plasma, (S) 0.125 M sucrose. Different superscripts indicate statistical difference between respective values of compared groups (P<0.05).

In our work we have investigated the following five treatments groups (Figure 13):

Group 1: Cortland® medium only (frozen control)

Group 2: Cortland® medium+ 1% BSA

Group 3: Cortland® medium+ 1% BSA + 0.125M sucrose

Group 4: Cortland® medium+ 1% BSA + 40% seminal plasma

Group 5: Cortland® medium+ 1% BSA + 40% seminal plasma + 0.125M sucrose.

The vitrification / warming of rainbow spermatozoa was proceeded as following:

Sperm samples were centrifuged at 300 g for 10 min at 4 °C. The seminal plasma (supernatant) was retained and the sperm suspension diluted with Cortland® medium to a concentration of 40×10^6 spermatozoa/ml. Five equal 500-μl aliquots from each preparation were placed in individual 1ml tubes for vitrification. Twenty microliters of sperm suspension from each tube was dropped directly into liquid nitrogen, during which the droplet adopted a spherical form approximately 3mm in diameter. After 5min, the solidified droplets were placed into 2-ml cryovials pre-cooled in liquid nitrogen with precooled tweezers. After storage for at least 24 h in liquid nitrogen, the samples were warmed by plunging the droplets into a 15ml tube containing 5ml Cortland® medium supplemented with 1% BSA at 37°C with intense agitation. After warming (one droplet/tube), the tubes were maintained at 37°C for 5–10 min prior to evaluation of spermatozoa quality.

The spermatozoa quality was tested according the following parameters:

- Motility - (percent of motile spermatozoa detected during 30 s after warming) was performed using phase contrast microscope (Carl Zeiss Jena, Jena, Germany);
- Cytoplasmic membranes integrity was assessed using the LIVE/DEAD Sperm Viability Kit (SYBR-14 dye; Invitrogen Inc., Eugene, OR, USA) and propidium iodide. The analysis of spermatozoa was carried out under an epifluorescence microscope (Axiolab drb KT 450905, Zeiss) at 400x magnification (Figure 14).
- Mitochondrial membranes integrity (mitochondrial activity) was assessed due to relative levels of M ΔΨ using the fluorescent cationic dye, JC-1 (5,50,6,60-tetrachloro-1-10,3,30-tetraethylbenzamidazolocarbocyanin iodide, according to the manufacturer's protocol (MIT-E-_TM, BIOMOL® International LP, Plymouth Meeting, PA, USA) and observed under epifluorescent optics (Axiolab drb KT 450905) at 400x magnification at room temperature. For the MIT-E-Ψ reagent, an excitation/emission filter of 488/490nm was used. The monomeric dye structure emits at 527nm (Green FITC channel), whereas J-aggregates indicative of high potential of undamaged mitochondria emit at 590nm (red, RITC channel). If ΔΨm is above a certain mV threshold, JC-1 monomers multimerize into a crystalline-like state that shifts the emission spectrum to higher frequencies (orange-red) and it is the multimerization of the monomer that is potential sensitive, the paracrystals can be stable after formation, which is why we can still get red fluorescence in JC- 1 treated cells after collapsing ΔΨm with inhibitors, which is why staining is done after inhibitor treatment (Figure 15).

To investigate these cold sensitive (Holt, 2000; O'Connell et al., 2002) organelles of spermatozoa were necessary because the retention of plasma membrane integrity and mitochondrial function after cryopreservation is too important with regard to fertilization capacity of both spermatozoa and oocytes (Gao et al., 1997; de Lamirande et al., 1997). For all species, normal mitochondrial function is a key factor in the fertilizability of spermatozoa and for fish it is especially critical to maintain mitochondrial activity because high motility normally lasts for only 30 s to few minutes. Spermatozoa of rainbow trout have only one mitochondrion to produce sufficient ATP to drive this transient high motility, and damage during cryopreservation will certainly lead to decrease of motility and as a result, fertilization ability (Maisse, 1996). In this case the stability of mitochondrion during cryopreservation can be used as a specific test for applicability of a any investigated cryopreservation protocol (Meseguer et al., 2004; O'Connell et al., 2002).

The results of this investigation showed that the proportion of sperm showing normal, high motility varied between 82% and 95% in fresh samples. In Groups 1, 2, 3, 4, and 5, motility in these solutions was 86%, 71%, 79%, 81%, and 82%, respectively (Figure 13).

The percent of spermatozoa with intact cytoplasmic membrane after thawing was similar between the 5 experimental groups, ranging from 81.8% to 90%, as shown in Figures 13 and 14. Nevertheless, the integrity of mitochondrial membrane potential of spermatozoa (Figures 13 and 15) in Groups 1, 2, 3, 4, and 5 was decreased significantly compare to non

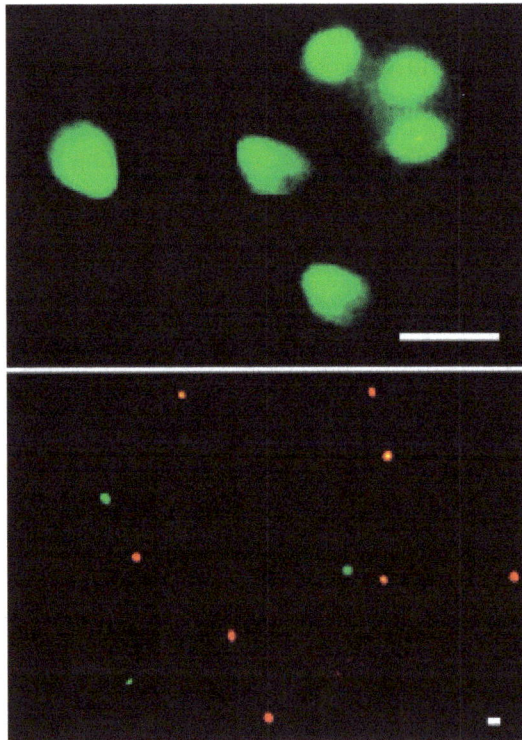

Fig. 14. Example of rainbow trout spermatozoa with non-damaged (green) and damaged (red) cytoplasmic membranes. Bar = 8μm.

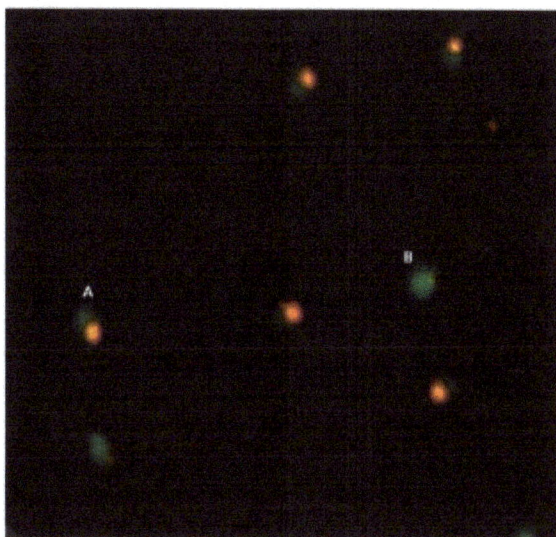

Fig. 15. Example of rainbow trout spermatozoa with (A) non-damaged and (B) damaged mitochondria. Bar = 2.5μm.

treated spermatozoa (5.5%, 49.8%, 37.1%, 54.7%, and 34.4%, respectively). As possible to see from our results they can have the following potential question. How spermatozoa can have a high level of motility with low level of the integrity of mitochondrial membranes? Especially this question is actual taking into account that fish spermatozoa have only one mitochondrion. A lower M $\Delta\Psi$, as reflected by green fluorescence, simply can have the following explanation. Activity of the mitochondria, including ATP production, is reduced when compared to their red counterparts. Cell often have "green" and "red" mitochondria with shifts between all green or all red governed by a variety of external and internal conditions, and this is normal. Individual mitochondria constantly shift from red to green and back to red in response to rapid changes in local conditions, including calcium levels and pH (Vanblerkom, personal communication). While the plasma membrane is known to be sensitive to cryopreservation (Cabrita et al., 2001; Aitken and Baker, 2006; Muller et al., 2008), our results shows that it is cryostable in rainbow trout sperm, as indicated by ~90% of non-damaged plasmatic membrane in sperm vitrified in culture medium only i.e., without permeable cryoprotectants and additional proteins). This is similar to levels reported after conventional freezing with permeable cryoprotectants. We suggest that the vitrification technique described here which associated with high rate of cooling allows to avoid the formation of large extracellular water crystals. Sucrose is well known to have a beneficial influence on the plasma membrane of cells subjected to cryopreservation (Anchordoguy et al., 1987; Rodgers and Glaser, 1993). For human spermatozoa, the drop-wise technique of vitrification is a major technical advance because it includes a mixture of non-permeable cryoprotectants such as serum albumin (Isachenko et al., 2008). However, we report that the inclusion of sucrose in the vitrification solution was ineffective for rainbow trout spermatozoa. According to Lahnsteiner (2007), lipoproteins in the seminal plasma of rainbow trout likely maintain the lipid composition of the plasma and may increase the cryostability of spermatozoa. Our results support this point of view and we suggest that the method of sperm vitrification described here could also be applied to other species. As a rule, carbohydrates are used for sperm cryopreservation to compensate for the decrease in osmotic pressure caused by the permeable cryoprotectant glycerol, which works as an additional dissolvent and has the ability to decrease the medium's osmotic pressure. Based on this evidence, we investigated whether sucrose had a similar cryoprotective effect on fish spermatozoa during freeze–thaw. We found that its inclusion in vitrification medium has no visible protective effect on mitochondrial membrane integrity nor does it provide significant protection for spermatozoa when compared to other vitrification mediums containing BSA or BSA + seminal plasma. Indeed, the addition of these non-permeable cryoprotectants did not increase either the motility or plasma membrane integrity of rainbow trout spermatozoa. However, described here technology of cryopreservation of fish spermatozoa by direct plunging into liquid nitrogen has big disadvantage because did not protect the biological material against direct contact with liquid nitrogen. In this connection in the future investigation it would be necessary to find a synthetic substitute for seminal plasma to avoid the possible microbial contamination. In fact, any technology in reproductive biology, and especially in a therapeutic medical approach, must guarantee the full protection of cells from microorganisms that might survive in liquid nitrogen temperatures (Gardner, 1998; Bielanski et al., 2003), and it has been suggested that liquid nitrogen can be contaminated by microorganisms (Tedder et al., 1995). The problem of potential microbial contamination of spermatozoa during cryopreservation, especially by the virus of Infectious Salmon Anemia is significant in the fish industry, especially in Latin America (Ellis, 2007; Fortt and Buschmann, 2007; Sommer, 2009). In spite of that the results of our experiments conformed that for fish spermatozoa the developed method of cryopreservation by direct plunging into liquid nitrogen (vitrification) without permeable cryoprotectants is potentially significant for this industry, but the development of "aseptic" methods, in which the spermatozoa suspension is enclosed in capillaries or straws to prevent direct contact of sperm with liquid nitrogen, will need to be considered. Filtration or ultraviolet treatment of liquid nitrogen cannot guarantee the absence of contamination of biological material by viruses. For example, Tedder et al. (1995) reported the contamination of blood probes by hepatitis virus during the storage of probes in liquid nitrogen. Different types of viruses, such as hepatitis virus, papova virus, vesicular stomatitis virus and herpes virus, which are simple and very cryostable structures, may increase their virulence after direct plunging and storage in liquid nitrogen (Hawkins et al., 1996; Charles and Sire, 1971; Schaffer et al., 1976; Jones and Darville, 1989).

6. General conclusion

Data presented in this review shown that the technique of cryopreservation by direct plunging into liquid nitrogen (vitrification) in absence of permeable cryoprotectants has a great perspective. This technique allows significantly protect the important physiological parameters of mammalian and fish spermatozoa against cryo-injures.

7. References

Aitken RJ and Clarkson JS. Cellular basis of defective sperm function and its association with the genesis of reactive oxygen species by human spermatozoa. J Reprod Fertil 1987;81:459-469.

Aitken RJ, Baker MA. Oxidative stress, sperm survival and fertility control. Mol Cell Endocrinol 2006;250:66–69.

Anchordoguy T, Rudolph A, Carpenter J, Crowe J. Mode of interaction of cryoprotectants with membrane phospholipids during freezing. Cryobiology 1987;24:324–331.

Andrews FC. Colligative properties of simple solutions. Science 1976;194:567-571.

Bailey JL, Bhur MM. Cryopreservation alters the Ca2 + flux of bovine spermatozoa. Can J Anim Sci 1994;74:45–51.

Billard R. Ultrastructure of trout spermatozoa: Changes after dilution and deep-freezing. Cell Tissue Res 1983; 228:205-218.

Barbas JP, Mascarenhas RD. Cryopreservation of domestic animal sperm cells. Cell Tissue Bank 2009;10:49–62.

Bielanski A, Bergeron H, Lau PCK, Devenish J. Microbial contamination of embryos and semen during long term banking in liquid nitrogen. Cryobiology 200346:146–152.

Blaxter J H S. Sperm storage and cross-ertilization of spring and autumn spawning herring. Nature 1953;172:1 189-1 190.

Cabrita E, Ane L, Herraez M. Effect of external cryoprotectants as membrane stabilizers on cryopreserved rainbow trout sperm. Theriogenology 2001;56:623–635.

Cervera R, Garcia-Xime´nez F. Vitrification of zona-free rabbit expanded or hatching blastocysts: a possible model for human blastocysts. Hum Reprod 2003;18:2151–2156.

Charles GN, Sire DJ. Transmission of papova virus by cryotherapy applicator. J Am Med Assoc 1971;218 :1435.

Chen S L, Ji XS, Yu GC, Tian YS, Sha ZX. Cryopreservation of sperm from turbot (Scophthalmus maximus, and application to large-scale fertilization. Aquaculture 2004; 236: 547-556.

Chen S, Lien Y, Cheng Y, Chen H, Ho H, Yang Y. Vitrification of mouse oocytes using closed pulled straws (CPS) achieves a high survival and preserves good patterns of meiotic spindles, compared with conventional straws, open pulled straws (OPS) and grids. Hum Reprod 2001;16:2350–2356.

Chohan KR, Griffin JT, Carrell DT. Evaluation of chromatin integrity in human sperm using acridine orange staining with different fixatives and after cryopreservation. Andrologia 2004;36:321–326.

Conget P, Fernfindez M, Herrera G, Minguell JJ. Cryopreservation of rainbow trout (Oncorhynchus mykiss) spermatozoa using programmable freezing. Aquaculture 1996;143:319-329.

Cormier N, Bailey JL. A differential mechanism is involved during heparin- and cryopreservationinduced capacitation of bovine spermatozoa. Biol Reprod. 2003; 69:177–185.

Cosson J, Groison A-L, Suquet M, Fauvel C, Dreanno C, Billard R. Marine fish spermatozoa: racing ephemeral swimmers. Reproduction 2008;136 277–294

Critser JK, Huse-Benda AR, Aaker D, Arneson BW, Ball GD. Cryopreservation of human spermatozoa. III. The effect of cryoprotectants on motility. Fertil Steril 1988;50:314–320.

de Lamirande E, Leclerc P, Gagnon C. Capatitation as a regulatory event that primes spermatozoa for the acrosome reaction and fertilization. Mol Hum Reprod 1997;3:175–194.

Deppe M, Ortloff C, Salinas G, Bravo D, Sánchez R. Efecto de la temperatura de incubació́n y adició́n de glicerolsobre la preservació́n del acrosoma en espermatozoides humanos. Rev Invest Clin 2004;56:477–482.

Didion BA, Dobrinsky JR, Giles JR, Graves CN. Staining procedure to detect viability and the true acrosome reaction in spermatozoa of various species. Gamete Res 1989;22:51–57.

Dreanno C, Suquet M, Quemener L, Cosson J, Fierville F, Normant Y, Billard R. Cryopreservation of turbot (Scophthalmus Maximus) spermatozoa. Theriogenology 1997; 48: 589-603.

Drokin S, Stein H, Bartscherer H. Effect of cryopreservation on the fine structure of spermatozoa of rainbow trout (Oncorhynchus mykiss) and brown trout (Salmo trutta F. fario). Cryobiology 1998;37:263–270.

Ellis S. Risk and factors salmon chilean. Assessing infectious disease emergence potential in the U.S. Aquacult Industr 2007,1-111. www.scribd.com/doc/1446078/USDA-aquaculture23.

Fabbrocini A.S, Lavadera L, Rispoli S, Sansone G. Cryopreservation of Seabream (Sparus aurata) Spermatozoa. Cryobiology 2000;40: 46-53.

Fraga CG, Motchnik PA, Shigenaga MK, Helbock HJ, Jacob RA, Ames BN. Ascorbic acid protects against endogenous oxidative DNA damage in human sperm. Proc Nat Acad Sci USA. 1991;88:11003-11006.

Franks F. Biological freezing and cryofixation. J Microsc 1977;111:3-16.

Fortt Z, Buschmann AR. Residuos de tetraciclina y quinolonas en peces silvestres en una zona costera donde se desarrolla la acuicultura del salmón en Chile. Rev Chil Infect 2007;24:8–12.

Gao D, Mazur P & Critser J 1997 Fundamental cryobiology of mammalian spermatozoa. In Reproductive Tissue Banking, 4 edn, pp 263–328. Eds AM Karow & JK Critser. London: Academic Press

Gardner DK. Development of serum-free media for the culture and transfer of human blastocysts. Hum Reprod 1998;13:218–225.

Gilmore JA, Liu J, Gao DY, Critser JK. Determination of optimal cryoprotectants and procedures for their addition and removal from human spermatozoa. Hum Reprod 1997;12:112–118.

Glander HJ, Schaller J. Hidden effects of cryopreservation on quality of human spermatozoa. Cell Tissue Bank 2000;1:133-42.

Gorczyca W, Traganos F, Jesionowska H, Darzynkiewicz Z. Presence of DNA strand breaks and increased sensitivity of DNA in situ to denaturation in abnormal human sperm cells; analogy to apoptosis of somatic cells. Exp Cell Res 1993;207:202–205.

Green CE, Watson PF. Comparison of the capacitation-like state of cooled boar spermatozoa with true capacitation, Reproduction 2001;122: 889–898.

Hammadeh ME, Askari AS, Georg T, Rosenbaum P, Schmidt W. Effect of freeze-thawing procedure on chromatin stability, morphological alteration and membrane integrity of human spermatozoa in fertile and subfertile men. Int J Androl 1999;22:155-162.

Hawkins AE, Zuckerman MA, Briggs M, Gilson RJ, Goldstone AH, Brink NS, Tedder RS. Hepatitis B nucleotide sequence analysis: linking an outbreak of acute hepatitis B to contamination of a cryopreservation tank. J Virol Methods 1996;60:81-88.

Henkel R, Ichikawa T, Sánchez R, Miska W, Ohmori H, Schill WB. Differentiation of ejaculates showing reactive oxygen species production by spermatozoa or leukocytes. Andrologia 1997;29:295-301.

Henkel RR, Schill WB. Sperm preparation for ART. Reprod Biol Endocrinol 2003;1:108.

Henkel R, Kierspel E, Stalf T, Mehnert C, Menkveld R, Tinneberg HR, Schill WB, Kruger TF. Effect of reactive oxygen species produced by spermatozoa and leucocytes on sperm functions in non-leucocytospermic patients. Fertil Steril 2005;83:635-642.

Henkel R, Bastiaan HS, Schüller S, Hoppe I, Starker W, Menkveld R. Leucocytes and intrinsic ROS production may be factors compromising sperm chromatin condensation status. Andrologia 2010;42:69-75.

Hoagland H, Pincus G. Revival of mammalian sperm after immersion in liquid nitrogen. J Gen Physiol 1942;25:337-344.

Holt W, North RD. Cryopreservation, actin localization and thermotropic phase transitions in ram spermatozoa. J Reprod Fertil 1991;91:451-461.

Holt WV. Fundamental aspects of sperm cryobiology: the importance of species and individual differences. Theriogenology 2000;53:47-58.

Isachenko E, Isachenko V, Katkov II, Dessole S, Nawroth F. Vitrification of mammalian spermatozoa in the absence of cryoprotectants: from past practical difficulties to present success. Reprod Biomed Online 2003;6:191-200.

Isachenko, E., Isachenko, V., Katkov, I.I., Rahimi, G., Schondorf, T., Mallmann, P., Dessole, S., Nawroth, F. (2004a). DNA integrity and motility of human spermatozoa after standard slow freezing versus cryoprotectant-free vitrification. Hum. Reprod. 19, 932-939.

Isachenko, V., Isachenko, E., Katkov, I.I., Montag, M., Dessole, S., Nawroth, F., van der Ven, H. (2004b). Cryoprotectant-free cryopreservation of human spermatozoa by vitrification and freezing in vapor: effect on motility, DNA integrity, and fertilization ability. Biol. Reprod. 71, 1167-1173.

Isachenko V, Isachenko E, Montag M, Zaeva V, Krivokharchenko A, Nawroth F, Dessole S, Katkov I, Van der Ven H. Clean technique for cryoprotectant – free vitrification of human spermatozoa. Reprod Biomed Online 2005a;10:350-354.

Isachenko V, Montag M, Isachenko E, van der Ven H Vitrification of mouse pronuclear embryos after polar body biopsy without direct contact with liquid nitrogen. Fertil Steril 2005b; 84:1011-1016.

Isachenko, V., Katkov, I.I., Yakovenko, S., Lulat, A.G., Ulug, M., Arvas, A., Isachenko, E. (2007). Vitrification of human laser treated blastocysts within cut standard straws (CSS): novel aseptic packaging and reduced concentrations of cryoprotectants. Cryobiology 54, 305-309.

Isachenko E, Isachenko V, Weiss JM, Kreienberg R, Katkov II, Schulz M, Lulat AG, Risopatron MJ, Sanchez R. Acrosomal status and mitochondrial activity of human spermatozoa vitrified with sucrose. Reproduction 2008;136:167-173.

Isachenko E, Isachenko V, Sánchez R, Katkov II, Kreienberg R. Cryopreservation of spermatozoa: old routine and new perspective. In: Donnez J, Kim SS, eds. Principles and Practice of Fertility Preservation. Cambridge, UK: Cambridge University Press, 2011a:177-198.

Isachenko V, Isachenko E, Petrunkina A.M., Sánchez R. Human spermatozoa vitrified in the absence of permeable cryoprotectants: birth of two healthy babies. Reprod Fertil Dev. 2011b, 24:323-326.

V. Isachenko, R. Maettner, A. M. Petrunkina, K. Sterzik, P. Mallmann, G. Rahimi, R. Sánchez, J. Risopatrón, I. Damjanoski, E. Isachenko. Vitrification of Human ICSI/IVF Spermatozoa Without Cryoprotectants: New Capillary Technology. J Androl., 2011c. In press

Isachenko V, Maettner R, Petrunkina AM, Sterzik K, Mallmann P, Rahimi G, Sánchez R, Risopatrón J, Hancke K, Damjanoski I, Kreienberg R, Isachenko E. Cryoprotectant-free Vitrification of Human Spermatozoa in Large (to 0.5 mL) Volume: Novel Technology. Clin Lab. 2011d, 57:643-650.

Jahnel F. Resistance of human spermatozoa to deep cold, Klinische Wochenschrift 1938;17:1273-1274.

Jones SK, Darville JM. Transmission of virus-particles by cryotherapy and multi-use caustic pencils: a problem to a dermatologist? Br J Dermatol 1989;121:481-486.

Karlsson JOM, Cravalho EG. A model of diffusion-limited ice growth inside biological cells during freezing. J Appl Phys 1994;75:4442-55.

Katkov II, V Isachenko, E Isachenko, M S. Kim, A G-M.I. Lulat, A M. Mackay, F Levine. Low- and high-temperature vitrification as a new approach to biostabilization of reproductive and progenitor cells. Int J Refrigeration 2006;29:346–357.

Kay VJ, Coutts JR, Robertson L. Effects of pentoxifylline and progesterone on human sperm capacitation and acrosome reaction. Hum Reprod 1994;9:2318–2323.

Koshimoto C, Gamliel E, Mazur P. Effect of osmolality and oxygen tension on the survival of mouse sperm frozen to various temperatures in various concentrations of glycerol and raffinose. Cryobiology 2000;41:204-31.

Kumar R, Venkatesh S, Kumar M, Tanwar M, Shasmsi MB, Kumar R, Gupta NP, Sharma RK, Talwar P, Dada R. Oxidative stress and sperm mitochondrial DNA mutation in idiopathic oligoasthenozoospermic men. Indian J Biochem Biophys. 2009;46:172-177.

Koshimoto CH, Mazur P. Effects of cooling and warming rate to and from -70°C, and effect of further cooling from -70 to -196°C on the motility of mouse spermatozoa. Biol Reprod 2002;66:1477-1484.

Kuleshova LL, MacFarlane DR, Trounson AO, Shaw JM. Sugars exert a major influence on the vitrification properties of ethylene glycol-based solutions and have a low toxicity to embryos and oocytes. Cryobiology 1999;38:119- 130.

Lahnsteiner F, Berger B, Horvath A, Urbanyi B, Weismann T. Cryopresevation of spermatozoa in cyprindid fishes. Theriogenology 2000;54:1 477-1 496.

Lahnsteiner F, Weismann T, Patzner RA. Fine structural changes in spermatozoa of the grayling, *Thymallus thymallus* (Pisces: Teleostei), during routine cryopreservation. Aquaculture 1992;103: 73-84.

Lahnsteiner F, Berger B, Wiesmann T, Patzner R. Changes in morphology, physiology, metabolism, and fertilization capacity of rainbow trout semen following cryopreservation. Prog Fish Cult 1996;58:149-159.

Lahnsteiner F. Characterization of seminal plasma proteins stabilizing the sperm viability in rainbow trout (Oncorhynchus mykiss). Anim Reprod Sci 2007;97:151–164.

Larson JM, McKinney KA, Mixon BA, Burry KA, Wolf DP. An intrauterine insemination-ready cryopreservation method compared with sperm recovery after conventional freezing and post-thaw processing. Fertil Steril 1997;68:143-8.

Levin RL. A generalized method for the minimization of cellular osmotic stresses and strains during the introduction and removal of permeable cryoprotectants. J Biomech Eng 1982;104:81-86.

Lozina-Lozinski LK. Action of cooling on cells and the body as a multifactorial process. Tsitologiya 1982;24:371-390.

Luyet BJ, Hoddap A. Revival of frog's spermatozoa vitrified in liquid air. Proc Meet Soc Exp Biol 1938;39:433–434.

Maisse G. Cryopreservation of fish semen: a review. In: Proceedings of Refrigeration and Aquaculture, Bordeaux, France, 1996March, pp. 443–466.

Mazur P. Kinetics of water loss from cells at subzero temperatures and the likelihood of intracellular freezing. J Gen Physiol 1963;47:374–369.

Maxwell WMC, Welch GR, Johnson LA. Viability and membrane integrity of spermatozoa after dilution and flow cytometric sorting in the presence or absence of seminal plasma. Reprod Fertil Dev 1997;8:1165–1178.

McLaughlin EA, Ford WC, Hul MG. A comparison of the freezing of human semen in the uncirculated vapour above liquid nitrogen and in a commercial semi-programmable freezer. Hum Reprod 1990;5:724-8.

Merino O, Risopatrón J, Sánchez R, Isachenko E, Figueroa E, Valdebenito I, Isachenko V. Fish (Oncorhynchus mykiss) spermatozoa cryoprotectant-free vitrification: Stability of mitochondrion as criterion of effectiveness. Animal Reproduction Science Animal Reproduction Science 2011;124 :125–131.

Merino O, Sánchez R, Risopatrón J, Isachenko E, Katkov II, Figueroa E, Valdebenito I, Mallmann P, Isachenko V. Cryoprotectant-free vitrification of fish (Oncorhynchus mykiss) spermatozoa: first report. Andrologia 2011;1–6.

Meseguer M, Garrido N, Martinez-Conejero JA, Simon C, Pellicer A, Remohi J. Role of cholesterol, calcium, and mitochondrial activity in the susceptibility for cryodamage after a cycle of freezing and thawing. Fertil Steril 2004;82:514–515.

Moreno D, Fuentes JL, Sánchez A, Baluja L, Prieto E. Conversio'n y reversio'n ge'nica en saccharomyces cerevisiae. Un modelo para el estudio del dan͂o producido por radiacio'n gamma. Rev Cubana Invest Biomed 2004;23:80–86.

Muller K, Muller P, Pincemy G, Kurz A, Labbe C. Characterization of sperm plasma membrane properties after cholesterol modification: consequences for cryopreservation of rainbow trout spermatozoa. Biol Reprod 2008;78:390–399.

Nakagata N, Takeshima T. High fertilizing ability of mouse spermatozoa diluted slowly after cryopreservation. Theriogenology 1992;37:1263–1291.

Nakagata N, Takeshima T. Cryopreservation of mouse spermatozoa from inbred and F 1 hybrid strains. Jikken Dobutsu 1993;42:317–320.

Nawroth F, Isachenko V, Dessole S, Rahimi G, Farina M, Vargiu N, Mallmann P, Dattena M, Capobianco G, Peters D, Orth I, Isachenko E. Vitrification of human spermatozoa without cryoprotectants, Cryo Lett 23;2002:93– 102.

Ngamwuttiwong T, Kunathikom S. Evaluation of cryoinjury of sperm chromatin according to liquid nitrogen vapour method (I). J Med Assoc Thai 2007;90:224–228.

Ogier de Baulny B, Labb C, Maisse G. Membrane integrity, mitochondrial activity, ATP content, and motility of the European catfish *(Silurus glanis)* testicular spermatozoa after freezing with different Cryoprotectants. Cryobiology 1999;39:177-184.

Paasch U, Sharma RK, Gupta AK, Grunewald S, Mascha EJ, Thomas AJ Jr, Glander HJ, Agarwal A. Cryopreservation and thawing is associated with varying extent of activation of apoptotic machinery in subsets of ejaculated human spermatozoa. Biol Reprod 2004;71:1828–1837.

Perez LJ, Valcarcel A, Delasheras MA, Moses D, Baldassarre H. Evidence that frozen/thawed ram spermatozoa show accelerated capacitation in vitro as assessed by chlortetracycline assay. Theriogenology 1996;46:131–140.

Pérez-Sánchez F, Cooper TG, Yeung CH, Nieschlang E. Improvement in quality of cryopreserved human spermatozoa by swim-up before freezing. Int J Androl 1994;17:115–120.

Parkes AS. Preservation of human spermatozoa at low temperatures, Br Med J 1945;2:212– 213.

Petrunkina AM, Volker G, Weitze KF, Beyerbach M, Töpfen-Petersen E, Waberski D. Detection of cooling-induced membrane changes in the response of boar sperm to capacitating conditions. Theriogenology 2005;63:2278-2299.

Petrunkina AM. Fundamental aspects of gamete cryobiology. J Reproduktionsmed Endokrinol 2007;4:78-91.

Quinn P, Warnes GM, Kerin JF, Kirby C. Culture factors affecting the success rate of in vitro fertilization and embryo transfer. Ann NY Acad Sci. 1985;442:195-204.

Rall WF, Fahy GM. Ice-free cryopreservation of mouse embryos at)196 _C by vitrification. Nature 1985;313:573–575.

Reed ML, Lane M, Gardner DK, Jensen NL, Thompson J. Vitrification of human blastocysts using the cryoloop method: successful clinical application and birth of offspring. J Assist Reprod Genet 2002; 19:304-306.

Risopatrón J, Catalan S, Miska W, Schill WB, Sànchez R. Effect of albumin and polyvinyl alcohol on the vitality, motility and acrosomal integrity of canine spermatozoa incubated in vitro. Reprod Domest Anim 2002;37:347-351.

Robertson L, Watson PF. Calcium transport indiluted or cooled ram semen. J Reprod Fertil 1986;77:117–185.

Robertson L, Watson PF, Plunner JM. Prior incubation reduces calcium uptake and membrane disruption in boar spermatozoa subjected to cold shock. Cryo Lett 1988; 19:286–293.

Roca J, Rodríguez M, Gil M, Carvajal G, García E, Cuello C, Vásquez J, Emilio A. Survival and in vitro fertility of boar spermatozoa frozen in the presence of superoxide dismutase and/or cabalase. J Androl 2005;26:15–24.

Rodgers W, Glaser M. Distributions of proteins and lipids in the erythrocyte membrane. Biochemistry 1993;32:12591–21298.

Samper JC. Reproductive anatomy and physiology of the breeding stallion. In: Current Therapy in Large Animal Theriogenology. Younquist RS (ed). WB Saunders, Pennsylvania, 1997; pp. 3–12.

Sánchez R, Schill WB. Influence of incubation time/ temperature on acrosome reaction/sperm penetration assay. Arch Androl 1991;27:35–42.

Sánchez R, Isachenko V, Petrunkina AM, Risopatrón J, Schulz M, Isachenko E. Live Birth after Intrauterine Insemination with Spermatozoa from an Oligo-Astheno-Zoospermic Patient Vitrified Without Permeable Cryoprotectants. J Androl 2011a, [Epub ahead of print]

Sánchez R, Risopatrón J, Schulz M, Villegas J, Isachenko V, Kreinberg R, Isachenko E. Canine sperm vitrification with sucrose: effect on sperm function. Andrologia 2011b;43:233-41.

Santiani A, Risopatrón J, Sepúlveda N, Sánchez R. Efecto de la lisofosfatidilcolina en la reaccio´n del acrosoma de espermatozoides caninos. Rev Cient FCV-LUZ 2004;14:311–316.

Sawetawan C, Bruns ES, Prins GS. Improvement of post-thaw sperm motility in poor quality human semen. Fertil Steril 1993;60:706-10.

Schaffner CS. Longevity of fowl spermatozoa in frozen condition, Science 1942;96:337.

Schuffner A, Morshedi M, Oehninger S. Cryopreservation of fractionated, highly motile human spermatozoa: effect on membrane phosphatidylserine externalization and lipid peroxidation. Hum Reprod 2001;16:2148-53.

Silva ME, Berland M. Vitrificacio´n de blastocitos bovinos producidos in vitro con el me´todo Open Pulled Straw (OPS): primer reporte. Arch Med Vet 2004;36:79–85.

Smiley ST, Reers M, Mottola-Hartshorn C, Lin M, Chen A, Smith TW, Steele GD Jr, Chen LB. Intracellular heterogeneity in mitochondrial membrane potentials revealed by a J-aggregate-forming lipophilic cation JC-1. Proc Natl Acad Sci USA 1991;88:3671–3675.

Sommer M. Acuicultura Insostenible en Chile (unsustainable aquaculture in Chile). Rev Electr Vet 2009;10:1–23, www.veterinaria.org/revistas/redvet/n030309.html.

Stanic P, Tandara M, Sonicki Z, Simunic V, Radakovic B, Suchanek E. Comparison of protective media and freezing techniques for cryopreservation of human semen. Eur J Obstet Gyn RB 2000;91:65-70.

Steer CV, Mills CL, Tan SL, Campbell S, Edwards RG. The cumulative embryo score: a predictive embryo scoring technique to select the opti-mal number of embryos to transfer in an in-vitro fertilization and embryo transfer programme. Hum Reprod 1992;7:117-119.

Trus-Cott B, Idler D, Hoyle R, Freeman H. Sub-zero preservation of Atlantic salmon sperm. J Fish Res Board Can 1968;25:363–372.

Tsutsui T, Tezuka T, Mikasa Y, Sugisawa H, Kirihara N, Hori T, Kawakami E. Artificial insemination with canine semen stored at a low temperature. J Vet Med Sci 2003;65:307– 312.

Tuset VM, Dietrich GJ, Wojtczak M, Słowińska M, de Monserrat J, Ciereszko A. Relationships between morphology, motility and fertilization capacity in rainbow trout (Oncorhynchus mykiss) spermatozoa. J Appl Ichthyol 2008;24:393–397.

Twigg JP, Irving DS, Aitken RJ. Oxidative damage to DNA in human spermatozoa does not preclude pronucleus formation at intracytoplasmic sperm injection. Hum Reprod 1998; 13:1864–1871.

Vadnais ML, Roberts KP. Seminal plasma proteins inhibit in vitro- and cooling-induced capacitation in boar spermatozoa. Reprod Fertil Dev. 2010;22:893-900.

Villlegas J, Schulz M, Vallejos V, Henkel R, Miska W, Sánchez R. Indirect immunofluorescence using monoclonal antibodies for the detection of leukocytospermia: comparison with peroxidase staining. Andrologia 2002;34: 69–73.

Wakayama T, Whittinhgam DG, Yanagimachi R. Production of normal offspring from mouse oocytes injected with spermatozoa cryopreserved with or without cryoprotection. J Reprod Fertil 1998;112:11-7.

Ward MA, Kaneko T, Kusakabe H, Biggers JD, Whittingham DG, Yanagimachi R. Long-term preservation of mouse spermatozoa after freeze-drying and freezing without cryoprotection. Biol Reprod 2003; 69:2100-2108.

Watson P, Plummer JM. The responses of boar sperm membranes to cold shock and cooling. In: Deep Freezing of Boar Semen. Johnson LA, Larsson K (eds). Swedish University of Agricultural Sciences, Uppsala, Sweden, 1985; pp 113–127.

Wheeler PA, Thorgaard GH. Cryopteservation of rainbow trout semen in large straws. Aquaculture 1991;93:95– 100.

World Health Organization. Laboratory Manual for the Examination of Human Semen–Cervical Mucus Interaction, 4th ed. Cambridge University Press,New York, 1999.

Yildiz C, Ottaviani P, Law N, Ayearst R, Liu L, McKerlie C. Effects of cryopreservation on sperm quality, nuclear DNA integrity, in vitro fertilization, and in vitro embryo development in the mouse. Reproduction 2007;133:585– 595.

Yin HZ, Seibel MM. Human sperm cryobanking. Use of modified liquid nitrogen vapor. J Reprod Med 1999;44:87-90.

Zamzami N, Marchetti P, Castedo M, Zanin C, Vayssiere JL, Petit PX, Kroemer G. Reduction in mitochondrial potential constitutes an early irreversible step of programmed lymphocyte death in vivo. J Exp Med 1995;181:1661– 1672.

Zhang, Y. Z., S. C. Zhang, X. Z. Liu, Y. Y. Xu, C. L. Wang, M. S. Sawant, J. Li and S. L. Chen, 2003. Cryopreservation of flounder (ParaBchthys olivaceus) sperm with a practical methodology. Theriogenology 60: 989-996.

2

Kinetic Vitrification of Spermatozoa of Vertebrates: What Can We Learn from Nature?

I.I. Katkov** et al.*

CELLTRONIX and Sanford-Burnham Institute for Medical Research,
San Diego, California,
USA

Dedicated to the memory of Father Basile J. Luyet (1897-1974)

1. Introduction

This as well as two other related Chapters, by Isachenko *et al.* and Moskovtsev *et al.*, open this Book neither accidentally nor by the Editor's preferences to his friends and collaborators; the reasons, in fact, lie quite deeper:

Why *sperm*? Cryobiology had actually started from freezing sperm. We will skip all those very early anecdotes but should mention the Spallanzani attempt to freeze frog semen in the 18th century [Spallanzani, 1780]. Cryobiology as a science started with revolutionizing work of Father Luyet and other scientists of the late 1930's and 1940's, who we can collectively call *"the pioneers of the cryobiological frontiers"* (see the following sub-Chapter). There were several reasons why sperm was chosen, which included easiness in obtaining the samples, clear evidence of viability (moving – not moving, though later it was figured that everything was not so easy in this sophisticated living *"cruise missile"*), and importance for the farming industry with the emergence of systematic selective breeding (especially in cattle) with a powerful tool – artificial insemination (AI). AI started with the revolutionary work of W. Heape, I.I. Ivanov and other scientists at the dawn of the 20th century and was further developed by V.K. Milovanov in the 1930's as a viable breeding technology (see [Foote,

* V.F. Bolyukh[2], O.A. Chernetsov[3], P.I. Dudin[4], A.Y. Grigoriev[5], V. Isachenko[6], E. Isachenko[6], A.G.-M. Lulat[7], S.I. Moskovtsev[7,8], M.P. Petrushko[9], V.I. Pinyaev[9], K.M. Sokol[10], Y.I. Sokol[2], A.B. Sushko[3] and I. Yakhnenko[1]

[1] *CELLTRONIX and Sanford-Burnham Institute for Medical Research, San Diego, California, USA*
[2] *Kharkov National Technical University "KhPI", Kharkov, Ukraine*
[3] *Animal Reproduction Center, Kulinichi, Kharkov region, Ukraine*
[4] *Raptor Restoration and Reintroduction Program, National Reserve "Galichya Gora", Voronezh region, Russia*
[5] *Kharkov Zoo, Kharkov, Ukraine*
[6] *Dept. Obstetrics and Gynecology, Ulm University, Germany*
[7] *CReATe Fertility Center, Toronto, Ontario, Canada*
[8] *Dep. Obstetrics and Gynecology, Toronto University, Toronto, Ontario, Canada*
[9] *ART Clinic, Kharkov, Ukraine*
[10] *Kharkov National Medical University, Kharkov, Ukraine*
** Corresponding Author

2002] and [Milovanov, 1962] for detailed history of AI). Whatever case(s) for such specific interest to freezing sperm had been, it was the first subject of systematic research in cryobiology. For a long time after the 1940's, cryopreservation (CP) of sperm would be overshadowed by successes in CP of other types of cells: peripheral blood, blood, embryos, cord blood, stem cells, and other cells, tissues and organs. However, the recent progress and rejuvenation of the old method of sperm vitrification (see following Chapters by Isachenko and Moskovtsev) makes us to believe that it can bring a new shift in the cryobiological paradigm, which we will discuss later in this Chapter.

Why *vitrification*? As we will discuss below, the only method of stable and long-term (practically infinite) preservation and storage of any perishable biological materials, particularly cells, (a.k.a. *"biostabilization"*) is to keep them in the glassy (vitreous) state. This was clearly understood by Father Luyet when he titled his pioneering work "The *vitrification* of organic colloids and of protoplasm" and "Revival of frog's spermatozoa *vitrified* in liquid air" [Luyet & Hodapp, 1938; Luyet, 1937]. He and other *"pioneers of the cryobiological frontiers"* clearly understood 70 years ago that only glassy state would insure stable and non-lethal preservation of cells. With time, we saw the development of a variety of biopreservation methods,such as slow freezing (which, as we will see below, is just a way of achieving glassy state inside the cells and within their close vicinity - cells cannot live neither within ice without a glassy border between cells and ice, or with ice within them). Another method is *equilibrium* vitrification with large amounts of exogenous thickeners (vitrification agents, or VFAs). Eventually, many cryobiologists, especially the new generation and many practitioners, have forgotten that all those methods are basically different ways of achieving vitrification of the intracellular milieu (or at least, without the formation of intracellular types of ice that kill the cells) and the cell's close extracellular vicinity. This has led to several common misconceptions:

- The fact that permeable substances such as glycerol, dimethyl sulfoxide (DMSO), ethylene glycol (EG), propylene glycol (PG or PrOH) and some other small permeable compounds play absolutely different roles during *slow freezing*, when they serve mainly as osmotic buffers and during *vitrification* (VF), when they play the role of thickeners so they increase viscosity and deplete growth and propagation of ice. As a result, in both cases, these substances are called "cryoprotective agents" (CPA's) across the board even though the concentrations used, the modes of addition and elution, and the mechanisms of action are very different for the cases of *slow freezing* (SF) *vs. equilibrium* vitrification (E-VF) and *kinetic* vitrification (K-VF) (we will explain the difference between E-VF and K-VF later). We prefer to distinguish these two roles and call 10% of DMSO used for *slow freezing* of stem cells as *"CPA"* and 40% of DMSO used for *equilibrium* VF of embryos of kidneys as *"VFA"*. As we can see however, for *kinetic* VF, even 10% of glycerol can help vitrify the cells and can be used as the vitrification agent (with some reservation).
- The second misconception that has an even larger implication and can be seen mainly in the work of practitioners is that slow freezing is often called *"cryopreservation"* and is contrasted to vitrification. It is *all* essentially cryopreservation, just by different methods. Moreover, it is actually *vice versa*: slow freezing (*"cryopreservation"* in their terms) is just a way of intracellular *vitrification* with ice being present in the extracellular compartment (see below for details). We can see such erroneous terminology in some Chapters of this Book (especially in Volume 2). The Editor, however, has decided to keep a *democratic* approach and not impose his point of view, thereby letting the reader understand their mistakes after reading this Chapter for future publications. It is the authors' choice to use

incorrect terminology, and as the result, to be a target of criticism in following publications.

- The drastic decrease in the rate of degradation at low temperatures is contributed not *only* (and not mainly) by the simple Arrhenius decreases of the rate of a chemical reaction at lower temperature as all molecules *per se* move slower at lower temperatures even in a vacuum or air as suggested in [Suzuki, 2006]. The practically infinite stability in the vitreous state is achieved *mainly* due to the enormous increase in *viscosity* of the surrounding milieu, which at the glass transition point is determined as $10^{13.6}$ Pa x sec. At such conditions, according to the Einstein-Stokes Law, the destructive molecules such as reactive oxygen species can reach a biomolecule in time that is longer than the age of the universe [Katkov & Levine, 2004]. This is true in the opposite way as well; the degradation of the sperm after freeze-drying at different temperatures, as observed by Suzuki, had occurred mainly because the cells were kept at some level above the crucial temperature of the glass transition (T_g): As higher the cells are kept above T_g, as more soft (rubbery) and later liquid the sample became, therefore the cells degraded more rapidly. We can judge from Fig. 1 in Suzuki's paper that intracellular T_g was above -80°C but below +4°C, a typical scenario for *lyophilization* of sperm and other cells.

Why *kinetic*? As we will also discuss below, the modern shift from Fahy's *equilibrium* back to Luyet's *kinetic* vitrification has brought not only clear technical advantages and better survival of oocytes and embryos. The resurrection and successful re-emergence by the Isachenkos of the Luyet's method in regards to the very subject he and other *"pioneers of the cryobiological frontiers"* attempted to preserve more than 70 years ago - the sperm, has not only brought a simple and convenient technique to the field of assisted reproduction (human spermatozoa first, then animal ones followed). As we can see later in this Chapter, both success of K-VF for some species of sperm and failure of the same method for the others would prompt us to a more general idea: the *"Universal Cryopreservation Protocol"*, which could have a much broader impact and if realized physically by a new type of cryogenic devices that would insure hyperfast cooling and warming, it would shift the whole cryopreservation paradigm. We feel that we will soon witness some sort of a *"Kinetic Vitrification Spring"* as to draw a political analogy, and that is why we have put these three Chapters at the spearhead of the Book.

In this Chapter, we summarize the basic thermodynamical and biophysical distinctions between K-VF , E-VF , slow freezing (SF), analyze present and predict future developments that will widen the K-VF niche, and hypothesize why K-VF of some species of sperm was more successful than the others. We then briefly explore our idea that with the development of a new generation of hyper-fast cooling devices (up to several hundred of thousand °C/min), we will witness the *"Race for the Pace"* for the *Universal Cryopreservation Protocol* without any exogenous VFA's that can be applicable to *any* cell type.

2. Brief history of kinetic vitrification of sperm and cryobiology in general related to the goal of this Chapter

2.1 Early attempts of *kinetic* vitrification of sperm and other cells

In the dawn of cryopreservation, vitrification of small samples by ultra-fast cooling (tens of thousands °C/min) without additional thickening and ice-blocking agents (VFAs), which is referred here as kinetic VF, had been considered as the major method of cryopreservation at that time [Graevsky, 1948a, b; Graevsky & Medvedeva, 1948; Hoagland & Pincus, 1942; Jahnel, 1938; Luyet & Hodapp, 1938; Luyet, 1937; Park *et al.*, 2004; Schaffner, 1942]. Note that some authors contributed the first understanding of the importance of vitrification for biopreservation to a an earlier work of Walter Stiles [Stiles, 1930], as it, for example, is done in [Fahy & Rall, 2007]; we think, the Stiles's notion however was vague and had had a marginal impact. It was Luyet's work, which would make cryopreservation a *science*. From the outset, he recognized that ice damage must be avoided and vitrification could be a method for long-term preservation of cell viability [Luyet, 1937]. In 1938 Luyet and Hodapp achieved survival of frog spermatozoa vitrified by plunging into liquid air [Luyet & Hodapp, 1938], and later several Western European groups reported their experiences with attempts in kinetic vitrification of fowl [Schaffner, 1942], human [Hoagland & Pincus, 1942; Jahnel, 1938; Parkes, 1945], and rabbit spermatozoa [Hoagland & Pincus, 1942] with varying success. While not directly related to the K-VF of sperm, a clear notion of vitrification as the only way of viable stabilization of cells has been expressed by Graevsky in USSR. As he worked with bacteria, it was natural to use a bacterial sample collection loop to freeze the cells in thin pellicles [Graevsky, 1948a, b]. A similar approach was used by Hoagland and Pincus in Germany in 1942 [Hoagland & Pincus, 1942], which seems a very natural approach for very fast K-VF. Yet, in the money-driven 21st century, the term *"Cryoloop"* is a registered as a trademark. Apparently, those early scientists would have infringed the trademark law now!

These early efforts of K-VF of sperm did not receive the recognition they deserved, hindered by the low repeatability and poor survival, as well as difficulties in communication due to various "iron walls" that existed between scientists of the Western Allies, Germany and USSR in the era of WWII followed by the Cold War.

2.2 The rise of slow freezing

The breakthrough came from an independent discovery of the protective role of a permeable CPA glycerol by two groups in 1948-49 [Polge *et al.*, 1949; Smirnov, 1949].The high permeability of glycerol to the sperm membrane in conjunction to relatively low toxicity seemed to be the crucial factor; both groups unsuccessfully tried either non-permeable sugars such as glucose (Parkes's group) or very permeable but very toxic lower alcohols such as ethanol or methanol (Smirnov). The high membrane permeability of glycerol and, thus, fast penetration inside the cells allowed to preserve the cells using slow (10-40 °C/min) freezing, and very moderate warming rates by direct thawing on air or in a water bath. It then became the mainstream of the cryopreservation methods, and a vast variety of cell species of different biological taxa have been preserved by slow (also called *equilibrium*) freezing. It revolutionized two very important fields: the cattle industry (with preservation of bovine sperm and later bovine embryos) and cryopreservation of blood components. It is worth noting that 12 years before the discovery of Parkes's and Smirnov's groups, Bernstein and Petropavlovski had reported the protective role of glycerol during the freezing of sperm [Bernstein & Petropavlovski, 1937] to -20°C, but that work had gone largely unnoticed.

With the development of Peter Mazur's equations and the 2-factor hypothesis of cryodamage [Mazur, 1963; Mazur *et al.*, 1972] and work of other cryobiologists on slow (equilibrium) freezing in 1960's, it became clear that a particular cell would need its own optimal cryopreservation protocol, which would largely depend on the cell cryobiological and physiological parameters as well as on the type of cryoprotective agents (CPA's) used.

Particularly, equilibrium CP of embryos would require much slower pace of cooling (0.3-1 ºC/min) so the whole cryopreservation process would take several hours.

Following cryopreservation of animal spermatozoa, the successful slow freezing of human sperm with glycerol followed, and the first   birth was reported by Sherman and colleagues 1964 [Perloff *et al.*, 1964]. It was then followed by the use of frozen spermatozoa for practically all assisted reproduction techniques (ART) mentioned above. Yet, since his first publications, Sherman had questioned the efficiency of glycerol as the ideal CPA for human spermatozoa [Sherman, 1963, 1964]. The addition and especially removal of permeable osmotically active cryoprotective agents (cryoprotectants) during freezing and warming can induce a lethal mechanical stress *per se*. Further problems include the chemical toxicity of cryoprotectants and the possible negative influence on the genetic apparatus of the mammalian spermatozoa [Gilmore *et al.*, 1997].

2.3 The emerging of *equilibrium* vitrification

On the other hand, Greg Fahy and colleagues [Fahy *et al.*, 1984] reported the vitrification of a whole organ--a rabbit kidney--using high pressure and *extremely high concentrations* of permeable vitrificants. We will call that approach, which for all intents and purposes will be clarified later, *equilibrium* vitrification. The needs of more quick and robust methods of cryopreservation of mammalian embryos had been clear, since Mazur and colleagues and Wilmut had obtained the "frozen mice" by SF in 1972 [Whittingham *et al.*, 1972; Wilmut, 1972].Plus, Fahy's initial report led to the collaboration between him and W. Rall (former Mazur's student, who specialized in freezing embryos) and the first successful vitrification of mouse embryos was reported a year after Fahy's first report [Rall & Fahy, 1985]. The first human baby from a vitrified embryo was reported in 1990 ?? [Gordts *et al.*, 1990]. Since then, vitrification has become an equally spread assisted reproduction technique (ART) as programmed slow freezing of embryos and, especially, oocytes for *in vitro* fertility (IVF) (see [Rezazadeh *et al.*, 2009] for examples and background).

For detailed state of the art of vitrification of reproduction cells, see several Chapters of this Book and Book 2, as well an excellent book by Tucker and Lieberman [Tucker & Liebermann, 2007]. Several Chapters in that book will be referred throughout this Chapter as well. Particularly, an interesting history and even possible natural occurrence of E-VF in nature is described in the Chapter 1A of that book by Fahy and Rall (*"Certain Alaskan beetles dehydrate sufficiently to generate concentrations of up to 10 mol/L of endogenous glycerol,26 which is enough to vitrify aqueous solutions under laboratory conditions"*) [Fahy & Rall, 2007]. Note, however that this particular Chapter 1A is substantially biased against K-VF in favor of E-VF, which we will address throughout the following sub-chapters, and toward the founder of the method, Father Luyet, including some far from diplomatic language escapades. That part will be addressed at the end of the Chapter.

2.4 Vitrification *of the majority* of reproductive cells is moving from *equilibrium* to *kinetic approach*

While slow freezing showed its limitations for certain cell types (e.g. oocytes), a new era started when Rall and Fahy vitrified mouse embryos [Rall & Fahy, 1985] using essentially the same high concentrations of vitrificants vitrified by Fahy *et al.* used in its original report [Fahy *et al.*, 1984]. However, such high concentrations (40-60 % v/v) of VFA's such as glycerol, DMSO, and PG are osmotically damaging and chemically toxic so they are intolerable for many cells such as oocytes and spermatozoa, many of which can withstand at best 10-15% DMSO or glycerol. As a result, researchers moved from *equilibrium* VF to much more rapid *kinetic* vitrification that requires much lower concentrations. It is especially clear for CP of oocyte, which cannot tolerate either slow freezing or equilibrium VF apparently due to their cytoskeletal osmotic fragility. To date, many methods and sample carriers have been designed for K-VF of oocytes and embryos, but they all require small sample volumes and precise timing, which makes them vulnerable to technical errors. We will further explore this aspect in a sub-Chapter below.

2.5 Resurrection and rise of *kinetic* vitrification of sperm: the Isachenkos' contribution

The true "second wind" of the *kinetic* VF was brought in with re-discovering of VF of human spermatozoa *without* any exogenous vitrificants (a.k.a., „*cryoprotectants*" even though they play a completely different role than in slow equilibrium freezing) by the Isachenkos and their colleagues. It came with two seminal appears and two presentations in 2002 and 2003, which, as one of the author remembers, stirred a pot and met a lot of resistance and denial from vitrification experts and other prominent "classical" cryobiologists. In 2002, Vladimir and Evegenia Isachenkos and their colleagues reported that human sperm can be vitrified without endogenous vitrificants ('cryoprotectants" as they called it). It worked with the same success or even better than slow freezing [Nawroth *et al.*, 2002], so the Isachenkos showed that it *did* work. Later, Igor Katkov joined the team and tried to explain *why* it actually worked in [Isachenko *et al.*, 2003] and gave a presentation in CRYO-2003 in Coimbra [Katkov *et al.*, 2003]). It was clearly emphasized that at least three factors might have played a crucial role in the successful K-VF of human sperm without exogenous permeating vitrificants: i) small size of the cells, ii) compartmentalization, and iii) high amount and *concentration* of endogenous natural vitrificants such polymers, sugars and nucleotides. We will explore those aspects later in some detail. This quite novel at the time notion is so *"well known"* now that does not even need mentioning the source (e.g., p.649 [Isachenko *et al.*, 2011]); however, it was not so *"obvious"* back in 2003. Here we want to emphasize that despite of skepticism, denial, or even open hostility towards publications and presentations faced by the Isachenkos (and by Katkov as their strong proponent), the method had matured into a *technology*, which proved to be robust and feasible for ART practitioners as well brought food for thoughts to those who works in the realm of basic cryobiology. Most importantly, the results led to the birth of healthy babies and at least one group has repeated the Isachenko method and has obtained good results completely independently- they report their data in Chapter 3 [Moskovtsev *et al.*, 2012]. The authors dedicate a separate Chapter 2 in this Book for summarizing their achievements [Isachenko *et al.*, 2012]. Below, we not only briefly explore progress of the method but also show that even as the staunchest opponent of the method (more precisely, interpretation of the results) as Dr. Fahy has also evolved in his perception of "legitimacy" of *kinetic* vitrification, which we had never doubted at the beginning.

The Isachenko group has recently expanded K-VF method to other mammalian species (dogs) and to an even more distance vertebrate taxon, the fish (see below and also a separate Chapter [Isachenko *et al.*, 2012]). However, our experiments on K-VF of sperm of rodents was not so successful, and attempts of kinetic VF of sperm of the polar bear and raptor birds (falcons and eagles) failed completely, which actually would prompt us to an even more interesting hypothesis of *"The Universal Preservation Protocol"* and prediction of the *"Race for the Pace"*.

3. Five basic methods of long-term cell biostabilization: *pro's* & *con's*

3.1 All five basic methods of long-term biostabilization cell requires vitrification of the intracellular milieu

We have defined 5 *major ways* of cell stabilization that all lead to low- or high-temperature VF of intracellular milieu as we outlined in [Katkov *et al.*, 2006], which are shown on a schematic phase diagram (**Fig. 1**) adapted from [Devireddy & Thirumala, 2011] with some corrections and additions.

Equilibrium (slow) freezing (points A-B' in green) allows to freeze-out the bulk of both extracellular and intracellular water (which escapes from the cell as the extracellular liquid phase becomes more and more concentrated) to ice. Finally, the cells are vitrified in the inter-ice "channels" that are surrounded by ice but always make a connected network (due to barometric restrictions) and surrounded by ice. Yet, the glass transition temperature in those channels is still low so the cells must be stored in LN_2 at -196 ℃, in nitrogen vapor, or in industrials freezers at -130℃ and for a limited time at higher temperature than the T_g of water (around -136℃), for example in more accessible -80℃ freezers. This is the mainstream conventional cryopreservation, which in the majority of cases requires the use of permeable and impermeable cryoprotective agents (CPAs).

Ice-free *equilibrium* vitrification (E-VF) of cell suspensions, tissues, and organs at very low temperatures and moderate to high rates of freezing (points E-F in red). This method requires the use of high concentrations of vitrificants, which elevates the viscosity of the milieu and prevents the ice formation during cooling and de-vitrification (sometimes called re-crystallization, which is not exactly the same) during warming. Some researchers [Fahy & Rall, 2007] refer to this method as *"vitrification proper"*, and in its "pure form" (see below) has had very limited success in preserving animal oocytes, embryos, some tissues and *one* organ, as well as some plant specimens.

Intracellular ice-free *kinetic* vitrification of a bulk solution by very fast (abrupt) plunging into a cooling agent such as liquid nitrogen (points G-H in purple). The extremely high rate of cooling (10^4–10^6 ℃/min) and practically instant warming prevents ice formation inside the cells (the ice still can be formed outside but it has no time to cause any osmotic damage to the cells as K-VF occurs in fractions of a second). As the result, it does not require the use of potentially toxic high concentrations of "CPAs" (vitrificants) or no permeable exogenous vitrificants at all, it is referred to as *"CPA-free vitrification"* by the Isachenkos in regards to sperm. We deliberately include in this method cooling of sperm at much lower rates because the very high T_g of the intracellular milieu does not require such high rates. This is one of the major themes of this chapter.

Slow freezing to moderately low (around -40 ℃ -- -60℃) temperatures, which comprises two steps; i) primary drying - sublimation of the bulk of ice at very high vacuum (points A-D, and ii) secondary drying of the 'cake' at elevated (up to +30-40 ℃) temperatures (points D-C). This method is called *lyophilization* and it is widely used in food production, microbiology and in the pharmaceutical industry; but so far it has had very limited applications in the preservation of *animal* cells and higher plants.

High temperature vitrification of a highly dehydrated sample (desiccation) and its stabilization by air/vacuum drying at temperatures above ^{o}C is so no ice is formed (points A-C in orange). In some sources, it also erroneously called *"lyopreservation"* [Chakraborty *et al.*, 2011], which is incorrect as *"lyo"* implies sublimation (Greek *luien*- loosing of ice during sublimation (http://dictionary.reference.com/browse/lyo-). In contrast, the Greek word *xero* means *"dry"* (http://dictionary.reference.com/browse/xero-), thus *"xerophile organisms"*, or even the *Xerox* machine! Subsequently, *xeroportective* agents such as trehalose are often used to prevent damage associated with high levels of dehydration when it is used in secondary drying during freeze-drying, and during the whole desiccation cycle. Note that the temperature of drying T_{dr} is *always* above the glass transition temperature of the sample T_g (blue curve) for both methods on definition (otherwise, neither sublimation nor evaporation will occur due to extremely high viscosity). For stable storage on another hand, the temperature of storage T_{st} must be *below* T_g, so the conditions of stable drying are following $T_{st} < T_g < T_{dr(f)}$ (final tememperature of drying). Many papers on drying of biologicals report T_g above T_{dr}, which is incorrect (see [Katkov & Levine, 2004] for details and possible explanation of such *"paradox"*). It can explain instability of samples at long storage [Suzuki, 2006] that are often claimed to have T_g +60-70 ^{o}C while in fact they barely exceed $0^{o}C$ or fall within the negative range and cannot be long-term stored at ambient temperatures.

The first three methods imply the low temperature and thus, are in the scope of these two books *("cryo"* means cold). Biostabilization above $0^{o}C$ is often considered as a part of the preservation science and traditionally reported on the cryo-meeting and published in the specialized journals such as *"Cryobiology"*, *"CryoLetters"* and *"Problems of Cryobiology and Cryomedicine "*(a bilingual journal of the Institute for Cryobiology in Kharkov, Ukraine). We deliberately excluded those topics from the scope of our Books as they need special consideration; nonetheless, we will briefly discuss some aspects below.

3.2 At present, desiccation and especially lyophilization can *not* be considered as major approaches for biostabilization of *viable* cells

Despite the reports of "successful" xeropreservation and lyophilization of live vertebrate cells from time to time by many groups including prominent cryobiologists since the end of 1940's, with three notable reports of Meryman and the birth of a cow called *"Desicca"*(see [Suzuki, 2006] for an excellent mini-review on the topic), it turned out that neither of the methods to date have proven to produce *stable* and *viable* cells that could be stored for *long* periods of time. It mainly contributes to the fact that even for such good vitrificants, such as proteins, achieving a true high T_g coincides with very low water content (in a range of 0.3 g H_2O per 1 g dry weight), which apparently is not sustainable by vertebrate cells insofar. Whether the very recent reports by Devireddy and Thirumala [Devireddy & Thirumala, 2011] and by the Mehmet Toner's group [Chakraborty *et al.*, 2011]will change the situation, or they will fade away as all the proposed methods have so far needs to be seen. Our approach is expressed in [Katkov & Levine, 2004; Katkov *et al.*, 2006; Katkov, 2008]. The discussion of what has been done wrong so far, and what could and should be done, would need a separate Chapter, and as we said before, is out of the scope of this Book. However, there are two things that should be mentioned.

3.2.1 Freeze-drying/desiccation of spermatozoa has not produced *motile and viable cells* but it fits for intracellular sperm injection (ICSI) as it stabilizes the nucleus

First, we have to remember that *"successful* (i.e., bringing offspring) freeze-drying or drying of spermatozoa" is a confusing and actually misleading statement. The *properly* freeze-dried/desiccated spermatozoa are dead, they are never motile, and neither do they have intact acrosome (in the majority of cases). It is the genetic apparatuses, which include such excellent endogenous vitrificants as proteins (e.g., histones), and at lesser extent, DNA that indeed can be stabilized at high temperatures (above 0°C) for long time by xeropreservation (preferably) or by lyophilization (if secondary drying is done properly). Naturally, if intracellular sperm injection (ICSI) is performed, both methods can and do bring offspring (see [Suzuki, 2006] for references).

3.2.2 On reasoning of creating *"xerobanks"* of dried genetic material

Secondly, as the nucleus of somatic cells can be kept intact after desiccation, it (theoretically) can be used for cloning by somatic-cell intracellular nuclear transfer (SCNT). So, those two aspects, ICSI and SCNT raise the question whether the xerobanks of both gametes and somatic cells should be created for human, model (laboratory), agricultural and wildlife species. We personally believe (though it might change with the time) that except for xerobanks of sperm of laboratory animals, such as transgenic mice and rats, for which both ICSI and SCNT have been well established [Katkov, 2008], the other types of xerobanks are not a necessity, and people should focus their resources and money (which are often scarce in this field) on the methods that have been proven to produce viable cells (i.e. on cryopreservation). In situations where the cold chain is not as easily available (for example, for the preservation of a genome of species that are on the verge of extinction), drying could be considered as the last resort, but for now, it should not be considered as an alternative to cryobanks. That might change where ICSI and SCNT become routine for many species, but so far we should concentrate on CP. And again, it is gametes, embryos and other reproductive cells that should be preserved first to save genetic material of endangered species even after their death [Maksudov *et al.*, 2009] while, for example, the CP of stem and other somatic cells should be kept as the last resort when the reproductive cells are unavailable. Note, that some other authors of this Book are much more optimistic on that matter of both drying (e.g., the Chapter by Joseph Saragusty ([Saragusty, 2012] sub-chapters 2.3 and 4.2), and cryobanks of stem cells for restoration of species ([Saragusty, 2012], sub-chapter 4.3).

3.3 Slow freezing: Still the mainstream of cryopreservation but...

As we mentioned above, the discovery of "enigmatic glycerol" [Polge *et al.*, 1949; Smirnov, 1949] led to the explosion of methods of cryopreservation and types of species cryopreserved and development of the first cryobanks that marked the 1950's. It revolutionized first the cattle industry, than blood transfusion and many others followed. However, while many of them being successful, the method *per se* remained semi-empirical. However, it has changed with introduction of the 2-factor hypothesis and the equations for the equilibrium slow freezing (minimal intracellular ice formation) by Peter Mazur [Mazur, 1963; Mazur *et al.*, 1972]. Using this truly fundamental approach, Mazur and colleagues in USA and Ian Wilmut in UK were be able to cryopreserve the mouse embryo [Whittingham *et al.*, 1972; Wilmut, 1972]. Since then, slow freezing has been the mainstream of modern cryobiology, and while VF is an emerging method that will one day replace SF for many types of cells it has not been done yet: right now SF is an imperative for the majority of cell types.

With the development of Peter Mazur's equations and the 2-factor hypothesis of cryodamage and work of other cryobiologists on slow (equilibrium) freezing in 1960s, it became clear that a particular cell would need its own optimal cryopreservation protocol, which would largely depend on the cell cryobiological and physiological parameters as well as on the type of cryoprotective agents (CPAs) used. Particularly, equilibrium freezing of embryos would require very slow pace of cooling (0.3-1 $^{\circ}$C/min) so the whole cryopreservation process would take several hours. In contrast, for small oblate (flat) ellipsoids such as the red blood cells (RBCs) with an excellent surface-to area ratio, which would allow them to lose water very quickly, the optimal freezing rate of cooling would be in the range of several thousand $^{\circ}$C/min. Thus, if we consider an intermediate cooling rate, say 10 $^{\circ}$C/min, it would kill oocytes at a very fast rate due to the intracellular ice formation (IIF). But the same cooling rate is too slow for RBC's so they will be dead, due to excessive shrinkage and prolonged degradation ("solute effects"). Yet, for lymphocytes, which are intermediate between oocytes and RBC's that rate would be optimal.

Addition of a CPA shifts the survival curve toward the lower rate and higher survival, which indicates that the CPA protects mostly during suboptimal cooling (see Fig. 2) acting as an osmotic buffer that prevents excessive shrinkage and other "solute effects"[Lovelock, 1953; Mazur, 1970, 1984; Mazur & Koshimoto, 2002]. The effect of protective action of the CPA is much more pronounced for larger bone marrow cells while small erythrocytes perfectly survive the absence of CPA if cooled fast enough. The optimal concentration, however, is in the same magnitude of 1- 2 M. Note that the mechanism of cryoprotective action of CPAs such as glycerol or DMSO at slow sub-optimal cooling rates is absolutely different and works at much lower concentrations than their role as vitrificants ("thickeners") that elevate the viscosity during vitrification (VF). From that standpoint, they should NOT be called "CPAs" but rather "VFAs" in case of VF.

Thus, the optimal ("*maximum maximorum*") concentration of the CPA (more precisely, the combination of concentration of CPA *and* rate of cooling) are unique for a particular type of cells. These two concepts are illustrated in Figures 2 and 3. The bottom line is that SF often needs elaborate multi-step protocols, whichrequires special equipment, and it can do exceptionally well, especially if combined with other "tricks" that are specific to the particular species of cells. A good, recent example is the CP of human pluripotent (embryonic and induced alike) stem cells (hESc's and iPSC's respectively). Introduction of i) multi-step freezing, ii) ROCK inhibitors in combination with full cell dissociation, and iii) freezing pluripotent SC's in adherent stage as they are prone to *anoikis* (cell death after cell are detached from extracellular matrix, [Wagh *et al.*, 2011]) have dramatically increased survival and functionality of human pluripotent cells after SF ([Katkov *et al.*, 2011; Li *et al.*, 2008; Martin-Ibanez *et al.*, 2009; Mollamohammadi *et al.*, 2009; Stubban *et al.*, 2007; Ware & Baran, 2007], see Chapter by Martin-Ibanez [Martin-Ibanez, 2012] in this Book). It now highly supersedes various vitrification techniques proposed from time to time [Beier *et al.*, 2011; Reubinoff *et al.*, 2001; Zhou *et al.*, 2004] despite what is claimed otherwise by the authors.

However, the strengths of SF freezing can be its weaknesses as well: it indeed needs *elaborative* protocols that have to be developed for each new species of cells separately. Secondly, it is difficult to implement for CP of large chunks if tissues, and especially if we are talking about CP of a whole organ. Yet, the methods and equipment are being developed, see a chapter by Butler in our Book [Butler & Pegg, 2012].

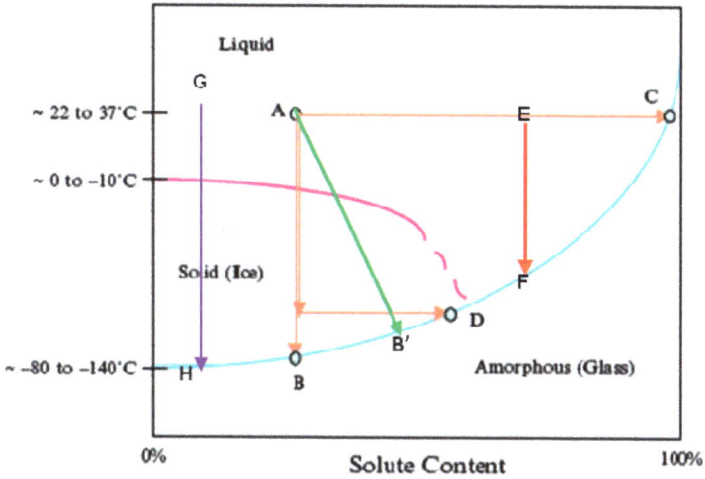

Fig. 1. Five ways of vitrification: A corrected and supplemented phase diagram adapted from [Devireddy & Thirumala, 2011].

Light Blue line represents the glass temperature T_g curve of the sample.

1. Points **A-B'** (**green**): *slow equilibrium freezing*, often called *cryopreservation per se*. Note, that the solute concentration is dynamically changing during freezing of extracellular ice so the original authors' line **A-B** (**orange**) is substituted by **A-B'** (**green**).

2. Points **E-F** (**red**): *equilibrium vitrification* (often referred as VF per se). The very viscous solution of the permeable vitrificant (solute) prevents the formation and/or growth of both intracellular and *extracellular* ice the sample can vitrify without the ice phase practically at *any rate* of cooling and warming (the E-F is locate at higher concentration than the line of freezing (heterogeneous ice nucleation) shows in a sketchy form in magneto color crosses the T_g line at lower concentrations and only two phases, amorphous and liquid, exist on the right side of the x-axis.

3. Points **G-H** (**purple**): *kinetic vitrification* that occurs *intracellularly* at a much lower concentration of the vitrificant or even without permeable VFA. This however, requires much higher rates of cooling and warming so the damaging ice crystals cannot be formed during rapid cooling and re-crystallization (de-vitrification) will be blocked and, and thus, will not damage the cells during very fast warming.

See also other set of phase diagrams in the Fig. 4 and explanation in the text.

4. Points **A-D-C** (**orange**): *freeze-drying (lyophilization)* (not **A-D**, as originally is stated in [Devireddy & Thirumala, 2011]). **A-D** (**orange**) represents freezing and sublimation of ice (primary drying) followed by elevation of temperature of drying above ^{O}C (secondary drying) **D-C**.

5. Points **A-C** (**orange**): *desiccation (xeropreservation)* is either vacuum or air/humidity chamber drying where the temperature of drying is always above ^{O}C so no freezing phase is present. Note that the temperature drying T_{dr} is *always* above the glass transition temperature of the sample T_g (blue curve) on definition (otherwise, evaporation will not occur due to extremely high viscosity), while for stable storage, the temperature of storage T_{st} must be *below* Tg, so the conditions of stable drying are following $T_{st} < Tg < T_{dr}$ (final). Many papers on drying of biological reports T_g above T_{dr}, which as incorrect, see [Katkov and Levine, T_g] for details and possible explanation of such "paradox". It can explain instability of samples at long storage [Suzuki, 2006] that are often claimed to have T_g +60-70 ^{O}C.

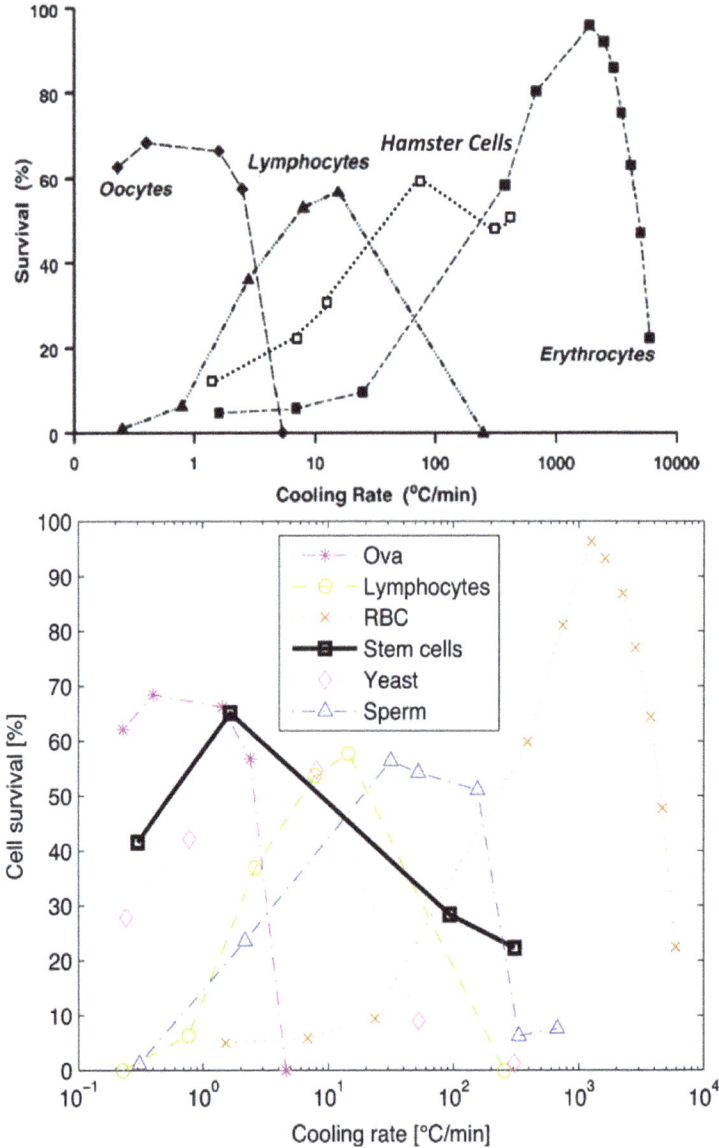

Fig. 2. The two-factor hypothesis of the cryoinjury by Peter Mazur: survival of cells of different size (oocytes >> lymphocytes > hamster cells >> erythrocytes) as function of the cooling rate. **Top**: Mazur's original graph, adapted from [Mazur *et al.*, 1972; Mazur *et al.*, 2008]. Bottom: Updated for stem cells (large size), yeast moderate) and sperm (slow) in [Cipri *et al.*, 2010]. Note that "slow (sub-optimal) and "fast" (supra-optimal) freezing in this case largely depends on the cells size: e.g., the rate of cooling 10 °C/min is very fast for oocytes (lethal IIF), very slow for erythrocytes ('damage due to the "solute effects") and close to the optimal for lymphocytes.

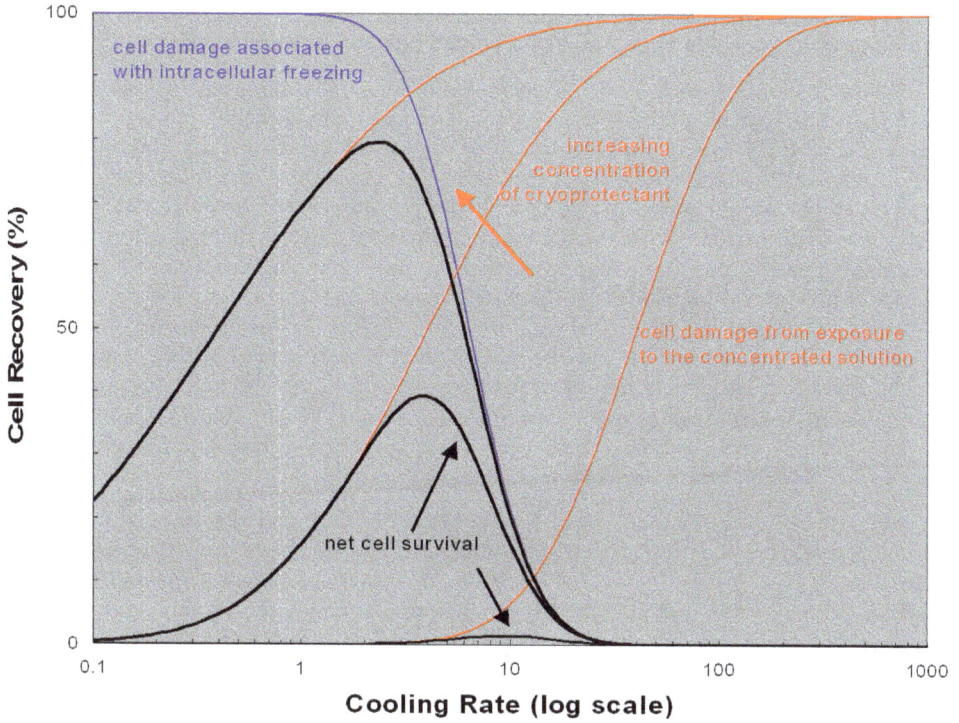

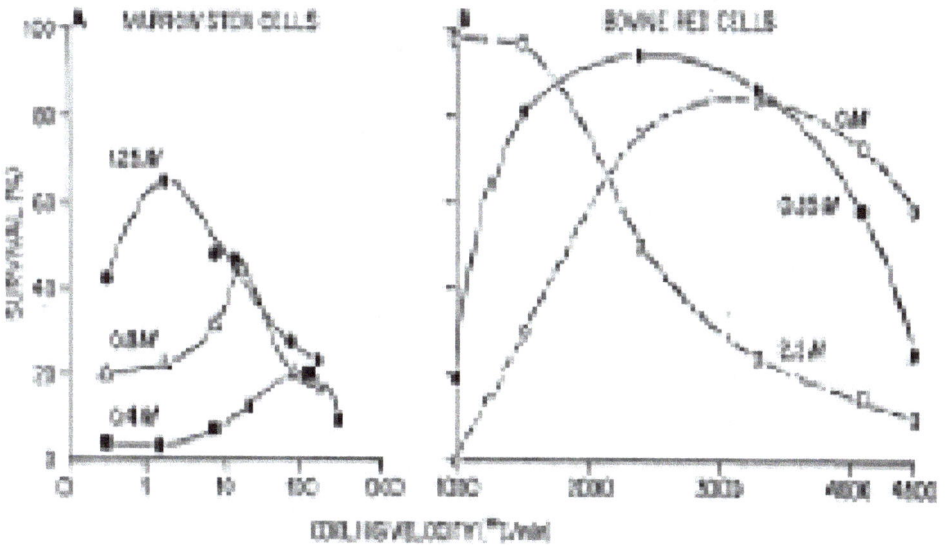

Fig. 3. The role of a cryoprotective agent (CPA) at slow freezing: Survival of cells of different size (marrow cells, the left panel >> erythrocytes on the right) as function of CPA concentration and cooling rate. Adapted from [Mazur, 1970].

3.4 Equilibrium vitrification and *"magic"* ice blockers: True 21st century medicine or *"Fahy's tyranny"* and the spearhead of cryonics pseudo-science?

On the other hand, Greg Fahy and colleagues [Fahy *et al.*, 1984] reported vitrification of the whole organ (i.e. kidney), and later report E-VF of mouse embryo by Bill Rall and Greg Fahy [Rall & Fahy, 1985]. Since, the fate of VF of these two types of cells and fields split dramatically: E-VF of the whole organ has been essentially *stuck in the rut*, with *very few progress*, that has been reported mostly by the Fahy's group *per se* [Fahy *et al.*, 2009] despite of 25 years of research and substantial amount of financial support that the author received from many sources including taxpayers money. For example, accordingly the Fahy's company with a promising name *"21st Century Medicine"(21 CM)*, posted on their Wikipedia site *"In 2004 21CM received a $900,000 grant from the U.S. National Institutes of Health (NIH) to develop solutions and processes to improve human heart transplantations"* [Wikipedia, 2011a]. Since 1+ million dollars (including previous Phase I) in funding and eight years after that announcement, we have not found any progress report or reliable publication on that topic from the company's scientists in scientific peer-reviewed journal. The vitrification of a heart (even an animal one) is not even close to realization apparently.

The company and its scientific team heavily rely on so called *"ice-blockers"*, chemical substances that block the propagation of ice in big samples cooled very slowly, thus helping vitrification. The company has made progress in the development a pipeline of such reagents. However, they are used mostly as "helpers" to lower the osmotic and chemical toxicity of the enormous concentrations of "common" vitrificants that are necessary for equilibrium (slow vitrification). Whether that approach will ever meet real progress in the remaining 88 years of the *21st century medicine*, needs to be seen.

Nonetheless, Dr. Fahy has been very proactive in promoting *equilibrium* vitrification and denying *kinetic* one whenever and wherever it is possible. He basically ignores and calls it *"quasi vitrification"*(e.g., in [Fahy & Rall, 2007]), and in doing so he contradicts himself within three pages of his own review [Fahy & Rall, 2007]! In Fig. 1A.1, he placed the start of citations on vitrification of cells and organs. Of course, he starts counting from his publication 1984 totally ignoring the *earlier* work of Luyet, Boutron, Farrant and other scientists, the very work that Fahy is discussing is a couple of sub-chapters later. Yet, it was he, *"the world's foremost expert in cryopreservation by vitrification"* (http://en.wikipedia.org/wiki/Twenty-First_Century_Medicine), who "truly" vitrified cells *first*. We will come back to this attitude a bit later when we compare E-VF and K-VF. Now, we only say that while his chapter in that book is #1A, the majority of the next 19 Chapters in fact describe various *kinetic* vitrification techniques with small size and fast cooling and warming, a typical pattern of K-VF. Few people even mentioned the term ice-blockers, fewer used it in reproduction practice, mainly as "helpers" (see our explanation above).

Who has been truly benefitting from Fahy's and his colleagues work? The people that have been engaged in a pseudo-scientific activity called *'cryonics'* (http://en.wikipedia.org/wiki/Cryonics) They freeze deceased people, or sometimes even just their heads or brains (as did Saul Kent (http://en.wikipedia.org/wiki/Saul_Kent), the founder of the "21st CM" (http://www.biomarkerinc.com/saul_kent_page.html) in the hope that one day the dead will be *"resurrected"* (?!), or even that the brain can be somehow *'translocated'* into a new body. This is at least science fiction and naive beliefs (a type of *"transhumanism"*) and at most a charlatanic snake oil scheme aimed *"to skim off big bucks"* from the human tragedy so it has as much in common with cryobiology as astrology with astronomy or alchemistry with chemistry. Not surprisingly, cryonicists are banned from publication in all scientific cryobiological journals and from the membership in the cryo-societies as their activities have nothing to do with real scientific premises. Yet, they skillfully wrap their messages, post some valid statements, and add some useful websites, for example on physics of glass transition (apparently, they have good physicists among their "disciples") to make cryonics seem like a *legitimate* science And of course, they cite Fahy's and Brian Wowk's work wherever they can. They actually admit that they buy those ice-blockers from the *"21 CM"*. They are very active in Wikipedia so we can see all biographies of prominent cryonicists, and even much less prominent and rather obscure ones like a former bookkeeper Danila Medvedev in Russia [Wikipedia, 2011c], which the company *"KrioRus"* proudly announces how many bodies and other parts of humans (including some brains, which they call *"Neurovitrification"*!), dogs, cats and birds they "vitrified" [Wikipedia, 2011b]. Of course, you can find in Wiki also the biographies of Greg Fahy, Brian Wok, and a detailed description of the *"21 CM"* company. None of these scientists has ever claimed any of the cryonics beliefs openly (they value their scientific carriers as well as an ability to apply for NIH money, for example, which considers cryonics as a pseudo-science), and we don't imply that those cryobiologists and the *current* "21 CM" management are "hidden cryonicists". Moreover, the Company's website clearly distances itself from cryonics (http://www.21cm.com/cryobiology.html). However, the fact that cryonics companies and organizations heavily rely on E-VF and ice-blockers as the major method of preservation and future *resurrection*, their connection, both *"ideological*, (e.g., hiring a *very* controversial Ukrainian scientist and former 21 CM employee Yuri Pichugin) and financial (being presumably valued customers of the *"21 CM"* by buying those ice-blockers) is self-evident (http://www.cryonics.org/century.html,   http://en.wikipedia.org/wiki/Cryonics, http://cryonics.org /yuri.html).

3.5 Modern methods of vitrification of reproductive, stem and other germplasm cells are in the realm of *kinetic* vitrification, but still many questions remain

The fate of the second direction of vitrification, which was initiated with the paper by Rall and Fahy [Rall & Fahy, 1985] on E-VF of embryos, was completely different: it definitely has not been stuck in the rut but rather quite opposite. The use of vitrification for cryopreservation of reproductive cells and tissues has boomed over the last 20+ years since that seminal paper was published. However, the modern methods of VF of oocytes, embryos, sperm, ovarian and testicular tissue are in fact the varieties of *kinetic* vitrification. Elaborative multi-step protocols of the addition of VFA's before VF and elution of them after warming have been developed to decrease the toxic and osmotic effects of vitrification. Some of those methods are covered by other Chapters or our Books 1 and 2 and by the above-mentioned excellent book by Tucker and Lieberman [Tucker & Liebermann, 2007]. Up to now, a vast variety of carriers has been developed as well. They are summarized in an excellent review by Saragusty and Arav [Saragusty & Arav, 2011] and is reproduced on Fig. 4.

While there is still a debate over what is better for a particular cell type or species, slow freezing or kinetic vitrification, the latter one is gaining ground, particularly for VF of oocytes, thanks to ART scientists and practitioners such as Kuwayama, Vajta, Sheldon, Liebermann, Tucker and many others. Note however, that sometime that "cold war" may erupt and evolve into a "hot war" when it comes to which set of VF media, the protocol, and the carrier are better. Thus, while being faster and simpler than slow freezing (though much farther from automation and "full proof"), vitrification at this moment has been struggling basically with the same problem as the SF has been plagued with: each type of cells, the carrier, and VF media need own protocol, and very often a VF media that work for open carries are too diluted for closed carriers, while using open carriers raises concern of contamination etc. The bottom line is that *kinetic* vitrification as it is now, offers a vast variety of the methods, that have to be checked and adjusted when a new type of cells of/or new species of animals are in consideration.

As we can see later, our experience with vitrifcation led us to conclude that it might change soon, but before moving further, we have to look in more detail at the distinction and principal differences between equilibrium and kinetic vitrification from the standpoint of thermodynamics. In other words, we have to look at the supplemental phase diagram, or as we call it here, the "Fahy's diagram" as it was first published and explained in detail from the cryobiological audience by Fahy and colleagues [Fahy *et al.*, 1984].

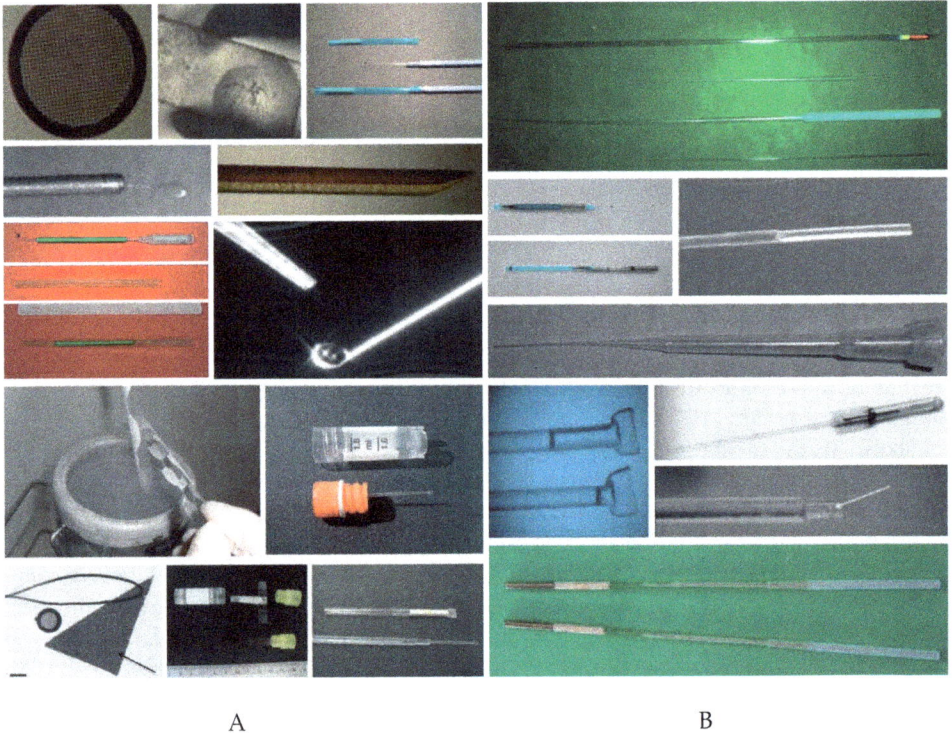

A   B

Fig. 4. Vitrification carrier systems [Saragusty & Arav, 2011].
A: surface carriers. **First raw:** electron microscope grid, minimum drop size; Cryotop; **Second raw:** Cryoloop, Hemi-straw; **Third raw:** Cryoleaf, fiber plug, **Fourth raw**: direct cover VF, VF

spatula; **Fifth raw**: nylon mesh - arrow points at the nylon mesh, plastic blade, Vitri-Inga.
B: tubing carriers. **First raw** (top to the bottom): 0.25-mL mini-straw, 0.25 ml mini-straw,
Open-pulled straw (OPS), Superfine OPS (SOPS), Flexipet-denuding pipette (170 μm end
hole); **Second raw**: CryoTip (opena and loaded), high-security vitrification device; **Third
raw**: pipette tip, **Fourth raw**: sealed pulled straw (left), (Cryopette (top right), Rapid-I
(right-bottom); **Fifth raw**: JY Straws.
See [Saragusty & Arav, 2011] for more details and references.

4. Equilibrium *vs.* kinetic vitrifcation; Evolution of the "Fahy's" phase diagram

This sub-chapter discusses in detail the phase diagram (*"Fahy-Rall"* vitrification diagram).
We will also discuss using this diagram the two basic and reciprocal ways of achieving VF,
which can be done: i) by cooling and warming at relatively moderate rates but very high
concentrations of exogenous (and often toxic) vitrification agents/enhancers (VFAs), which
is defined as *equilibrium* VF and ii) by increasing the rate of cooling with a few or not at all
exogenous VFAs present, which we refer as *kinetic* VF. We will also emphasize that the
border of *"non-achievable"* and *"achievable"* VF that was once set up by Fahy is arbitrary and
largely depends on the currently achievable rates of cooling and warming.

Fig. 5A depicts the *original* diagram published by *Fahy* et al in **1984** [Fahy *et al.*, 1984]. The
diagram is divided in 4 distinctive zones. **Zone IV** is the *equilibrium VF*, when it occurs at
any practical rates of cooling and warming as it lies to the right of the junction of T_m (i.e., no
ice forming during cooling) and T_d (no de-vitrifcation during warming). It is basically the
zone where the line E-F on **Fig. 1** is drawn but with the notion that T_g in the Fahy's diagrams
(apparently, for glycerol) lies substantially lower than in Devireddy's diagram (T_g of a fully
dehydrated sample is well above 0 °C while T_g of glycerol is in range of -90°C and below
[Pouplin *et al.*, 1999]. For T_g's of some popular vitrificants see Table 1 in [Katkov & Levine,
2004]. **Zone III** is the zone when vitrification occurs. The left border is the junction of T_h
(showed in dotted line as it is hardly to estimate T_h of very viscous samples) and glass
transition curve T_g and it occurs at concentrated C_v's, the minimal concentrations where
equilibrium vitrification during cooling occurs at practically any speed. However, such
concentration still may produce de-vitrification during re-warming as the devitrification
curve T_d crosses the melting (equilibrium warming) curve at the critical concentration of
devitrification C_{dv}. Thus, this Zone III is the zone where warming must be done fast.

Zone II, called by Fahy and colleagues at that time (!) *"doubly unstable"* lies at concentration
below C_v. At those concentrations, both cooling and warming must be done fast to avoid ice
formatting and devitrification respectively. That is what we call kinetic vitrification as it
deals with the speed of cooling and re-warming rather than with the equilibrium values. It
means that the border between that Zone II, where vitrification is achievable with the **Zone
I**, where successfully vitrification is impossible at any "reasonable speed" of cooling and
warming largely depends on the rate of that cooling and warming: it is reciprocal to the C_v
and C_{dv} so they move to the left into the area of the lower concentrations.

Thus, **there are 2 basic and reciprocal ways of achieving VF**: i) by raising the
concentration, and as result, the viscosity of the intra- and extracellular milieu at
relatively moderate and even slow rates of cooling but very high concentrations VFAs,

which is defined as *equilibrium VF* and ii) by increasing the rate of cooling with a few or not at all exogenous VFAs present so deleterious intracellular ice formation is not achieved due to lack of time for growing ice crystal nuclei (*kinetic* VF). Note that the border between "*non-achievable*" and "*achievable*" VF (Zones I and II) that was once set up by Fahy is arbitrary and as we said, largely depends on the currently achievable rates of cooling and warming.

The position of the borders between the zones also depends on the glass transition temperature of the solute (T_g curve). As we mentioned above, Fahy *et al.* had considered a permeable vitrificant with very low T_g in range of -90°C (glycerol). In the paper on vitrification of sperm that we published in 2003, we hypothesized that T_g of the intracellular milieu could *be much higher*, so the T_g of the intracellular exogenous solute would go much above of T_g of glycerol. We published a review in 2006 [Katkov *et al.*, 2006] (the abstract was presented much earlier in 2003 [Katkov *et al.*, 2003]) with that concept superimposed onto Fahy's diagram. This concept and an explanation as to why we could vitrify sperm at much less or no exogenous vitrificants at all is shown in **Fig 5B**. We emphasized that the T_g of a internal vitrificants can go very high, so the border between Zones I and II can be shifted substantially to the left so the successful vitrification (straight blue line) is achievable even the extracellular milieu has no glycerol.

This explanation (and at lesser extent the experimental data published at that time *per se*) had been dismissed both by Fahy and by other prominent cryobiologists, most notable of them would be Stan Leibo. They either called it "*quasi-vitrification*" or ignored that such work was published (due to the fact that it was not referred at PubMed, even though the leading author had distributed its copies among numerous scientists in the field). The striking example is the Fahy's chapter [Fahy & Rall, 2007] where he spent a great deal of time bashing kinetic vitrification, giving intricate details as to how its first scientists had failed lately to implement K-VF in practice. In regards to our work, he simply ignored that paper even existed, even though it had been sent it to him and we discussed it with him [Dr. Fahy] in meetings.

So, the "*contemporary vitrificators*" ignored mentioning the paper and notion of kinetic vitrification at very low concentration of the extracellular solute. That however, did not mean they had not learned or gained from it. Neither could they ignore the booming success of K-VF in the assisted reproduction field, which we mentioned above. The whole set of innovation was aimed to cool and warm cells faster, which allowed the ART practitioners to move away from the humongous concentrations of DMSO, EG, PG or glycerol, which would be necessary for equilibrium VF. So, that was actually reflected in the very same chapter published in **2007** [Fahy & Rall, 2007] as it presented a Fahy's diagram with some interesting and *key changes* in comparison to the publication in year **1984** (**Fig. 5C**). We deliberately superimposed those changes on the original Fahy's diagram [Fahy *et al.*, 1984].

First, the words "*doubly unstable*" have been eliminated and it made sense because 99% of publications on vitrification of oocytes and embryos have been done at concentrations that exactly represented the "*doubly unstable*" Zone. That would not make all people who have successfully and *stably* frozen their precious happy to realize that they actually worked in the "*double instability*" zone (and which *is not correct* anyway: the vitrified cells *are stable* at temperatures below of their glass transition, i.e., well below the LN$_2$ temperature).

Secondly, the authors of the chapter shifted the border between *"achievable"* and *"non-achievable"* vitrification (Zones I and II) to the left. It was a small concession to the reality, people have been successfully vitrifying oocytes and other "watery' cells with substantially less concentrations than40%, but yet it reflects the general *"drift to the left"*, so to speak.

Most importantly in our opinion is that the notorious *"killing darts"*, which marked the "unsuccessful vitrification" in the Zone I on the original Fahy's diagram in 1984... suddenly disappeared in a newer version. It means that the authors allowed (at least theoretically) the

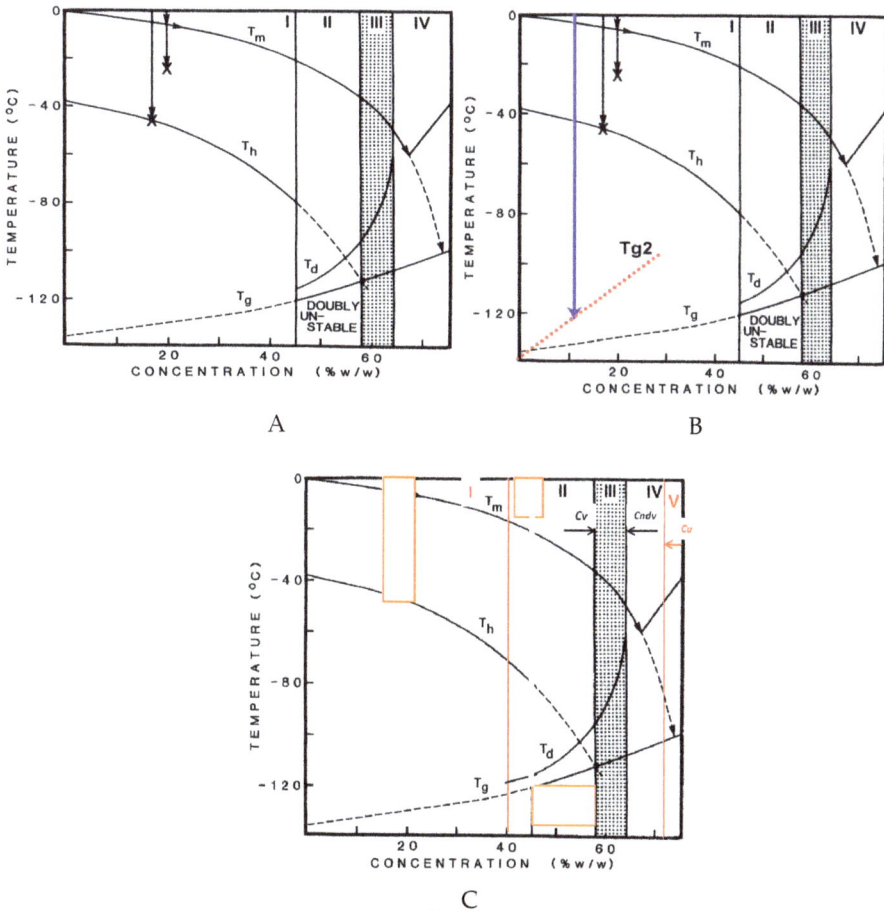

Fig. 5. **A (top):** original diagram published *Fahy* et al. in **1984** [Fahy *et al.*, 1984]. The diagram is divided in 4 distinctive zones. **Zone IV** is the *equilibrium VF*, when it occurs at any practical rates of cooling and warming as it lies to the right of the junction of T_m (i.e., no ice forming during cooling) and T_d (no de-vitrifcation during warming) it is basically the zone

when the line **E-F** on **Fig. 1** is drawn but with the notion that T_g in the Fahy's diagrams (apparently, for glycerol) lies substantially lower than in Devireddy's diagram (T_g of fully anhydrated sample is well above ^OC while for glycerol T_g depicted on this diagram is in range of -90^OC and below [Pouplin *et al.*, 1999]. For T_g's of other popular vitrificants see Table 1 in [Katkov & Levine, 2004]. **Zone I** is the zone of "non-achievable" VF, **Zone II** is *kinetic* VF marked as *"doubly unstable"*, and **Zone III** is an intermediate zone where devitrification must be avoided while VF is achievable at slow rates. Note two *"killing darts"* in Zone I that indicate ice crystallization (vitrification is not achieved).

B (middle): Fahy's diagram supplemented by us in **2006** [Katkov *et al.*, 2006] with the notion that the border between Zones I (unsuccessful VF) and II (successful *kinetic* VF) in diagram **A** is arbitrary and can be moved far left to the area of very low concentrations of external VFA's (or no VFA not at all as in case of human sperm). The blue line indicates successful *kinetic* VF, it is analogous with the **G-H** line on **Fig. 1**.

C (bottom): Fahy's diagram, version **2007** depicted in [Fahy & Rall, 2007] but superimposed by us on the original diagram **A**. Note the following notable changes: i) "disappearance of words "doubly instable"; ii) shifting the border between zones I and II to the left; iii) disappearance of the *"killing darts"* in Zone 1; in addition of **Zone V** (E-VF achievable even with a introduction of exogenous ice: propagation of the ice is stopped).
See the major text for further explanation.

blasphemous idea that vitrification could occur at *any* concentration of the solute, however low it might be. And it is true, even the pure water can also be vitrified, though the rate of vitrification must be in range of tens to hundreds of millions ^OC/min [Angell, 2004]. We can only speculate where all those Zones would go in *that* case. Apparently, they would all disappear! Finally, Fahy and Rall made two crucial concessions in their text (probably, insisted by Bill Rall taking to the account his vast experience and knowledge of the ART field), which we cite below in full:

- *"For the small samples often used in reproductive cryobiology, it becomes important that T_h, T_g, and T_d are all rate dependent (i.e,. T_h will go down, T_g will go up, and T_d will go up as the rate of change of temperature increases) because extremely high cooling and warming rates are feasible."*

- *"However, Figure 1A.5 [depicts relationships between the critical cooling rate of vitrification and concentration of the vitrificant, I.I.K.] is based on pure permeating cryoprotectants in water and does not take into account the effects of the carrier solution (see below) or additives such as serum or sucrose, nor does it take into account the effect of concentrated intracellular protein in shrunken cells or the naturally low water content of cells like spermatozoa"* [of course, no mentioning of our work whatsoever!]

Those two citations exactly explain how *kinetic* vitrification works without even mentioning it! While we are quite accustomed to the that style of ignoring "inconvenient" publications from several prominent cryobilogists and pushing their explanation aside the facts that "adjusted" (with the reality) Fahy's curve together with the two statements above clearly indicate that even as the staunchest orthodox proponents of equilibrium ("right") vitrification as Dr. Fahy could not ignore the facts and explanation why and how the *kinetic* one is working and dominating the scene now. Apparently and evidently they learned from our publication, as well as from the publications of others.

There are other peculiar similarities between that chapter and some of our *earlier* papers, such as use of the equation for determination of the viscosity of the solute near T_g [Katkov & Levine, 2004] and storage below and above T_g of the sample [Katkov *et al.*, 2006]; we would encourage our readers to compare our work and the Fahy's review with the notion that WLF relationship for viscosity near T_g in our work is substituted by an equivalent VTF equation in the Fahy's chapter (see **Appendix 1**).

In conclusion, of these sub-chapters, it is evident that the *kinetic* way dominates the present art of vitrification and all efforts are moving to the direction of increasing speeds and decreasing concentrations (see *"Race for The Pace"* below). On the other hand, the future of *equilibrium* vitrification even in the field where it cannot be substituted by K-VF such as organ CP (but can be done with precision SF as described in a Chapter by Butler and Pegg in this Book [Butler & Pegg, 2012]), remains largely unclear.

7. Kinetic vitrification of sperm: why some species have while others have not been vitrified?

Now, as we are fully equipped to discuss the core topic of the Chapter, let us refresh the turn of (relatively) recent events related to the *kinetic* VF of spermatozoa.

7.1 A turn of the helix: The Isachenkos' experiments on vitrification of human sperm

As we mentioned in the Introductory sub-chapter, 1, after earlier attempts to vitrify sperm with contradictory results, the findings of the cryoprotective role of glycerol and other CPAs at slow freezing moved the field of cryopreservation of spermatozoa from early attempts of K-VF toward E-SF. It has been successfully applied to many types of sperm, yet somewhere in 1990s, the data started accumulating that suggested that glycerol, DMSO and other permeable CPAs might adversely affect the genetic and especially epigenetic fabric of spermatozoa. At the same time, several Japanese groups had successful CP of very sensitive mouse spermatozoa without any permeable CPA but with 18% of impermeable raffinose (a 3-ring sugar) and a mixture of proteins (skim milk) [Okuyama *et al.*, 1990; Tada *et al.*, 1990; Yokoyama *et al.*, 1990]. It worked so exceptionally well, that the Mazur's group, which had originally cryopreserved mouse sperm with glycerol [Mazur *et al.*, 2000] (though found that it can be indeed chemically toxic to the sperm [Katkov *et al.*, 1998]) finally also reported that fast immersion of mouse spermatozoa into liquid nitrogen without any CPA worked perfectly [Koshimoto *et al.*, 2000]. In any case, those data had inspired Evgenia and Vladimir Isachenko to freeze human sperm in tiny pellicles by plunging those "cryogenic loops" without any CPA whatsoever. They published their findings in 2002, and a year later, the explanation why it worked was followed [Isachenko *et al.*, 2003; Nawroth *et al.*, 2002]. That marked the *"second wind"* in the kinetic VF of spermatozoa. The history of the development is described in numerous papers [Isachenko *et al.*, 2004a; Isachenko *et al.*, 2008; Isachenko *et al.*, 2004b; Isachenko *et al.*, 2005] and several reviews by the authors [Isachenko *et al.*, 2007; Isachenko *et al.*, 2010; Katkov *et al.*, 2007] and briefly touched in this Book in Chapter 2 [Isachenko *et al.*, 2012]. The method has been involved from a cryo loop (pellicle) through droplets in LN_2 to quite elaborated "aseptic technology". Some of the carriers used by the Isachenkos at different stages are shown in **Fig. 6**.

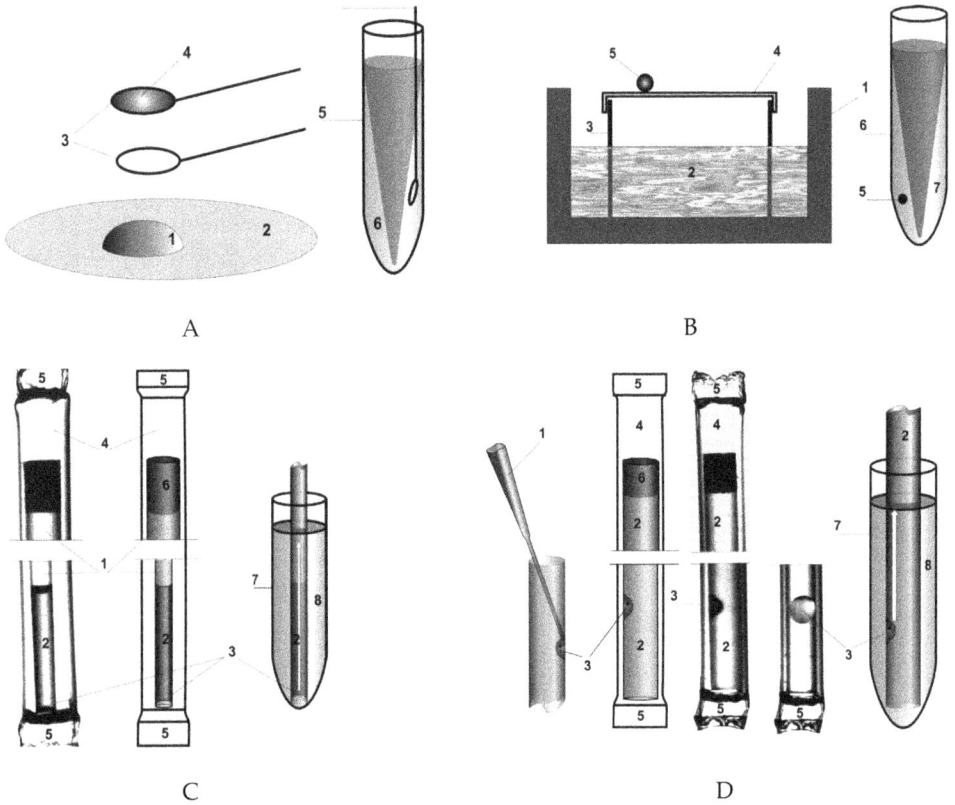

Fig. 6. Different techniques for kinetic vitrifcation of human sperm developed by the Isachenkos and colleagues: A: copper or nylon cryo loop (pellicle); B: a modification of a droplet technique; C; open-pulled straw; D; straw-in- straw. See [Isachenko *et al.*, 2007] for details.

The method and the scientist themselves were first dismissed, than ignored, than … ignored again. **Table 1** represents just two examples when the Isachenkos published their paper, and other cryobiologists, who came to the same conclusions, namely:

-   There is no proof of the *absence* of vitrification inside the sperm even at quite slow cooling [Morris, 2006]; the role of intracellular ice in the death of fast cooling mouse sperm is also questioned in [Mazur & Koshimoto, 2002].
-   Some cells can be vitrified in *"diluted"* solutions at relatively slow rate of cooling but very fast warming is essential for kinetic VF [Mazur & Seki, 2011]

Thus, both cryobiologists have reported similar findings as the Isachenkos observations, but they unfortunately fell short of mentioning Isachenkos in their own publications and presentations (e.g., in Cryo-2010 in Bristol), which might have made looking their observations (that were solid, of course) for an unfamiliar reader as "pioneering" or even as *"a **new** paradigm for cryopreservation by vitrifcation"* [Mazur & Seki, 2011]. The argument *"that paper by the Isachenkos et al. was not citable because the effect of the warming rates was not*

V. & E. Isachenkos and I.I. Katkov:	John Morris:
Factors that may enhance intracellular vitrification of human sperm: ...*Cells naturally contain high concentrations of proteins, which help in vitrifcation... this would enhace both the viscosity and Tg of the intracellular cytosol of permatozoa* . [Isachenko et al., 2003].	- *"It is generally assumed* that the intracellular environment of sperm has a low water content coupled with high protein levels. These data demonstrates that it heterogeneous nucleation sites are absent in that intracellular vitrification can occur:" (CRYO-2010). - "We demonstrate that the high intracellular protein content together with the osmotic shrinkage associated with extracellular ice formation leads to intracellular vitrification of spermatozoa during cooling" (Morris et al, 2011, *Cryobiology*, 64:71-80). - "The results described in this article suggest that it is now appropriate for new models to be developed that exclude the formation of intracellular ice" [Morris, 2006].
V. & E. Isachenkos and I.I. Katkov: Crucial role of fast warming *... As a result, we can speculate that we were able to achieve intracellular vitrification of the human spermatozoa even at such a low range cooling rate. ... However, as we discuss below, our method of* **instant thawing** *seemed to prevent cell damage even after relative slow freezing in liquid nitrogen vapor.* [Isachenko et al., 2004b]	Peter Mazur and Shinsuke Seki: *WARMING rate is much more critical in "diluted" vitrification solutions than the cooling rate* (CRYO-2010), [Mazur & Seki, 2011]

Table 1. Comparison of statements published by Isachenkos et al in 2003-4 and by other scientists reported in 2010 on Cryo-2010 in Bristol, UK and other publications.

thoroughly investigated" makes sense in the matter of describing a particular technique/ protocol but it does not hold water when the claim of a *"new paradigm"* in vitrification was put on the table seven years after the Isachenkos' paper, with essentially the same claim that had been published [Isachenko *et al.*, 2004b]. That new paradigm was indeed established but it was done in 2002-4, not in 2010-11!

Note that the role of ultrafast warming during kinetic vitrification had been known at some extend before so neither of the authors (the Isachenkos or Mazur & Seiko) can claim the absolute priority. In case with the crucial role of endogenous proteins and other high molecular weight components for the intracellular *kinetic* vitrification of spermatozoa, Katkov and colleagues clearly presented this idea (and indirect proof of it) in 2003. Therefore, any attempts to completely ignore that fact by Morris and colleagues and to position themselves as *"pioneers"* of this idea much later can be considered as blunt plagiarism.

7.2 Kinetic vitrification of sperm of *other* vertebrates: history of success and stories of failure

As we have mentioned, kinetic VF of human sperm in all its varieties shown on Fig. 6 seemed to be working equally well; however, when we tried the "droplet method" described in [Isachenko *et al.*, 2008] (20 µL droplets of swam-up washed sperm supplemented by 0.25 M sucrose) on model animals (rodent spermatozoa, the results (Figs. 7) were not so pronounced. So, while it worked well for human sperm, the droplet kinetic VF did not work so well for mouse sperm, and it worked poorly (at a much lower survival rate than conventional slow freezing) on rat sperm. Note that both rats and mice sperm have larger and apparently more watery heads.

But still, the Isachenkos' method worked in general so Celltronix and Kharkov Zoo launched 2 field expeditions (with the participation of a Moscow Zoo's specialist) for freezing polar bear (Ursus maritimus) sperm (in a distant Russian zoo) and sperm of gyrfalcons (Falco rusticolus), golden eagles (*Aquila chrysaetos*) and Eastern imperial eagles (*Aquila heliaca*) in the Russian Raptor Breeding Center Galichya Gora near Voronezh.

For the polar bear, for which sperm, to our knowledge, had not been frozen yet, the basic slow protocol developed for spectacled bear (*Tremarctos ornatus*) [Erokhin *et al.*, 2007] was used. That protocol worked quite poorly, vitrification protocol was even worse.

For the all 3 raptor species, gyrfalcon, the golden and the imperial eagles, slow freezing (using the slow freezing protocol in [Blanco et al., 2000]) worked very poorly in our hands despite the fact that artificial insemination with fresh sperm is a routine and successful procedure in that Center. And finally, kinetic vitrification using the Isachenko's "droplet" method failed completely.

After several sleepless nights of thinking what went wrong besides our insufficient experience with freezing the raptor sperm (my counterpart had frozen crane sperm before), difficulties related to small volume and a lot of fecal particles and urine in sperm, I.I. Katkov realized that the sperm of those species was fundamentally different in geometry from that of human sperm: their heads were much larger, and they looked much more watery, less condensed than the compact human portions. The rodent sperm heads were also relatively large, but those species, where we failed, the heads apparently contained much more water and presumably less so called "inactive osmotic volume", which means the concentration of internal proteins, sugars, nucleotides and other "internal endogenous vitrificants" was much lower in the polar bear spermatozoa, and especially in the sperm of the raptors. And according to the thermodynamic of the glassy state, as lower concentration of vitrificants the faster you need to cool the cells. From the personal communication with the Isachenkos several years before, it was known that kinetic vitrification of oocytes and embryos without cryoprotectants had been failed completely even with the smallest drops. And those cells have the ratio internal vitrificants: water about 7-9 lower than in human sperm as its osmotically active volume (i.e. water per se) is about 75%, while in human sperm it is only 25%. That meant that we just did not have sufficient cooling speed to vitrify those species!! That crystallized the hypothesis that if we would cool it fast enough, faster than the critical rates of cooling and warming for the most watery cells, we can vitrify all cells with the same protocol. That is how the concept of the Universal Cryopreservation Protocol was born (published first in the "Embryomail" in the spring of 2010).

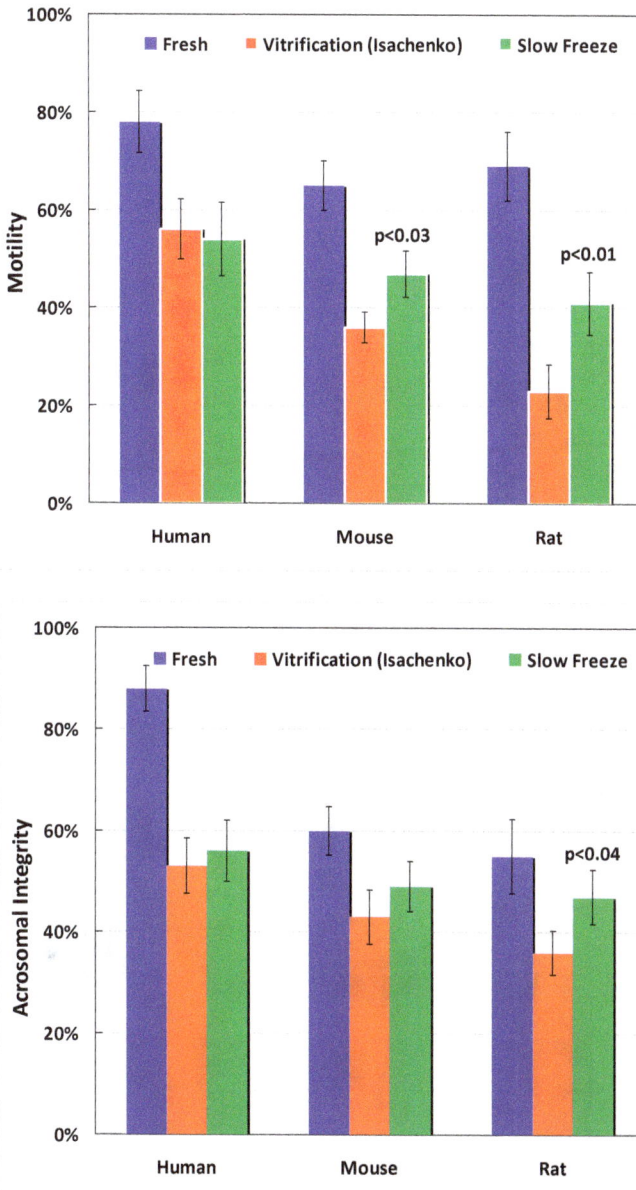

Fig. 7. Progressive motility (**left**) and acrosomal integrity (**right**) of 3 species of sperm frozen by conventional slow-freezing protocols in the media customized for different species (**green**) and by an identical protocol of vitrification [Isachenko et al., 2008] by quenching droplets in PBS containing 20% human serum albumin (HSA) and 0.25 M sucrose (**red**) directly into LN_2. Two methods of cryopreservation are compared.

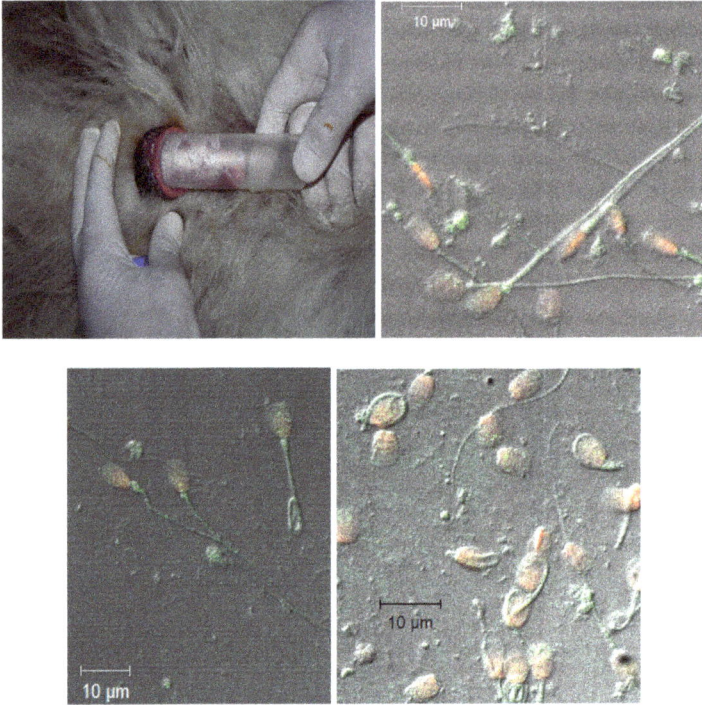

Fig. 8A. An attempt to vitrify sperm of the polar bear, **from left to right**: sperm retrieval; fresh; slow frozen; and vitrified sperm. Slow frozen sperm protocol worked poorly, *kinetic VF failed*.

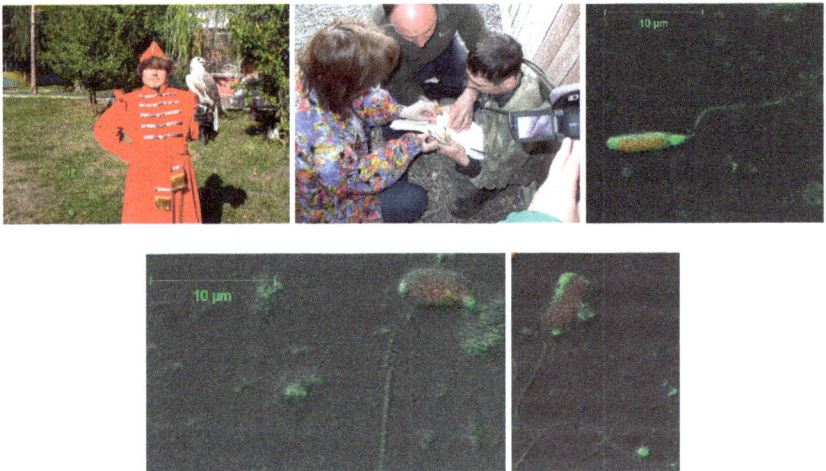

Fig. 8B. An attempt to vitrify sperm of the gyrfalcon, **from left to right**: I.I.K. with the bird; sperm retrieval process; fresh; slow frozen; and vitrified sperm. Slow frozen sperm protocol worked poorly, *kinetic VF failed*.

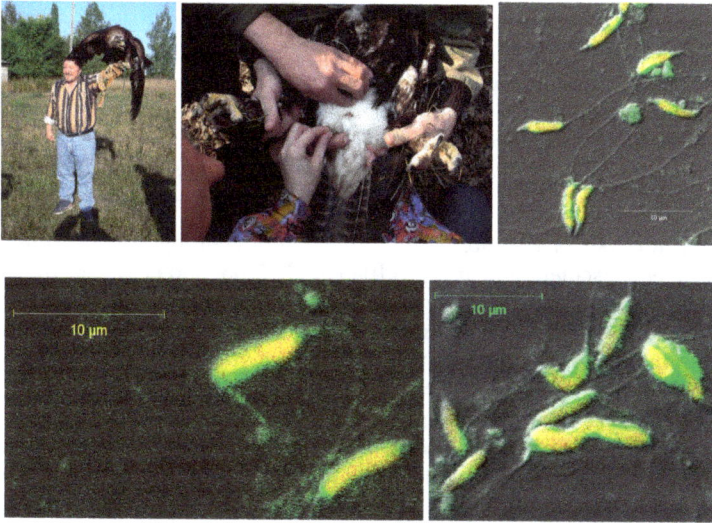

Fig. 8C. An attempt to vitrify sperm of the golden eagle, **from left to right**: I.I.K. with the bird; sperm retrieval process; fresh; slow frozen; and vitrified sperm. Slow frozen sperm protocol worked poorly, *kinetic VF failed*.

8. Conclusion: *"Race for the Pace"*: Is the universal cryo-protocol possible?

The universal cryoprotocol, that would fit *all* types of cells, at least if they are in suspension on make a thin layer, would be the *Holy Grail* of cryobiology. Here is our hypothesis for consideration [Katkov, 2010]:

1. Every cell has its own critical rates of cooling and thawing, at which and higher the cell can be vitrified during cooling (B_{cr_cool}) and will not devitrify during warming (B_{cr_warm}) *without* any external "cryoprotectants" (they must be called "vitrificants" in this case). Or it might be just that non-lethal ice (i.e. cubical vs. hexagonal "killer ice") is formed during cooling and its transformation (recrystallization) to hexagonal type is precluded during warming. In any case, at rates higher than those two B_{cr}'s, the cell will survive without any exogenous compounds.

2. Those rates are substantially *lower* than predicted by the contemporary theories (Fahy and Rall, Boutron's work, Cravalho 's school: Toner, Karlsson, *et al*). We will not go into the details of the thermodynamics of the glassy state but the three main reasons are: i) presence of the internal cell vitrificants with high T_g; ii) small compartmentalized intracellular milieu; iii) no "true" extracellular VF is needed for survival as the cell has no time and shrink at such fast time. In any scenario, the cell *survives* if the pace of cooling and warming is higher than those two B_{cr}'s, and that is what matters.

3. The distribution and average values of those B_{cr}'s 's depend on the species of cells, particularly on the abundance of endogenous vitrificants, how "watery" those cells are, the level of compartmentalization, the size of the compartments, etc. It may well be that the same species (such embryos) might have *very* different B_{cr}'s at different stages of their development.

4. We predict that while those speeds are relatively high for the majority of cell species (in range 200,00-1,000,000 ^{O}C/min), we have already achieved those critical rates in one well-documented case, namely humans as well as some other species of vertebrate sperm, thanks to the early work in the 1930s and by the Isachenkos in this century.
5. We believe that those high speeds are achievable for *all* species but that needs entirely new cryogenic equipment. Such rates if they are high enough to surpass the highest B_{cr}'s would be *universally* applicable to *any* type of suspendial and single cells so we will be witnessing the *"Race for the Pace"* very soon, some groups have been already actively working on it now.

Thermodynamic analysis of the most recent attempts of creating novel systems for kinetic VF such as cryogenic oscillating heat pipes [Jiao et al., 2006; Jiao et al., 2009], nano-droplets [Demirci & Montsesano, 2007], quartz capillaries [Risco et al., 2007], and some others approaches that claim *"ultra-fast"* rates (see a comprehensive review by Criado in this Book), which in our opinion, do not produce the rates fast enough to reach the majority of B_{cr}'s without using exogenous permeable (and thus, potentially toxic) vitrificants. Thermodynamical considerations that prove this statement are not in the scope of this Chapter and will be done elsewhere. In fact, the *hyper*-fast rates of cooling and warming will be needed, and there is about of an order or two of magnitude difference between *"ultra-"* and *"hyper-"* (cf. *ultrasonic* and *supersonic* speed of flight as an example).

Introduction of such a *"Universal Kinetic Vitrification Protocol"* applicable for *all* cells (at least for those that are in suspension or make a thin attached layer) would shift the whole paradigm in cryopreservation of germplasm (and other types of suspendial cells) and in cryobiology as a science. It will require both new equipment for realization of hyper-fast rates (on which we are working now) and new methods of measurements. For example, it is not clear how T_h, T_m, and T_d curves on the Fahy diagram would behave at speeds of cooling and warming in order of thousands ^{O}C/min, and how that could be measured: they may disappear completely! In any case, it will open not only the possibility of development of a uniform protocol and equipment for all existing and (which especially important) *new* types of cells and species, but it will also bring new, very challenging but exciting horizons for basic cryobiology as well.

Epilogue: *"In Defence of the Genius"*(Editor's Reflection)

"In Defense of the Genius"

Dr. Gregory Fahy called in his Chapter #1A [Fahy & Rall, 2007] the Luyet method and his promotion of kinetic VF as *"Lyuet's tyranny"* so my choice of words in this Chapter is just as a *"symmetrical response "*to that stye, nothing personal is intended. Greg spent a great deal of time in that and other numerous reviews and lectures describing the Luyet's unsuccessful attempts to implement K-VF in late 1930s -beginning of 1950's. However, we have not spent *so* much time in *this* Chapter on describing the *failure* to realize the promised potential of equilibrium VF and ice blockers for organ vitrification since 1984, even though Dr. Fahy and his colleagues have had in order of magnitude more resources, knowledge, and time than that of Father Luyet had had in 1939-1954. We would call this situation as being *"stuck in the rut"* and *"the promise is not fulfilled"*.

At the same time, we completely agree with a statement that Fahy and Rall made in a sub-chapter titled *"The ghost of Luyet"*:

*"Here we can only note the irony that, having been launched by breaking free of Luyet's tyranny of ultrarapid cooling, **vitrification methods have now essentially turned back closer to Luyet's original idea of cooling as quickly as possible with minimal intracellular exposure to cryoprotectants** [i.e., kinetic vitrifcation VF, I.I. K.],, albeit this time using at least marginally adequate concentrations of intracellular solutes. The ghost of Luyet lives on in the form of this ongoing methodological evolution, and we think he would have been pleased to see how his ideas about vitrification ultimately related to the now widespread use of vitrification as a practical and successful method of cryopreservation long after he, himself, had abandoned this approach."* [Fahy & Rall, 2007]

As a cryobiologist, who has been working in the same field, the Editor (at the Eves of 2012) might announce the following "resolutions":

-   I wish cryonics would have been a *real* science;
-   I wish ice-blockers would have been *worked* for the human body ;
-   I wish cryonics would have *preserve* not only the "ghost" (we, actually, preserve SPIRIT of his science), but his body as well;
-   I wish he could have been vitrified successfully in 1974 and re-warmed alive in 2012. Had that all happened, Father Luyet would have been indeed thrilled to see how his method has been spread and are opening new horizons!
    And for me, it is better to live under "tyranny" of the genius than under "ochlocracy" of the ignorant or "democracy" of the arrogant, and seeing as Wikipedia and other internet resources have been invaded (and infested) by cryonics "experts", and are full of their biographies, cryonic companies' descriptions, etc, while a reader can find neither biographies of Father Lyuet nor other prominent cryobiologists, both who passed away and live and in good health today. Unfortunately, Wikipedia has failed to be fair and balanced on this matter, but hope it'll change with time.
    But that would be a topic of our other story, here we must stop and say just only that: kinetic vitrification of sperm, the early child of Father Luyett and the other *pioneers of the cryobiological frontiers*, is very much alive and on the march! And our own success and failures, honestly described in this Chapter, have only strengthened the position K-VF as a viable (not marginal!) and very promising method of cryopreservation.

Appendix 1. Some peculiar similarities between [Fahy & Rall, 2007] and our earlier papers, which are not cited there

We invite the readers to compare the physical description of vitrification in that chapter by Fahy and Rall, particularly sub-chapter *"The kinetic basis of vitrification"*, and the first part of the *"Optimal storage below Tg"* with our preceding publications (pp. 71 and 75 in [Katkov & Levine, 2004] and pp. 353-4 *"6. Storage at temperatures higher than T_g of water"* in [Katkov et al., 2006] respectively). The only substantial difference is that *"the most widely used"* WLF equation (25) in our paper [Katkov & Levine, 2004].

$$\eta(T) = \eta(T_g)10^{\frac{-C_1(T-T_g)}{C_2+(T-T_g)}} \equiv \eta(T_g)e^{\frac{-2.303C_1(T-T_g)}{C_2+(T-T_g)}} \tag{1}$$

is replaced by the VTF equation in [Fahy & Rall, 2007] as following:

$$\eta(T) = \eta(T_g)e^{\frac{B}{T-T_v}}, \tag{2}$$

where $\eta(T)$ is the viscosity at temperature T *above* the glass transition temperature T_g, $\eta(T_g)$ is the viscosity at the glass transition tem temperature, T_v is the "Vogel temperature" [Zhai & Salomon, 2011] and C_1, C_2 and B are empirical constants.

If the assumption of a linear relationship between the fractional free volume and temperature holds (free volume theory for WLF), VTF equation can be transformed into the WLF equation so the equations (1) and (2) must be combined as an equality, i.e.:

$$\frac{-2.303C_1(T - T_g)}{C_2 + (T - T_g)} \equiv \frac{B}{T - T_v} , \tag{3}$$

which is true at *any* T and T_g in the range of being considered. Equalizing the numerators and denominators separately, the relationships between the WLF and VTF constants can be found as following:

$$C_1 \equiv \frac{B}{2.303(T_g - T_v)} \tag{4}$$

$$C_2 \equiv T_g - T_v , \tag{5}$$

which coincides with [Zhai & Salomon, 2011].

Beside this substitution of WLF with VTF, which accordingly to [Zhai & Salomon, 2011] *"has a more profound physical meaning that relates both thermodynamic and kinetic concepts"*, there are definite similarities between [Fahy & Rall, 2007] and [Katkov & Levine, 2004; Katkov et al., 2006], which of course might be purely accidental (with a notion that those two papers had been sent to G.F. by I.I.K. well before 2007), so we will follow the spirit of the Open Access, namely *"Let's the readers to decide"*.

Appendix 2. On the recent paper by the Isachenkos on *"vitrification in large volumes"*

Recently, the Isachenkos group published a report on vitrification of 500 µL of human sperm vitrified with 0.25 M sucrose Recently, the Isachenkos group published a report on vitrification of 500 µL of human sperm vitrified with 0.25 M sucrose [Isachenko *et al.*, 2011]. Unfortunately, our two other teams that have co-authored this Chapter were not able to repeat the method: both human and bovine spermatozoa sperm survived vey poorly (single alive spermatozoa were observed) after vitrification in 0.5 mL straws. Interestingly enough, morphology of the sperm was practically intact (**Fig. 9**). This, together with the failure of the method even for 25 µL droplets (another Isachenkos modification) to vitrify spermatozoa of polar bear and 4 raptor bird species described in the major text of this Chapter, indicates that the Isachenko method of *"cryoprotectant"*-free cryopreservation" works satisfactory for some species of sperm [Merino *et al.*, 2011a; Merino *et al.*, 2011b; Sanchez *et al.*, 2011] but not for the others, and it is completely inapplicable to big and watery "oocytes and embryos. And without further clarification, the method of kinetic VF in *"large volumes"* have not been able to be repeated independently even for human and bovine sperm. Some of the questions that have been raised from other co-authors of the present Chapter in regards to that paper are:

A (human native)   B (human, slow freezing)   C (human, VF 0.5 mL)

D (bull native)   E (bull, slow freezing)   F (bull, VF 0.5 mL)

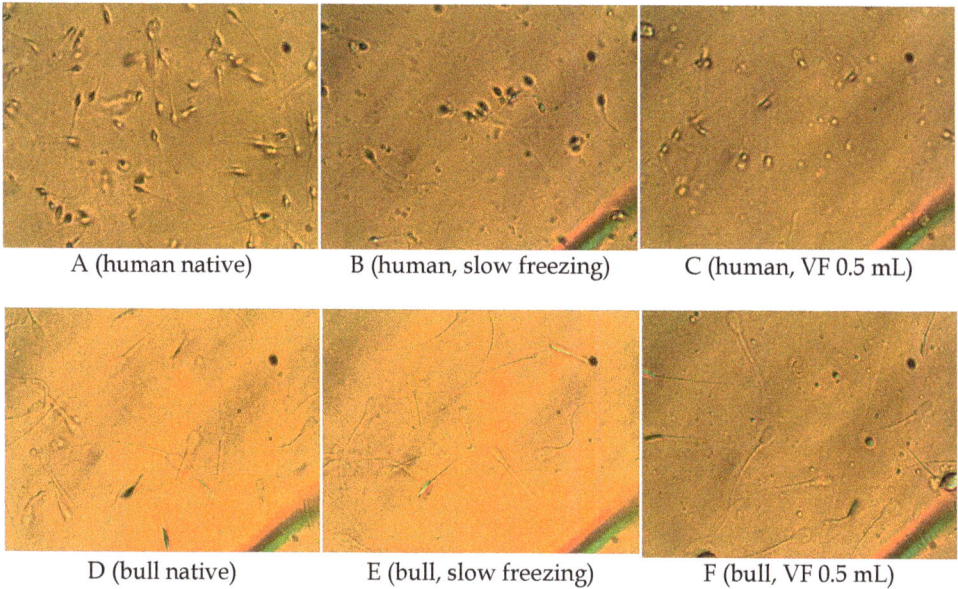

Fig. 9. Attempts to vitrify human (**A-C**) and bovine (**D-F**) spermatozoa using "large volume"(500 µL) method [Isachenko *et al.*, 2011]

A and **D** – native sperm, **B** and **E** spermatozoa frozen slowly with glycerol, **C** and **F**- sperm vitrified accordingly the Isachenko protocol.

Approximately 50% of human and bull sperm survived slow freezing,. The vitrified cells are not visibly damaged but no motile spermatozoa were observed for both species

- Sucrose is considered as "natural" CPA while glycerol is not. See examples given in [Fahy & Rall, 2007] about Alaskan beetles than can cumulate 10 Mole/L of glycerol!

- Glycerol cannot *"dilute"* intracellular osmolites - it can only add additional osmotic pressure inside the cell, thus, preventing more extensive shrinkage during slow freezing; that is exactly how it works as the CPA (not to be confused with its role in VF). Apparently, the authors confused it with the glycerol action as a plasticizer of the intracellular milieu, thus, lowering its T_g inside the cells, but this is a completely different topic.

- Sucrose per se, as an impermeable solute, cannot directly penetrate internal organelles such as mitochondria; the role of sucrose, apparently, is to dehydrate the cells and make intracellular vitrification easier using lower cooling and warming rates. That in theory would allow us to vitrify larger volumes.

- It is not clear whether *"It is known that human spermatozoa contain large amounts of proteins, sugars, and other components that make the intracellular matrix highly viscous and compartmentalized and may act as natural cryoprotectants"* for the authors (the idea had been originated form the paper published in 2003 [Isachenko et al., 2003] but that that source is NOT referred) or it is a "common knowledge" (as the explanation of the

possibility of kinetic VF of sperm) that had been recognized for many year prior to that publication in 2003. Note that the authors confused vitrificants with CPAs: intracellular proteins do not help slow freezing.
- Pre-cooling of the vitrified group is not detailed and it is not clear whether that grouped was pre-cooled at all.

Thus, there is a disparity in experimental verification of the method between several groups that have been contributed to this Chapter, which we feel should be clarified.

10. References

Angell C. A. (2004). "Amorphous water." Annu Rev Phys Chem 55: 559-83.

Beier A. F., Schulz J. C., Dörr D., Katsen-Globa A., Sachinidis A., Hescheler J. &Zimmermann H. (2011). "Effective surface-based cryopreservation of human embryonic stem cells by vitrification." *Cryobiology* 63: 175-85.

Bernstein A. D. &Petropavlovski V. V. (1937). "[Influence of non-electrolytes on viability of spermatozoa]." *Buleten' Eksperimentalnoi Biologii i Medicini (Rus.)* III(4225): 21-25.

Blanco J. M., George Gee G., Wildt D. E. &Donoghue A. M. (2000). "Species Variation in Osmotic, Cryoprotectant, and Cooling Rate Tolerance in Poultry, Eagle, and Peregrine Falcon Spermatozoa." *Biol Reprod* 63,: 1164–1171.

Butler S. &Pegg D. (2012). Precision in cryopreservation - equipment and control. In: *Cryopreservation / Book 1 (ISBN 979-953-307-300-1)*. Eds:I. I. Katkov, InTech. 1: *in press.*

Chakraborty N., Chang A., Elmoazzen H., Menze M. A., Hand S. C. &Toner M. (2011). "A Spin-Drying Technique for Lyopreservation of Mammalian Cells." Ann Biomed Eng 39: 1582-1591.

Cipri K., Lopez E. &Naso V. (2010). "Investigation of the use of Pulse Tube in cell cryopreservation systems." *Cryobiology* 61: 225-30.

Demirci U. &Montsesano G. (2007). "Cell encapsulating droplet vitrification." *Lab Chip* 7: 1428-1433.

Devireddy R. &Thirumala S. (2011). "Preservation protocols for human adipose tissue-derived adult stem cells." Methods Mol Biol 702: 369-394.

Erokhin A. S., Shishova N. V. &Maksudov G. Y. (2007). "[A comparison of two diluents for the cryopreservation of ejaculated spermatozoa in spectacled bear (*Tremarctos ornatus*)]." [Reproduction in Domestic Animals] Rus. 42: 95-96.

Fahy G. M., MacFarlane D. R., Angell C. A. &Meryman H. T. (1984). "Vitrification as an approach to cryopreservation." *Cryobiology* 21(4): 407-26.

Fahy G. M. &Rall W. F. (2007). Vitrification: an overview. In: *Vitrification in Assisted Reproduction: A User's Manual and Troubleshooting Guide.* Eds:M. J. Tucker and J. Liebermann. London, UK, Infroma Healthcare. Ch.1A: 1-20.

Fahy G. M., Wowk B., R. P., Chang A., Phan J., Thomson B. &Phan L. (2009). "Physical and biological aspects of renal vitrification." Organogenesis 5(165-175).

Foote R. H. (2002). "The history of artificial insemination: Selected notes and notables." J Anim Sci 80(E-Suppl. 2): E1-E10.

Gilmore J. A., Liu J., Gao D. Y. &Critser J. K. (1997). "Determination of optimal cryoprotectants and procedures for their addition and removal from human spermatozoa." *Hum Reprod* 12(1): 112-8.

Gordts S., Roziers P., Campo R. &Noto V. (1990). "Survival and pregnancy outcome after ultrarapid freezing of human embryos." Fertil Steril 53(3): 469-72.

Graevsky E. Y. (1948a). "[Living matter and low temperatures]." *Priroda* (Rus.) 5: 13-25.

Graevsky E. Y. (1948b). "[Glassy state of protoplasm in conditions of deep cooling]." *Uspehi Sovr. Biol.* (Rus.) 14: 186-202.

Graevsky E. Y. &Medvedeva Y. A. (1948). "[Causes of damage of protoplasm during deep freezing]." *Zh. Obschey Biologii* (Rus.) 9: 436-469.

Hoagland H. &Pincus G. G. (1942). "Revival of mammalian sperm after immersion in liquid nitrogen." *J Genet Physiol* 25: 337-344.

Isachenko E., Isachenko V., Katkov I. I., Dessole S. &Nawroth F. (2003). "Vitrification of mammalian spermatozoa in the absence of cryoprotectants: from past practical difficulties to present success." *Reprod Biomed Online* 6(2): 191-200.

Isachenko E., Isachenko V., Katkov I. I., Rahimi G., Schondorf T., Mallmann P., Dessole S. &Nawroth F. (2004a). "DNA integrity and motility of human spermatozoa after standard slow freezing versus cryoprotectant-free vitrification." *Hum Reprod* 19(4): 932-9.

Isachenko E., Isachenko V. &Katkov I. I. (2007). Cryoprotectant-Free Vitrification of Spermatozoa. In: *Vitrification in Assisted Reproduction: A User's Manual and Troubleshooting Guide*. Eds:M. J. Tucker and J. Liebermann. Oxon, UK, Infroma Healthcare Medical Books: 87-105.

Isachenko E., Isachenko V., Weiss J. M., Kreienberg R., Katkov I. I., Schulz M., Lulat A. G., Risopatron M. J. &Sanchez R. (2008). "Acrosomal status and mitochondrial activity of human spermatozoa vitrified with sucrose." Reproduction 136(2): 167-73.

Isachenko E., Isachenko V., Sanchez R., Katkov I. I. &Kreienberg R. (2010). Cryopreservation of spermatozoa: Old routine and new perspectives. In: *Principles and Practice of Fertility Preservation*. Eds:J. Donnez and S. S. Kim. Cambridge, UK,, Cambridge University Press: Ch. 14, pp 177-99.

Isachenko E., Mallmann P., Isachenko V., Rahimi G., Risopatron J. &Sanchez R. (2012). Vitrification Technique- New Possibilities for Male Gamete Low-Temeprature Storage. In: *Cryopreservation / Book 1 (ISBN 979-953-307-300-1)*. Eds:I. I. Katkov, inTech. 1: Chapter 2 (in press).

Isachenko V., Isachenko E., Katkov I. I., Montag M., Dessole S., Nawroth F. &Van Der Ven H. (2004b). "Cryoprotectant-free cryopreservation of human spermatozoa by vitrification and freezing in vapor: effect on motility, DNA integrity, and fertilization ability." Biol Reprod 71(4): 1167-73.

Isachenko V., Isachenko E., Montag M., Zaeva V., Krivokharchenko I., Nawroth F., Dessole S., Katkov I. I. &van der Ven H. (2005). "Clean technique for cryoprotectant-free vitrification of human spermatozoa." Reprod Biomed Online 10(3): 350-4.

Isachenko V., Maettner R., Petrunkina A. M., Mallmann P., Rahimi G., Sterzik K., Sanchez R., Risopatron J., Damjanoski I. &Isachenko E. (2011). "Cryoprotectant-free vitrification of human spermatozoa in large (up to 0.5 mL) volume: a novel technology." *Clin Lab*(57): 643-650.

Jahnel F. (1938). "[Resistance of human spermatozoa to deep cold]." Klinische Wochenschrift 17: 1273-1274.

Jiao A., Han X., Critser J. K. &Ma H. (2006). "Numerical investigations of transient heat transfer characteristics and vitrification tendencies in ultra-fast cell cooling processes." Cryobiology 52(3): 386-92.

Jiao A. J., Ma H. B. &Critser J. K. (2009). "Experimental investigation of cryogenic oscillating heat pipes." Int J Heat Mass Transf 52(15-16): 3504-3509.

Katkov I. I., Katkova N., Critser J. K. &Mazur P. (1998). "Mouse spermatozoa in high concentrations of glycerol: chemical toxicity vs osmotic shock at normal and reduced oxygen concentrations." Cryobiology 37(4): 325-38.

Katkov I. I., Isachenko E., Isachenko V. &Nawroth F. (2003). "Why can we vitrify mammalian spermatozoa without cryoprotectants? Physicochemical considerations." Cryobiology 47: 267.

Katkov I. I. &Levine F. (2004). "Prediction of the glass transition temperature of water solutions: comparison of different models." Cryobiology 49(1): 62-82.

Katkov I. I., Isachenko V., Isachenko E., Kim M. S., Lulat A. G.-M. I., Mackay A. M. &Levine F. (2006). "Low- and high- temperature vitrification as a new approach to biostabilization of reproductive and progenitor cells." Int J Refrigeration 29(3): 346-57.

Katkov I. I., Isachenko V. &Isachenko E. (2007). Vitrification in Small Quenched Volumes with a Little Amount of or without Vitrificants: Biophysics and Thermodynamics. In: Vitrification in Assisted Reproduction: A User's Manual and Troubleshooting Guide. Eds:M. J. Tucker and J. Liebermann. London, UK, Infroma Healthcare: 21-32.

Katkov I. I. (2008). High temperature stabilization by drying and storage at ambient temperatures as the emerging alternative to transgenic animal sperm cryobanking. Int-I Kirpichnkov's Conf. "Genetics, Selection, and Breeding of Fish", Saint-Petersburg, Russia.

Katkov I. I. (2010). "Race for the pace: is the universal cryoprotocol a dream or reality?" Cryobiology 61: 374-375.

Katkov I. I., Kan N. G., Cimadamore F., Nelson B., Snyder E. Y. &Terskikh A. V. (2011). "DMSO-free programmed cryopreservation of fully dissociated and adherent human induced pluripotent stem cells. 2011:981606.." Stem Cells International 2011: 981606.

Koshimoto C., Gamliel E. &Mazur P. (2000). "Effect of osmolality and oxygen tension on the survival of mouse sperm frozen to various temperatures in various concentrations of glycerol and raffinose." Cryobiology 41(3): 204-31.

Li X., Krawetz R., Liu S., Meng G. &Rancourt D. E. (2008). "ROCK inhibitor improves survival of cryopreserved serum/feeder-free single human embryonic stem cells." Hum Reprod.

Lovelock J. E. (1953). "The mechanism of the protective action of glycerol against haemolysis by freezing and thawing." Biochim Biophys Acta 11(1): 28-36.

Luyet B. &Hodapp A. (1938). "Revival of frog's spermatozoa vitrified in liquid air." Proc Meet Soc Exp Biol 39: 433-434.

Luyet B. E. (1937). "The vitrification of organic colloids and of protoplasm." Biodynamica 1: 1-14.

Maksudov G. Y., Shishova N. V. &Katkov I. I. (2009). Ch. 8, In the Cycle of Life: Cryopreservation of Post-Mortem Sperm as a Valuable Source in Restoration of Rare and Endangered Species. In: Endangered Species: New Research. Eds:A. M. Columbus and L. Kuznetsov. Hauppauge, NY, NOVA Science Publishers: 189-240.

Martin-Ibanez R., Stromberg A. M., Hovatta O. &Canals J. M. (2009). "Cryopreservation of dissociated human embryonic stem cells in the presence of ROCK inhibitor." Curr Protoc Stem Cell Biol Chapter 1: Unit 1C 8.

Martin-Ibanez R. (2012). Cryopreservation of human pluripotent stem cells: are we going in the right direction? In: *Cryopreservation / Book 1 (ISBN 979-953-307-300-1)*. Eds:I. I. Katkov, InTech. 1: *in press*.

Mazur P. (1963). "Kinetics of Water Loss from Cells at Subzero Temperatures and the Likelihood of Intracellular Freezing." *J Gen Physiol* 47: 347-69.

Mazur P. (1970). "Cryobiology: the freezing of biological systems." Science 168(934): 939-49.

Mazur P., Leibo S. P. &Chu E. H. Y. (1972). "2-factor hypothesis of freezing injury - evidence from chinese-hamster tissue-culture cell." *Experimental Cell Res.* 71: 345-349.

Mazur P. (1984). "Freezing of living cells: mechanisms and implications." *Am J Physiol* 247(3· Pt 1): C125-42.

Mazur P., Katkov I. I., Katkova N. &Critser J. K. (2000). "The enhancement of the ability of mouse sperm to survive freezing and thawing by the use of high concentrations of glycerol and the presence of an Escherichia coli membrane preparation (Oxyrase) to lower the oxygen concentration." Cryobiology 40(3): 187-209.

Mazur P. &Koshimoto C. (2002). "Is intracellular ice formation the cause of death of mouse sperm frozen at high cooling rates?" *Biol Reprod* 66(5): 1485-90.

Mazur P., Leibo S. P. &Siedel G. E. (2008). "Cryopreservation of the germplasm of animals used in biological and medical research: Importance, impact, status, and future directions." *Biol Reprod* 78: 2-12.

Mazur P. &Seki S. (2011). "Survival of mouse oocytes after being cooled in a vitrification solution to -196°C at 95° to 70,000°C/min and warmed at 610° to 118,000°C/min: A new paradigm for cryopreservation by vitrification." *Cryobiology* 62: 1-7.

Merino O., Risopatron J., Sanchez R., Isachenko E., Figueroa E., Valdebenito I. &Isachenko V. (2011a). "Fish (Oncorhynchus mykiss) spermatozoa cryoprotectant-free vitrification: stability of mitochondrion as criterion of effectiveness." Anim Reprod Sci 124(1-2): 125-31.

Merino O., Sanchez R., Risopatron J., Isachenko E., Katkov I. I., Figueroa E., Valdebenito I., Mallmann P. &Isachenko V. (2011b). "Cryoprotectant-free vitrification of fish (*Oncorhynchus mykiss*) spermatozoa: first report." *Andrologia: in press*.

Milovanov V. K. (1962). [Reproductive Biology and Artificial Insemination of Animals] Rus. Moscow, Nauka.

Mollamohammadi S., Taei A., Pakzad M., Totonchi M., Seifinejad A., Masoudi N. &Baharvand H. (2009). "A simple and efficient cryopreservation method for feeder-free dissociated human induced pluripotent stem cells and human embryonic stem cells." Hum Reprod.

Morris G. J. (2006). "Rapidly cooled human sperm: no evidence of intracellular ice formation." *Human Reprod.* 21(8): 2075-83.

Moskovtsev S. I., Lulat A. G. &Librach C. L. (2012). Cryopreservation of Human Spermatozoa by Vitrification vs. Slow Freezing: Canadian Experience. In: *Cryopreservation / Book 1 (ISBN 979-953-307-300-1)*. Eds:I. I. Katkov, inTech. 1: Chapter 3 (in press).

Nawroth F., Isachenko V., Dessole S., Rahimi G., Farina M., Vargiu N., Mallmann P., Dattena M., Capobianco G., Peters D., Orth I. &Isachenko E. (2002). "Vitrification of human spermatozoa without cryoprotectants." *CryoLetters* 23(2): 93-102.

Okuyama M., Isogai S., Saga M., Hamada H. &Ogawa S. (1990). "In vitro fertilization (IVF) and artificial insemination (AI) by cryopreserved spermatozoa in mouse." *J Fertil Implant (Tokyo)* 7: 116-119.

Park S. P., Lee Y. J., Lee K. S., Ah Shin H., Cho H. Y., Chung K. S., Kim E. Y. &Lim J. H. (2004). "Establishment of human embryonic stem cell lines from frozen-thawed blastocysts using STO cell feeder layers." Hum Reprod 19(3): 676-84.

Parkes A. S. (1945). "Preservation of human spermatozoa at low temperatures." *Br Med J* 2: 212-213.

Perloff W. H., Steinberger E. &Sherman J. K. (1964). "Conception with Human Spermatozoa Frozen by Nitrogen Vapor Technic." Fertil Steril 15: 501-4.

Polge C., Smith A. U. &Parkes A. S. (1949). "Revival of spermatozoa after vitrification and dehydration at low temperatures." *Nature* 164: 666-676.

Pouplin M., Redl A. &Gontard N. (1999). "Glass transition of wheat gluten plasticized with water, glycerol, or sorbitol." J Agric Food Chem 47(2): 538-43.

Rall W. F. &Fahy G. M. (1985). "Ice-free cryopreservation of mouse embryos at -196 degrees C by vitrification." *Nature* 313(6003): 573-5.

Reubinoff B. E., Pera M. F., Vajta G. &Trounson A. O. (2001). "Effective cryopreservation of human embryonic stem cells by the open pulled straw vitrification method." Hum Reprod 16(10): 2187-94.

Rezazadeh V. M., Eftekhari-Yazdi P., Karimian L., Hassani F. &Movaghar B. (2009). "Vitrification versus slow freezing gives excellent survival, post warming embryo morphology and pregnancy outcomes for human cleaved embryos." J Assist Reprod Genet 26: 347-354.

Risco R., Elmoazzen H., Doughty M., He X. &Toner M. (2007). "Thermal performance of quartz capillaries for vitrification." *Cryobiology* 55: 222-229.

Sanchez R., Risopatron J., Schulz M., Villegas J., Isachenko V., Kreinberg R. &Isachenko E. (2011). "Canine sperm vitrification with sucrose: effect on sperm function." *Andrologia*.

Saragusty J. &Arav A. (2011). "Current progress in oocyte and embryo cryopreservation by slow freezing and vitrification." *Reproduction* 141: 1-19.

Saragusty J. (2012). Genome Banking for Vertebrates Wildlife Conservation. In: *Cryopreservation / Book 1 (ISBN 979-953-307-300-1)*. Eds:I. I. Katkov, InTech. 1: *in press*.

Schaffner C. S. (1942). "Longevity of fowl spermatozoa in frozen condition." *Science* 96: 337.

Sherman J. K. (1963). "Questionable protection by intracellular glycerol during freezing and thawing." J Cell Comp Physiol 61: 67-83.

Sherman J. K. (1964). "Dimethyl Sulfoxide as a Protective Agent During Freezing and Thawing of Human Spermatozoa." Proc Soc Exp Biol Med 117: 261-4.

Smirnov I. V. (1949). "Preservation of domestic animal semen by deep cooling." *Sovetskaja Zootechnia* (Rus.). 4: 63-65.

Spallanzani L. (1780). "Dissertazioni di fisica animale e vegetale." *Modano.*

Stiles W. (1930). "On the cause of cold death of plants." *Protoplasma* 9: 459-468.

Stubban C., Wesselschmidt R. L., Loring J. F. &Katkov I. I. (2007). Cryopreservation of human embryonic stem cells. In: *Human Stem Cell Manual. A Laboratory Guide.* Eds:J. F. Loring, R. L. Wesselschmidt and P. H. Schwartz. London, Elsevier: 47-55.

Suzuki H. (2006). "Freeze-dried Spermatozoa and Freeze-dried Sperm Injection into Oocytes." J Mamm Ova Res(23): 91-95.

Tada N., Sato M., Yamanoi J., Mizorogi T., Kasai K. &Ogawa S. (1990). "Cryopreservation of mouse spermatozoa in the presence of raffinose and glycerol." J Reprod. Fertil 89: 511–516.

Tucker M. J. &Liebermann J., Eds. (2007). *Vitrification in Assisted Reproduction: A User's Manual and Troubleshooting Guide.* London, UK, Infroma Healthcare.

Wagh V., Meganathan K., Jagtap S., Gaspar J. A., Winkler J., Spitkovsky D., Hescheler J. & Sachinidis A. (2011). "Effects of cryopreservation on the transcriptome of human embryonic stem cells after thawing and culturing." Stem Cell Rev 7: 506-517.

Ware C. B. &Baran S. W. (2007). "A controlled-cooling protocol for cryopreservation of human and non-human primate embryonic stem cells." Methods Mol Biol 407: 43-9.

Whittingham D. G., Leibo S. P. &Mazur P. (1972). "Survival of mouse embryos frozen to -196 degrees and -269 degrees C." Science 178(59): 411-4.

Wikipedia (2011a). "21st Century Medicine." http://en.wikipedia.org/wiki/Twenty-First_Century_Medicine.

Wikipedia (2011b). "KrioRus." http://en.wikipedia.org/wiki/KrioRus.

Wikipedia (2011c). "Danila Medvedev." http://en.wikipedia.org/wiki/Danila_Medvedev.

Wilmut I. (1972). "The effect of cooling rate, warming rate, cryoprotective agent and stage of development on survival of mouse embryos during freezing and thawing." Life Sci II 22: 1071-1079.

Yokoyama M., Akiba H., Katsuki M. &Nomura T. (1990). "Production of normal young following transfer of mouse embryos obtained by in vitro fertilization using cryopreserved spermatozoa." Exp. Anim 39: 125–138.

Zhai H. & Salomon D. (2011). "Low Temperature Property Evaluation and Fragility of Asphalt Binders Using Non-Arrhenius Viscosity Temperature Dependency." http://www.technopave.com/publications/Fragility-of-Asphalt-Non-Arrhenius-Viscosity.pdf 05-0971.

Zhou C. Q., Mai Q. Y., Li T. &Zhuang G. L. (2004). "Cryopreservation of human embryonic stem cells by vitrification." Chin Med J (Engl) 117(7): 1050-5.

3

Cryopreservation of Human Spermatozoa by Vitrification *vs.* Slow Freezing: Canadian Experience

S.I. Moskovtsev[1,2], A.G-M. Lulat[1] and C.L. Librach[1,2,3]
[1]CReATe Fertility Centre
[2]Department of Obstetrics & Gynaecology, University of Toronto
[3]Division of Reproductive Endocrinology and Infertility,
Department of Obstetrics & Gynaecology, Sunnybrook Health Sciences Centre and
Women's College Hospital, Toronto, Ontario
Canada

1. Introduction

Many advances in reproductive medicine in the past five decades have made cryopreservation of human spermatozoa an invaluable tool for the clinical management of infertility and sperm banking. The advent of in vitro fertilization (IVF) and intracytoplasmic sperm injection (ICSI) with microsurgical sperm handling techniques along with advances in female gamete acquisition have resulted in an increased demand for the cryopreservation of semen and tissue samples, often containing a very limited number of spermatozoa. Sperm cryopreservation also makes it possible for cancer patients to preserve their fertility prior to gonadotoxic chemotherapy or radiation. Applications of sperm banking are not limited to cancer patients but extend to patients undergoing certain types of pelvic or testicular surgeries; those who suffer from degenerative illnesses such as diabetes or multiple sclerosis; spinal cord disease or injury; and persons in occupations where a significant risk of gonadotoxicity prevails. Sperm cryopreservation is also available to men undergoing surgical sterilization such as vasectomy, in the event that children may be desired in the future. Another use for semen cryopreservation is to allow donor semen samples to be quarantined while appropriate screening is performed to prevent the transmission of infectious pathogens during therapeutic donor insemination (TDI).

Although major improvements have been made in sperm cryopreservation, there are many unresolved technical issues. Since freezing protocols differ between types of cells, the ideal conditions for human sperm freezing and thawing need to be perfected. To add more complexity, samples with abnormal semen parameters, such as severe oligospermia or high seminal fluid viscosity, often require unique cryopreservation conditions. For example, the particular cryoprotectants can affect cooling rates. In addition, storage temperature can significantly influence cryopreservation outcome. Liquid nitrogen (LN_2) can offer long-term survival of spermatozoa due to essentially absent metabolic activity, such as chemical reactions, genetic modification or aging of cells (Mazur, 1984). A conventional slow freezing protocol has been in use for many years and very little has changed in terms of methodology and reagents. While freezing aims to preserve cells it can also easily destroy them if certain precautionary steps are not taken into consideration. During cryopreservation cells and tissue undergo dramatic transformation in chemical and physical characteristics as the temperature drops from +37 to -196°C. The cells can lose up to 95% of their intracellular water. The concentration of solutes increases considerably, triggering the possibility of osmotic shock. Moreover, potential intracellular ice crystallization and mechanical deformation by extracellular ice may cause significant injury leading to cell death. Furthermore, if cells survive freezing, they might sustain additional damage during the thawing process due to osmotic shock, uncontrollable swelling and ice re-crystallization (Woods, et al., 2004).

Recently scientists have begun to re-investigate the utility of ultra rapid freezing in the search for alternative methods of sperm cryopreservation. Slow freezing of sperm utilizes cooling rates of 1–10°C/min, while the rapid freezing, or vitrification, technique allows for cooling rates to reach more than 40-1000°C/min in order to avoid intracellular ice formation. As new techniques are perfected, there is a potential for sperm cryopreservation to greatly improve in the future.

2. Cryopreservation of human spermatozoa

2.1 History of human spermatozoa cryopreservation

Remarkably, the first reference of empirical sperm freezing dates as far back as the late 16th century, but it was only with the discovery in 1937 by Bernstein and Petropavlovski that glycerol can aid spermatozoa in surviving long term freezing, that sperm cryopreservation became practical. Expansion of artificial insemination for the dairy industry led to further important research in the field of cryobiology (E. Isachenko, 2003, as sited in Bernstein & Petropavlovski, Polge et al., 1949). Shortly after these practices were initiated with animals, the first pregnancies were reported in humans after insemination with frozen spermatozoa. The next milestone was the discovery of the possibility to store human spermatozoa in LN_2 at -196°C, resulting in superior recovery rates compared to storage at higher temperatures between -20 and -75°C. After the era of empirical freezing; cryobiology matured to its fundamental stage, focusing on the biophysical and biochemical principals of cryopreservation, further advancing the field (Mazur et al., 1972. A comprehensive review of the historical background of sperm freezing was recently published and is recommended for readers looking for more details (Katkov et al., 2006).

2.2 Biological aspects of freezing

Living cells have an isotonic condition with a melting point of their intracellular water of approximately -0.6°C. When cells are cooled below this standard freezing point, supercooling takes place and remains in a metastable state up to -5°C (Katkov et al., 2006; Mazur et al., 1972). Water crystallization and ice formation begin between -5 and -15°C, beginning with the formation of an ice nucleus (seed crystal) in the extracellular water. This 'nucleation' can be induced at a higher temperature by the planned external facilitation of ice formation, often referred to as "seeding". Prior to that stage, water remains unfrozen inside the cell as the membrane prevents ice crystals from intracellular penetration (Woods et al., 2004). Solutes are excluded from ice formation which results in rising concentrations of solutes within extracellular water. Due to the permeability of the plasma membrane, this chemical imbalance sets up the diffusion of solutes into the cell, forcing water out of the cell. Cells thus undergo excessive dehydration, losing up to 95% of their intracellular water content. This increases the intracellular concentration of solutes, resulting in denaturation of proteins, pH shifts and potential cell death.

Since velocity of cooling is crucial, inaccurate cooling rates can negatively affect sperm survival, motility, plasma membrane integrity and mitochondrial function (Henry et al., 1993). When cooling is slow enough, there is sufficient time for intracellular water efflux and balanced dehydration. If cooling is too slow, damage may occur due to exposure of cells to high concentrations of intracellular solutes. Extreme cellular dehydration leads to shrinkage of cells below the minimum cell volume necessary to maintain its cytoskeleton, genome-related structures, and ultimately cellular viability (Mazur, 1984). On the other hand, if cooling rates are too fast, external ice can induce intracellular ice formation and potential rupture of the plasma membrane and damage intracellular organelles. In addition, mechanical damage of cells is possible due to of extracellular ice compression and close proximity of frozen cells can result in cellular deformation and membrane damage (Fujikawa & Miura, 1986). In contrast, with ultra rapid cooling, the amount of ice formation is insignificant and the entire cell suspension undergoes vitrification. At this stage water transitions, ice formation slows, molecular diffusion and aging stops, and liquids turn into a glass-like condition (Katkov et al., 2006).

Despite the relative insensitivity of human sperm to freezing, optimal cooling rates are needed to ensure appropriate sperm recovery. Currently, there are two types of slow freezing, either static vapour phase freezing to a certain temperature, or the multistep approach using nonlinear controlled-rate freezers, followed by plunging into LN2. Most laboratories and sperm banks adapt simple static vapour phase cooling in order to avoid induction of ice nucleation by seeding. For this technique samples are lowered into a vapour phase just above the LN_2 level, allowing them to cool for 15-20 minutes before being plunged into LN_2. Alternately, controlled rate freezers can be used to cryopreserve human semen. Most of these protocols utilize a "no seeding" option where samples are cooled from room temperature to -4°C at the rate of 2°C/min, followed by an increase of the cooling rate to 10°C/min until -100°C is reached, and finally plunging into LN_2 (Morris et al., 1999). In contrast to these slow freezing techniques, single step ultra rapid cooling is used for the vitrification technique.

2.3 Cryoprotective agents (CPAs)

Most cells would not survive cryopreservation without CPAs, which can minimize cryo-injury of cells. CPAs are low molecular weight chemicals that serve to protect spermatozoa from freezing damage or ice crystallization by decreasing the freezing point of materials. There are two categories of CPAs, and they differ in their ability to penetrate the plasma membrane. Firstly, permeating CPAs such as dimethylacetaldehyde; dimethyl sulfoxide, glycerol, glycol, ethylene and methanol, stabilize cell plasma membrane proteins and reduce concentrations of electrolytes (Arakawa et al., 1990). In contrast, nonpermeating CPAs such as albumins, dextrans, egg yolk citrate, hydroxyethyl, polyethylene glycols, polyvinyl pyrollidone and sucrose, minimize intracellular crystallization by increasing viscosity of the sample. CPAs themselves can be toxic if used at high concentrations and spermatozoa are vulnerable to osmotic changes induced by these agents (Gao et al., 1993). Despite the use of CPAs, plasma membranes can still be damaged or ruptured due to the initial extensive dehydration followed by cell swelling and osmotic stress. Gradual introduction of CPAs to the cell suspension or stepwise increase in their concentration, with a limited waiting period prior to freezing, is utilized to minimize the potential negative effects of these agents (McGann & Farrant, 1976).

2.4 Biological aspects of thawing

While there are many risk factors associated with freezing of cells, thawing can also dramatically affect survival rates of spermatozoa. When frozen samples are returned to ambient temperature, a reversal of the freezing process takes place. Cells that were frozen by the slow method, are more vulnerable to rapid thawing, due to the fast influx of water into cells causing uncontrollable swelling and osmotic shock (Curry & Watson, 1994). If cells were frozen rapidly, intracellular ice crystals could re-crystallize and form larger crystals during a slow thaw. To minimize toxic effects, CPAs have to be promptly removed from the cell suspension by washing samples in isotonic solution. Therefore, the thawing process and CPAs removal technique utilized must take into account the original method that was used for freezing.

2.5 Cryopreservation of spermatozoa for assisted reproductive techniques (ART)

Sperm cryopreservation is widely used in combination with ART techniques such as intrauterine insemination (IUI), IVF and ICSI. Despite many years of research and the discovery of new CPAs, significant numbers of spermatozoa still do not survive cryopreservation (Morris, 1999). Both freezing and thawing can inflict irreversible injury on a proportion of human spermatozoa, marked by a significant increase in some apoptosis markers (Giraud et al., 2000). Lipid peroxidation can lead to a decrease in sperm velocity, motility, viability, and mitochondrial activity (Mossad et al. 1994; O'Connell et al. 2002). The recovery rates of intact spermatozoa are highly dependant on the pre-freezing sample quality (de Paula et al., 2006). Poor quality semen may be more prone to DNA damage and cell death after cryopreservation than normal semen samples and thus have lower fertilizing capacity (Borges et al., 2007). It has been shown that reactive oxygen species (ROS) production impacts membrane fluidity and the recovery of motile, viable spermatozoa after cryopreservation. As well, semen samples containing leukocytes may have higher DNA fragmentation. In addition, the cryopreservation process can diminish the antioxidant activity of the semen fluid making spermatozoa more susceptible to ROS-induced damage (Lasso et al., 1994). The occurrence of sperm DNA damage may also be associated with the thawing process. A rapid increase in post thaw sperm DNA fragmentation over time has been observed, with the highest rate of fragmentation occurring during the first four hours after thawing (Gosalvez et al., 2009).

Normozoospermic semen samples appear to be more resistant to damage induced by freezing and thawing compared with oligozoospermic or asthenozoospermic samples. It has been reported that motile spermatozoa can be recovered after five refreezing and thawing rounds in normozoospermic samples, but only after two rounds in cases of oligozoospermia (Verza et al., 2009). Spermatozoa of infertile men were also found to be less resistant to damage during cryopreservation compared with spermatozoa from fertile men (Donnelly et al., 2001). Optimization of both CPAs concentrations and cryopreservation protocols will maximize survival of spermatozoa and thus improve ART outcome.

2.6 Cryopreservation of epididymal and testicular spermatozoa

Couples with male factor infertility represent 30 to 40% of the infertile population. Azoospermia accounts for 10% of cases of confirmed male infertility, and often requires surgical retrieval of spermatozoa. Since the introduction of ICSI, many cases of severe male infertility can now be successfully treated. Cryopreservation of surgically retrieved spermatozoa is a valuable component in the effective management of male infertility, reducing the necessity of repeat surgeries. Diagnostic sperm retrieval prior to IVF has several benefits including the possibility of freezing spermatozoa for future use, or if none are retrieved, initiation of the IVF stimulation cycle can be postponed or avoided. Testicular spermatozoa have been utilized to achieve pregnancy in couples with severe male factor infertility, with reported pregnancy rates similar to ejaculated spermatozoa, according to a meta-analysis study (Nicopoullos et al., 2004). In the case of obstructive azoospermia, recovery of spermatozoa by aspirations varies from 45 to 97% (Craft et al., 1995; Lania et al., 2006). In cases of non obstructive azoospermia recovery depends on the degree of testicular pathology and varies from 0 to 64% (Schlegel et al., 1997; Hauser et al., 2006). A second or third surgery can increase the chance of complications including hematomas, inflammation, testicular devascularization, fibrosis and permanent testicular damage (Schlegel and Su, 1997). To avoid this, if pregnancy is not achieved during the first ICSI attempt, a repeat of the surgical procedure would not be required if a portion of the surgical specimen has been banked. Cryopreservation of surgically retrieved spermatozoa can also aid the coordination of oocyte retrieval and avoids the pressure of having the urologist available on the day of the ICSI procedure. Usually the number of spermatozoa obtained during a surgical procedure is limited, and in the case of testicular sperm they may not be fully matured. In the future, if no mature spermatozoa are recovered, spermatogonial stem cells or early germs cells could potentially be matured in vitro and used for fertility treatments (Hwang & Lamb, 2010).

There are significant technical challenges for successful cryopreservation of testicular tissue due to its complex structure and intracellular interactions. Different cells of testicular tissue will have dissimilar responses to cryopreservation and require different concentration of CPAs. Freezing larger pieces of tissue is not advisable as it would increase resistance of heat transfer and penetration of CPAs leading to variation in cooling rates within different parts of the tissue. In addition, seminiferous tubules capture liquid and increase chances of ice formation (Woods et al., 2004). To avoid these difficulties, cryopreservation of smaller tissue fragments or mincing tissue prior to freezing has been advocated (Hovatta, 2003).

2.7 Cryopreservation of low number or single spermatozoa

The idea of cryopreservation of low numbers or individual spermatozoa was introduced more than a decade ago (Cohen et al., 1997). While this approach remains very attractive, there are multiple biological and technical issues to overcome. Early attempts to freeze individual spermatozoa were performed by placing them in empty animal or human zona pellucida prefilled with CPAs (Walmsley et al., 1998). Data from these studies suggested lower recovery and fertilization rates with human zona in comparison to hamster, possibly due to the presence of the ZP3 binding protein and induced acrosome reactions when human zona were used (Cohen et al., 1997). While this method requires special skills, equipment, and is very labour-intensive; live births were reported using both human and hamster zona (Walmsley et al., 1998). Spermatozoa were also injected and frozen within spheres of Volvox Globator algae and recovered after thawing using an ICSI needle (Just et al., 2004). While all of these methods appear to be attractive for single spermatozoa cryopreservation, they have a number of limitations. Issues around the use of donor human zona pellucida as well as exposure of human gametes to animal or algae genetic materials present potential risks that restrict the use of such methods for human ART procedures. While in theory zona could be obtained from the female partner of men with severe male infertility, this would be unrealistic in the clinical setting, as it would require IVF egg retrieval and destruction of ooplasma to obtain empty zona pellucidae. An alternative proposal would be to use a non-biological carrier such as non-toxic polysaccharide alginate agarose to cryopreserve small numbers of sperm (Herrler et al., 2006, Isaev et al., 2007). In these studies spermatozoa were mixed with CPAs and added to the alginate before the gelatin stage and then frozen in small bead microspheres. After cryopreservation, they were dissolved in a sodium citrate solution. The residual alginic acid on the sperm membrane can reduce sperm motility with slow freezing (Herrler et al., 2006). Agarose microspheres were also frozen in 0.25 cc straws by vitrification with better recovery rates (Isaev et al., 2007). Another reported method was to divide the sample into several small aliquots of 15–20-µl and to freeze in 0.2-mm cryopreservation embryo straws cut into smaller sections, sealed on one end (Desai et al., 2004).Conventional and open-pulled straws containing 1 or 5 µl of sperm suspension frozen by vitrification has also been reported (V. Isachenko et al., 2005). However, individual spermatozoa could not be easily sequestered because of possible adherence to the walls of the straws. ICSI pipettes were suggested as a container to freeze individual spermatozoa by either the slow method or vitrification (AbdelHafez et al., 2009; Sohn et al., 2003). Cryopreservation of sperm in microdroplets containing 1 or 40 µl of spermatozoa in cryoprotectant placed on a cold surface or directly plunged into liquid nitrogen was also reported (Gil-Salom et al., 2000; Isachenko et al., 2005). Microdroplets covered by mineral oil in a plastic tissue culture dish placed in liquid nitrogen were also used to cryopreserve individual spermatozoa (Quintans et al., 2000; Sereni et al., 2008). A nylon cryoloop first introduced for embryo freezing was successfully used to cryopreserve small volumes of sperm suspension by both slow freezing and vitrification (Nawroth et al., 2002; Schuster et al., 2003; V. Isachenko et al., 2004; Desai et al., 2004). However direct placement of sperm into LN_2 without a container using an 'open system' such as cryoloop or unsealed culture dish increases the risk of cross-contamination and such techniques are discouraged by regulatory agencies such as the FDA and the European Tissue Directive on Sperm.

Overall reported recovery rates of a known number of frozen spermatozoa varied from 59 to 100% with reported survival rates of 8–85% and motility of 0 to 100%. The wide ranges of results depended on patient population, initial quality and number of frozen spermatozoa, as well as the type of cryopreservation device, type of cryoprotectant, and freezing and thawing protocols.

2.8 Sperm packaging and relation to the method of cryopreservation

Storage of frozen samples has to be in suitable freezing containers and at an optimal temperature to ensure long term survival. The packaging containers must meet several criteria. They must: 1)hold freezing temperatures without cracking or leaking, 2)have a large surface to enable a uniform cooling rate of the sample, 3)have proper heat exchange properties, 4)be easy to label and seal securely and 5)be available in small sterile units. When storage packaging is chosen, the possible risk of microbial or viral contamination must also be considered. The type of packaging also depends on the freezing protocol and sometimes on the quality of the sample. For conventional slow freezing, the two most common types of containers currently used are plastic screw-top vials or straws. Straws can be made of polyethylene terephthalate glycol (PETG) or ionomeric resin (CBS High Security Straws by CryoBioSystem, Paris, France). As described above, a low number or single spermatozoa have been experimentally frozen in empty animal or human zona pellucida, spheres of Volvox Globator algae, alginate agarose bead microspheres and ICSI pipettes (AbdelHafez et al., 2009; Herrler et al., 2006; Isaev et al., 2007; Just et al., 2004; Walmsley et al., 1998). For vitrification purposes different types of storing strategies have been suggested. These include: cryoloops, electron microscope copper grids, nylon meshes, open-pulled straws and standard open straws (V. Isachenko et al., 2005).

3. Cryopreservation of human spermatozoa by vitrification

3.1 Background on vitrification of spermatozoa

Vitrification is an alternative method of freezing based on the rapid coolling of water to a glassy state through extreme elevation of viscosity without intracellular ice crystallization (Fahy, 1986; Katkov et al., 2006). The relationship between the size of different cells, particularly, different spermatozoa species, and the ability of cells to be vitrified are discussed in details in the paper by Katkov (Katkov et al., 2006).

The earliest experiments on vitrification from the 1930s was not successful because critical rates of cooling were unachievable at that time. With the use of LN2 and the discovery of cryoprotectants, however, it became possible to vitrify many types of cells. The five basic ways to achieve vitrification have been described in details by Katkov et al.: equilibrium freezing-out of the bulk of water with the use of CPAs and storage at ultra low temperature; lyophilization using slow freezing to moderately low (-40 °C) followed by secondary drying at +30°C (mostly used in food and pharmaceutical industries); ice-free vitrification at high rates and high concentration of CPAs; ice-free vitrification at very fast rates without permeable agents ("CPAs-free vitrification"); high temperature' vitrification by air/vacuum drying at temperature above 0°C (Katkov et al., 2006).

However, until only recently, vitrification of spermatozoa was unsuccessful, possibly due to high concentrations of permeable CPAs (30-50% compared to 5-7% with slow freezing) and low tolerance of spermatozoa to permeable agents. Even brief exposure to a high concentration of CPAs can lead to toxic and osmotic shock and would be lethal for spermatozoa. One possible strategy to lower the concentration of CPAs could be to increase the speed of cooling and warming temperatures as higher rates of cooling and warming, require lower concentrations of CPAs; these conditions can help eliminate intracellular ice crystallization, and facilitate the formation of a glassy state (Katkov et al., 2006). Another option is to add non-permeable CPAs--such as carbohydrates--to permeable CPAs to minimize osmotic shock by decreasing osmotic pressure and stabilizing the nuclear membrane. Since the intracellular matrix of human spermatozoa contains large amounts of proteins and sugars, they can be successfully frozen in the absence of permeable CPAs using protein- and sugar-rich non-permeable agents (Koshimoto et al., 2000).

Successful vitrification of human spermatozoa was first reported by the Isaschenkos' group (Nawroth et al, 2002; E. Isachenko et al, 2003). The high viscosity of the intracellular milieu due to large amounts of proteins, nucleotides and sugars and low water in human spermatozoa content determines the ability of human sperm to be vitrified at relatively low cooling rates (Katkov et al., 2006). It was noted that human spermatozoon is one of the smallest germ cells among mammals, has almost no residual histones and has very compacted DNA (Holt, 2000), which indirectly confirms this hypothesis (see also Chapter by Katkov et al in this Book).

As we mentioned above, the major breakthrough in successful vitrification of *human* spermatozoa without the use of permeable CPAs was reported only recently by the Isachenko group (Nawroth et al., 2002), who actually re-invented the work of the "pioneers" in the 1930-40s mentioned above. The combination of extremely high rates of cooling/warming and utilization of vitrification media containing proteins and polysaccharides made it feasible to avoid de-vitrification during warming without use of toxic CPAs. The same group compared viability, survival rate and sperm DNA damage between slow freezing and vitrification and found that DNA integrity was independent from the mode of cooling and the presence of cryoprotectants in thawed spermatozoa (V. Isachenko et al., 2004). The acrosome reaction, capacitiation and mitochondrial activity of spermatozoa were compared vis-a-vis slow freezing and vitrification (E. Isachenko et al., 2008). The group reported that changes in the mitochondrial membrane potentials relate to the type of vitrification media with the best achieved results when both sugar and albumin were added to the media. To achieve high cooling rates the vitrification specimen volume needs to be kept to a minimum. Specially designed freezing carriers such as cryoloops and electron microscope copper grids have been suggested for vitrification of human spermatozoa (E. Isachenko et al., 2003; Nawroth et al., 2002). However, placing drops of semen directly into LN2 raises the issue of the potential risk of microbial or viral cross contamination during freezing and storage (Katkov, 2002). The development of aseptic techniques of vitrification allowing to freeze 5-10 µl of sperm suspension in open-pulled straws (OPS) or 1-2 µl of sample cut standard straw (CSS) placed inside of insemination straw further advanced the methodology of human sperm vitrification (V. Isachenko et al., 2005). The ultra-high freezing rates utilized for vitrification, via direct plunging of specimens into LN_2, leads to solidification of a solution by an intense increase in viscosity during cooling which avoids water crystallization and damaging ice formation (Katkov et al., 2006).

Most importantly, vitrified spermatozoa were successfully utilized in ICSI treatment with clinical pregnancy resulting in delivery of healthy twins (E. Isachenko et al., 2011). While only a small volume 0.2 to 40 µl of sample suspension was frozen in the past, recently larger amounts of spermatozoa (100 µl) were successfully vitrified using newly developed straw packaging system (SPS) made from cut in half 0.25 ml plastic straw (E. Isachenko et al., 2011). A first live birth was reported following intrauterine insemination of semen vitrified without permeable cryoprotectants from patient with oligoasthenozoospermia making this freezing technique even more attractive in clinical practice (Sanchez al., 2011).

3.2 Vitrification of human spermatozoa: Canadian experience

Encouraged by the findings of the German group, we have also looked at possibilities to utilize vitrification in our laboratory. We have compared sperm motility, kinetics and DNA damage between semen samples cryopreserved by standard vapour freezing verses vitrification protocols (Moskovtsev et al., 2011). Semen samples from 11 patients presenting for infertility were washed by density gradient centrifugation and evaluated by Computer-Aided Sperm Analysis (CASA). Subsequently kinematic parameters were assessed as previously described: sperm motility, average path velocity (VAP) curvilinear velocity (VCL), straight-line velocity (VSL), linearity (LIN), amplitude of lateral head displacement (ALH) (Moskovtsev et al., 2009). However, kinematic parameters are averages of values obtained from analyzing the entire motile fraction of cells in a sample and include absolute (actual) parameters (VAP, VCL, VSL, ALH) and relative (derived) such as LIN. When cryopreserved samples are evaluated after thawing, the CASA-paradox can take place, when despite of deterioration of semen samples after cryopreservation "pseudo-enhancement" of kinematics characteristics is observed (Katkov & Lulat, 2000). Modified Kinematic Parameters (MKP) were calculated as previously described: Kinematic Parameters (KP) x Motility/ 100% To account for this phenomenon, modifications of actual CASA-parameters are recommended and are incorporated into our data (Table 1). (Katkov& Lulat, 2000).

Parameters	Prior to freezing	Post vitrif.	Post vitrif. MKP	Post slow freezing	Post slow MKP	P (vitrif. vs. slow freezing)
Motility (%)	68.0 ± 10.79	25.4 ± 13.6		14.6 ± 10.2		< 0.05
Rapid motility (% of motile)	50.64 ± 4.52	19.45 ± 12.98	4.9 ± 3.3	10.45 ± 8.49	1.5 ± 1.2	< 0.05*
VAP (microns/sec)	60.73 ± 8.93	43.09 ± 14.24	10.9 ± 3.6	38. 00 ± 8.57	5.5 ± 1.2	< 0.05
VSL (microns/sec)	47.73 ± 9.39	35.45 ± 13.98	9.0 ± 3.5	31. 00 ± 8.27	4.5 ±1.2	< 0.05
VCL (microns/sec)	96.18 ± 3.68	81.18 ± 21.31	20.6 ± 5.4	68.73 ± 13.01	10.0 ± 1.9	< 0.05
ALH (microns)	4.42 ± 0.79	4.3182 ± 0.80	1.1 ± 0.2	3.67 ± 0.85	0.5 ± 0.1	< 0.05
LIN (ratio of VSL/VCL)	49.91 ± 0.55	42.45 ± 7.43	10.8 ± 1.9	45.91 ± 5.59	6.7 ± 0.8	< 0.05
TUNEL (%)	7.5 ± 5.5	9.6 ± 4.4		9.5 ± 5.1		NS

Note: * Statistically significant P values, when compared between MKP of samples frozen by vitrification vs. slow freezing.

Table 1. Comparison of CASA and TUNEL resuls between semen samples frozen by vitrification and slow freezing.

Our results indicate that sperm motility was significantly reduced for both types of frozen-thawed samples (P <0.03) (Table 1). Mean motility of vitrified samples was 25.4% ± 13.6 (a decrease of 36.4% compared to samples prior to freezing), which was almost two-fold higher compared to motility of samples frozen by standard slow vapour protocol (14.6% ± 10.2, decrease of 47.2% compared to samples prior to freezing), (P <0.05). Sperm kinematics such as VCL, VSL, and LIN were not significantly different between the two types of cryopreservation protocols without taking into account CASA- paradox. However, when MKP were calculated, it was revealed that indeed vitrified samples had superior recovery of sperm kinematic parameters in comparison to slow freezing.

Samples for slow vapour freezing were diluted 3:1 with commercial cryoprotectant medium and frozen by standard protocol in CBS. Aliquots of samples for vitrification were diluted 1:1 with a G-IVF medium (Vitrolife, Göteborg, Sweden) supplemented with 0.25M sucrose and 1% of LSPS (Life Global Protein Supplement, IVF Online, Guelph, ON, Canada). We have used 0.5 ml OPS and loaded 5 µl of vitrified sample in each straw by capillary; OPS were inserted into 0.5 CBS straws and sealed (Figure 1).

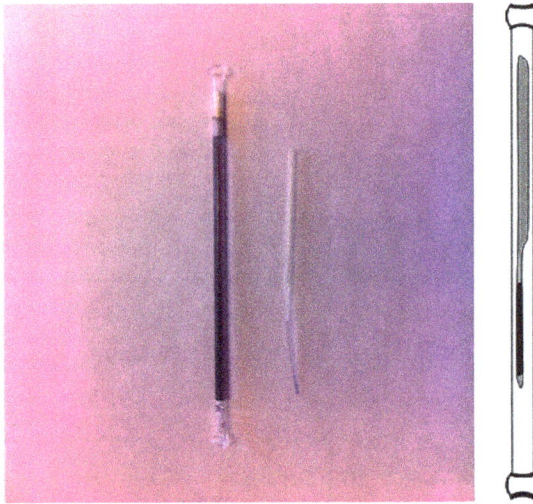

Fig. 1. Comparison of 0.5 ml CBS straw and 0.5 cc OPS straw and schematic of OPS inserted and sealed inside a CBS.

Samples were immediately plunged into LN2 and stored there for several days. For thawing procedure, OPS were rapidly removed from CBS straws, and plunged into 2 ml of the same medium used for vitrification at 37°C for 10 seconds.

We have evaluated the effect of cryopreservation on sperm DNA damage as the subject remains controversial. |While several reports indicate no negative effect of freezing on sperm DNA integrity (Duty et al., 2002; V. Isachenko et al., 2004). others have reported significant negative effect of sperm cryopreservation and DNA damage and chromatin stability (Hammadeh et al., 1999; Said et al., 2010). Significant increase in percentage of DNA fragmentation was associated with an increase in oxidative stress during cryopreservation

(Thomson et al., 2009). A slide-based technique for the assessment of sperm DNA was performed as previously described (TUNEL: TdT-mediated dUTP nick end labelling) (Moskovtsev et al., 2010) (Figure 2).

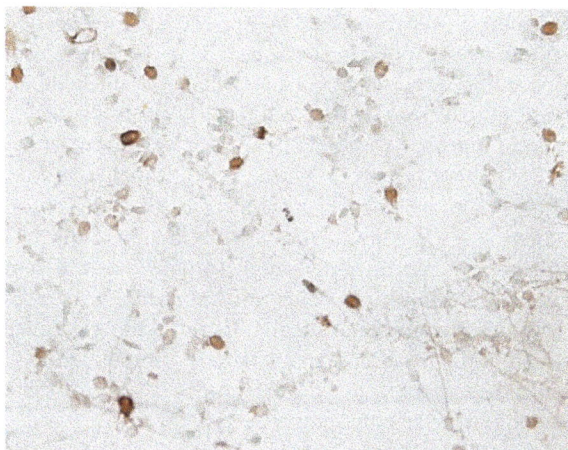

Note: Brown (TUNEL-positive): damaged DNA; gray-green (TUNEL-negative): undamaged cells.

Fig. 2. Sperm DNA damage assessment by TUNEL assay.

We found statistically significant increase in sperm DNA damage after both methods of sperm freezing (P < 0.05). However, the increase in DNA damage was minimal and to a degree probably irrelevant to clinical concerns. No significant differences were observed in sperm DNA damage between slow freezing and vitrification (9.6 ± 4.4 vs. 9.5 ± 5.1).

We can now confirm previous reports that human spermatozoa can be successfully vitrified without the use of potentially toxic cryoprotectants. The vitrification protocol showed significantly better results in preserving motility rates of spermatozoa when compared to slow vapour freezing. No significant differences were observed in post thaw sperm DNA damage in comparison to the standard slow freezing method. While our results are based on the freezing of a small volume of specimens, we are evaluating vitrification of larger volumes of spermatozoa with a proprietary mixture developed in our laboratory in CBS. We have achieved comparable results with both small volume (5 µl) and relatively large volume of 200 µl semen samples (unpublished data).

4. Sperm banking

4.1 Referring patients to a sperm bank

Human semen cryobanking can be divided into two broad categories: autologous banking for personal fertility preservation and donor sperm banking. Semen banking is useful in many situations and can be considered a safeguard against unforeseen future circumstances. These may include: prior to chemotherapy or radiation therapy; pre- vasectomy; before certain types of pelvic or testicular surgery; in cases of degenerative illnesses such as diabetes or multiple sclerosis, spinal cord disease or injury; high risk occupations or sports;

and preparation for future fertility treatment. Usually several steps are required prior to initiation of semen storage. Most sperm banks only accept patients referred by their physician, however sometimes self referral is possible. Patients have to be screened by collection of blood, urine and semen samples to ensure that their samples will not contaminate others with infectious diseases when placed into storage. Consent for freezing and storage of semen has to be signed and witnessed. It is important that patients have a clear understanding of the process of banking and that they provide clear instructions on who is responsible for the disposition of the specimens in the event of their demise. Similarly, costs associated with banking and long term storage need to be clearly defined and assigned in case the patient becomes incapacitated. When samples are either used for procreation, transported to another facility, or storage is to be discontinued, the patient or designate must complete specific consent forms allowing the bank to comply with their wishes. In most countries it is assumed or stated by law that the semen belongs to the individual who produced the sample, and it is this individual who must sign all consents, unless the sample was designated for donation (e.g., in the case of anonymous sperm donation). If possible, multiple donations are usually recommended to ensure an adequate sperm reserve for future procreation, depending on the quality of the sample, individual circumstances and reason for sperm banking.

4.2 Methods of sperm collection and retrieval

In most cases patients are able to produce an ejaculate by masturbation. Patients are given instructions to have minimum of 2 days and a maximum of 5 days of sexual abstinence, and to collect an entire ejaculate into a sterile specimen container. Condoms, creams or lubricants must not be used during collection as they can interfere with sperm motility and vitality due to the spermicidal properties of many products. Some patients would not be able to produce sample at the sperm bank by masturbation due to psychological, medical reasons or religious restrictions. In this case collection of samples by intercourse using a non-spermicidal condom, often referred to as a "semen collection device", is acceptable. Penile vibratory stimulation (PVS) can be helpful in spinal cord injury patients and those who are unable to produce semen by masturbation or intercourse (Brackett et al., 1998). However samples collected after PVS often exhibit relatively low motility (Hovav et al., 2002). PVS can help to produce a semen sample in most patients with spinal cord injuries (Brackett et al., 1998).

In some situations, collection of a retrograde sample is necessary. Retrograde ejaculation is a condition in which some or all of the sperm are not expelled through the urethra during ejaculation, but because of an incompetent bladder neck, the ejaculate refluxes back into the bladder. Reasons for this problem include organic conditions such as diabetes or multiple sclerosis, or pharmacological effects (eg. alpha adrenergic blocker use for hypertension). The general approach is to neutralize the urine pH and normalize the urine osmolarity by giving the patient sodium bicarbonate to alkalinize his urine. The urine sample is subsequently washed and used for insemination, in-vitro fertilization or cryopreservation. In patients with anejaculation, electro-ejaculation might be necessary to obtain semen, but this usually requires anaesthesia and has the associated risk of rectal injury. Additionally, samples produced by electro-ejaculation tend to be of relatively low quality (Denil et al., 1996).

Surgical sperm retrieval should be the last option in patients who can not produce a sample or patients who are diagnosed with azoospermia. Several methods of sperm retrieval are available depending on the etiology of the problem. To retrieve epididymal spermatozoa in cases of obstruction, percutaneous epididymal sperm aspiration (PESA) can be performed without surgical scrotal exploration, it is repeated easily, and does not require an operating microscope or expertise in microsurgery. Microsurgical epididymal sperm aspiration (MESA) is performed under the operating microscope and general anaesthesia. Individual tubules of the epydidymis are isolated and aspirated. Testicular sperm aspiration (TESA) is a needle biopsy of the testicle. It is an office procedure performed under local anaesthesia. Testicular sperm extraction (TESE) is the process of making a small incision and removing a small portion of tissue from the testicle under sedation or local anaesthesia. Microdissection (or sometimes referred to as microscopic or microsurgical) testicular sperm extraction (MicroTESE) is a very rigorous search for sperm under high magnification in cases of azoospermia or extremely low sperm production. MicroTESE is usually performed in the operating room under general anaesthesia utilizing an operating microscope. Cryopreservation of surgically retrieved epididymal and testicular spermatozoa is a valuable component in the effective treatment and management of these patients and it reduces the necessity of repeat surgeries when the intial procedure is unsuccessful or if additional children are desired.

4.3 Long term storage

Proper long term storage is usually achieved by placing specimens in LN_2 freezers, which have been safely used since the 1970s. Some automated systems are available and are capable of LN_2 autofilling, supplied with alarms and data recordings for all activities and are designed to minimize the chances of loss or damage of samples. Despite automation, quality control procedures must be implemented by sperm banks to ensure proper monitoring and safety of samples and staff. The LN_2 itself can be a source of microbial contamination so every available practical step has to be considered to reduce the risk of transmission (Fountain et al., 1997). In general cross-contamination of frozen samples by pathogens are extremely rare, and have not been reported in the setting of sperm banks. It is, however, theoretically possible, as often patients are not fully screened prior to freezing, as in cases of autologous sample cryopreservation or due to time constraints for oncology banking (Clarke, 1999). "Quarantine'" tanks are often used to separate samples with pending, unclear laboratory results or unscreened patients. However, contamination is still possible, as released samples that are moved to long term storage could have acquired pathogenic contamination from one of other "pending" samples in the quarantine tank. When samples are cryopreserved for patients with known infections such as HIV or hepatitis B and C carriers, separate tanks for each type of infection are required (Tomlinson & Sakkas, 2000). Cross-contamination can also be avoided by storage of samples in nitrogen vapour. However, in contrast to liquid nitrogen, there are some concerns that vapour has poor heat transfer rates, lower thermal capacity, and significant temperature fluctuation may exist within the vapour(Tomlinson & Saakas, 2000; Wood, 1999). Some older types of vapour storage systems could only guarantee the maintenance of temperature around -100°C and were not acceptable for long term sperm storage. Storage temperatures have to be maintained below -135°C to ensure a glass-like condition of frozen water and for secure long term storage of semen samples (Clarke, 1999). Newer types of high efficiency LN_2 vapour freezers and others that have a LN_2 "jacket"[1] provide working environments of below -160°C and are more suitable for sperm banks.

4.4 Quality control and quality assurance

Amongst the many government or professional organizations that require periodic inspection of frozen samples, the standards put forward by Health Canada are the strictest (Health Canada, 2000). Rigorous standards of operation are essential for sperm banks. Sperm banks must have specific requirements for screening, processing and quarantine of samples. Licensing is required in some countries and sperm banks are inspected in accordance with existing standards or regulations. While auditing is absolutely necessary, it might pose the risk of exposure of frozen specimens to room temperature while such inventory is performed. Straws thaw more rapidly than vials and can warm up to -80°C within 8 to 15 seconds at room temperature, dramatically increasing the possibility of damaging samples during inventory or verification of samples' identity (Tyler et al., 1996). Clear labelling systems to easily identify and link samples to a specific donor or patient must be in place to enable sample location and for performing inventory. The samples should remain in LN_2 during the duration of inventory performance and the audit must be performed by qualified and skilled staff.

Several facility and equipment-related quality control and risks factors must be considered for cryopreservation and storage of semen. Physical security of bank facilities and proper identification of sample location within freezers is crucial. Equipment must be appropriate and functional, with defined periodic service and maintenance schedules. Staff must be supplied with all necessary personal protective equipment. Adequate supplies of LN_2 gas must be guaranteed and spare LN_2 prefilled tank must be available in case of emergency. All staff involved in handling LN_2 must be properly trained by a certified organization. Standard operating procedures must be developed to clearly describe each step of the process of sample collection, processing, banking and handling. Annual reviews of both proper documentation and LN_2 training must be performed. Temperature of freezers chambers and LN_2 levels in tanks must be monitored on a continuous basis and all data logged in secure databases. Alarm systems and appropriate call procedures must be in place to attend to any emergencies. 24 hour monitoring and response is absolutely essential to safely maintain the integrity of the clinical samples in storage. Storage rooms must be monitored continuously for O_2 levels and staff activity in enclosed spaces must be monitored to avoid hypoxic injury. Backup power generation must be available in the event of a power failure. Each sample designated for storage has to be properly verified, assessed, processed, labelled, frozen and stored. Double-checking the identity of samples at each step is highly recommended. Some banks choose to divide samples from individual patients or couples and store them in different tanks or locations to minimize the risk of total loss of their biologic material (WHO, 2010).

5. Sperm donation

5.1 Applications for sperm donation

In cases of severe male infertility, single or lesbian women, the use of donor sperm is the only approach to address fertility issues (Botchan et al., 2001; Golombok, 2005). Advances in sperm cryopreservation have created opportunities for many families to achieve pregnancies through therapeutic donor insemination or IVF with donor sperm. Pregnancy rates are estimated to be around 10-12% per unstimulated cycle and can be achieved when at least 5×10^6 progressively motile spermatozoa inseminated into the lower cervical canal on 2-3 occasions during the ovulatory phase of menstrual cycle (Scott et al., 1990). At present, some 30,000 births per year worldwide are attributable to frozen donor sperm inseminations (Mortimer, 2004). While this seems like a large number, it may fall in the future, as the recruitment of sperm donors is increasingly difficult due to complicated and strict regulatory procedures, as well as lack of interest from potential donors.

The screening process for donor sperm is quite rigorous and includes obtaining a complete medical and sexual history, physical examination, psychological assessment and laboratory work-up on blood, urine and semen specimens to screen for pathogens including Hepatitis B, C, Human Immunodeficiency Virus (HIV 1&2), Human T-cell Lymphotrophic Virus (HTLV 1&2), Treponema pallidum (Syphilis), Cytomegalovirus, Chlamydia trachomatis and Neisseria gonorrhoea. Sperm banks perform genetic screening for heritable diseases based on the ethnic background of sperm donors (eg. Cystic Fibrosis for Caucasians). Donors must be retested after the required quarantine interval, and specimens may be released only if the results of repeat testing are negative. Specimens can only be used after they have been quarantined for a minimum of 180 days to avoid the risk of HIV transmission. Donor eligibility restrictions apply to employees of sperm banks, poor donors' health or quality of the semen and in some countries by sexual orientation of the donor, as gay or bisexual men are considered at higher risk for HIV and prohibited from being sperm donors in some countries (including Canada and USA). Many countries have age restrictions for sperm donation. The minimum age is usually 18 and the maximum 40 years of age (Health Canada 2000, American Society for Reproductive Medicine (ASRM), 2004)

5.2 Anonymous donors

Semen donors can be classified into two specific groups, anonymous and non-anonymous (known). Currently, with the establishment of many commercial sperm banks and the ability to safely transport samples even between continents, anonymous sperm donation is the method of choice for most recipients. The anonymity of the donor is maintained through the process. This is an important issue to both the recipient and the donor (Ernst et al., 2007). For fully anonymous sperm donation, the recipients would not be known to the donor and the donor offspring would have no future contact with the donor. The sperm donor gives up all legal rights over the biological children conceived from his samples donated to sperm bank. Anonymous donation allows parents, if they wish, to conceal the issue of infertility, or the fact of non-genetic parenting from the offspring. The motivation to hide this information most commonly is driven by pressure from other family members; fear of being rejected by the child or to protect children from the complicated psychosocial matters related to sperm donation. In many Western countries disclosure is encouraged by many counsellors, and if open disclosure is chosen by the parents, it is usually advised to disclose the method of conception to their children at an early age. Non-disclosure by parents of the biological origin of their children is viewed by some as misleading the child and could potentially affect trust between parents and their children, if their origin eventually becomes known to the child (Patrizio et al., 2001). However, it is ultimately the decision of the parents to disclose or not as in adoption cases.

There is some consensus that there should be limits on the number of offspring allowed from a given sperm donor. This is driven by possibility of accidental consanguinity between children from the same sperm donor. For example ASRM recommends a limit of 25 children per population of 800 000 for a single donor, but there are no federal or state laws limiting the number of sperm donation by a donor. In UK the number is limited to 10 different families, but does not apply if a genetically related sibling for an existing child is desired. Some countries limit the number of children to 4 in New Zealand; 5 in China; and 5 to 10 in Australia depending on the region; 25 in the Netherlands (Gong et al., 2009).

Recently, open-identity sperm donors have become available through many sperm banks. These donors have agreed to at least a single contact with any children born through use of their sperm, usually when the child reaches the age of consent (18 years old in most jurisdictions), for those individuals who wish to contact them (Gottlieb et al., 2000; Frith et al., 2007). In some cases audio interviews and pictures are available from these donors.

Two types of anonymous donor samples are usually available through sperm banks, prewashed or unwashed: Prewashed samples are obtained by processing the ejaculate by density gradient centrifugation for seminal fluid removal prior to freezing and can be directly inseminated into the uterine cavity after thawing (Larson et al. 1997). These samples are favoured by doctors' offices without access to an Andrology laboratory for post thaw processing. For processing unwashed samples, density gradient isolation is required to remove contaminants and CPAs after thawing the specimen prior to intrauterine insemination or for IVF. The removal of CPAs has to be performed step-by-step and gradually to minimize osmotic stress on spermatozoa. Drop-wise dilution of the sample with 1:10 sample to sperm wash medium ratio is recommended (Mortimer, 2004).

5.3 Non-anonymous sperm donors

Some donors and recipients choose to arrange donations privately and the donor in this case is known to the recipient(s). The donor may be a family member such as a brother or father or a friend. Most of these donations are done altruistically and acceptable only if all parties are in agreement. All participants involved in the donation process are generally required to attend a separate and a joint counselling session. An initial counselling interview with the donor and his spouse or partner (if applicable) is arranged to discuss the personal, social and legal aspects of donation. The known donor has to meet all requirements to be accepted for donation and undergo the same screening tests and laboratory evaluation as an anonymous donor, including 180 days quarantine for his frozen sample. Proper consent and declaration forms are required to be signed by known semen donor. Furthermore, a child conceived using donated semen is legally deemed to be the child of the recipient(s), and the donor has no legal rights or responsibilities regarding the child. Usually the donor may at any time prior to the use of his semen, vary or revoke his consents. Most clinics require a legal contract with all parties having received independent legal advice. As the use of the third party reproduction such as sperm and egg donation becomes more acceptable in many countries the ethical and legal aspects of these procedures become increasingly important. Issues of the donor's anonymity, financial compensation, religion and cultural acceptance, regulation of donor and prospective parent screening, as well as consideration of the welfare of children conceived with the use of donor sperm are widely discussed in the scientific literature and public media. While guidelines on the use of donated sperm come from government or professional organizations, they may also be influenced by religious institutions and they vary widely from country to country (Gong, 2009).

6. Social importance and psychological aspects around banking oncology patients, adolescent and young adults

When an individual is diagnosed with cancer almost every aspect of their physical and psychosocial well-being is altered. Quite often in clinical practice, the long term effects of cancer therapy on a patients' ability to have children in the future is not adequately addressed (Thaler-DeMers, et al., 2001). While the priority is to eliminate the cancer and save their life, fertility preservation especially among adolescent or young adults to ensure the potential of procreation with their own gametes after treatment, needs to be considered. Impaired spermatogenesis has been demonstrated before treatment in some patients with malignancies, depending on their location (eg. testicular cancer) or type (eg.Hodgkin's lymphoma) (Rueffer et al., 2001). Current treatment options such as surgery, chemotherapy and/or radiation can impair spermatogenesis and sexual function and lead to temporary or permanent infertility (Magelssen et al., 2006).

The scale of negative effects of cancer treatment on spermatogenesis depend on the specific gonadotoxicity of administered chemotherapeutic agents, number of chemotherapy treatment cycles, radiotherapy field location and dosage, type and stage of the cancer, and age of the patient. Considering combination cancer therapy, uncertainty in individual response to treatment and the large number of confounding variables, it becomes very challenging to assess the risk of iatrogenic infertility in many patients. The ability of cancer survivors to have their own biological offspring is very important for many oncology patients, especially at younger ages (Schover et al., 1999). Advances in early diagnostic investigation and treatments have led to increasing numbers of young cancer survivors. Unfortunately up to 30% of childhood cancer survivors are permanently sterile following cancer treatment (Tournaye et al., 2004). In Canada and the United States, cancer in patients 15 to 29 years of age who can benefit from sperm banking is nearly three times more common than in patients younger than 14 years (Bleyer et al., 2006). Early germ cells, (spermatogonia) are very sensitive to radiation and chemotherapy. Even low doses or a single dose treatment can potentially cause functional impairment of spermatogenesis. With increase in dosage or duration of the treatment, initially spermatocytes get damaged and as treatment progresses spermatids also become damaged. Radiation doses of less than 0.8 Gy can result in oligospermia and doses between 0.8 and 3 Gy can result in azoospermia (Rivkees & Crawford, 1988).

Cryopreservation of semen has changed the reproductive prospects for young patients diagnosed with cancer. Unfortunately, banking services continue to be underutilized since cancer patients and their families are not always informed about the potential fertility risks associated with cancer treatments, or the availability of banking. According to some surveys, less than 20% of patients undergoing chemotherapy or radiation treatment are informed about the adverse effects of such treatment on spermatogenesis or are offered sperm banking for fertility preservation. Cancer patients are usually under huge physiological and time pressure to make cryopreservation decisions while dealing with a life threatening situation. To complicate matters, some young patients are unable to produce semen samples by masturbation. In such cases, PVS or electro-ejaculation under general anaesthetic might be required. Surgical retrieval of testicular tissue may be an option for prepubertal boys who are not capable of producing mature sperm. Testicular tissue cryopreservation has been reported in boys with cryptorchidism to preserve fertility (Bahadur et al., 2000). Cryopreserved testicular tissues can be autografted to restore reproductive functions; however recurrence of neoplastic process is a concern in oncology patients and such procedures are still considered to be experimental (Hwang & Lamb, 2010). A multi-disciplinary team approach is important to ensure that patients have the opportunity to preserve their fertility potential if they elect to do so.

The posthumous use of semen is an entirely separate and complex ethico-legal subject. The ethical and legal aspects of posthumous assisted reproduction have been recently addressed by the European Society of Human Reproduction and Embryology Task Force on Ethics and Law (ESHRE, 2006).

7. Conclusions

Human spermatozoa can be successfully cryopreserved and utilized. Cryopreservation now plays an essential role in fertility preservation under the following scenarios:

- couples undergoing infertility treatment.
- cancer patients undergoing gonadotoxic chemotherapy or radiation.
- patients undergoing certain types of pelvic or testicular surgeries
- patients suffering from degenerative illnesses such as diabetes or multiple sclerosis; spinal cord disease or injury.
- men undergoing surgical sterilization such as vasectomy
- screening and quarantine of donor semen samples

Normozoospermic semen samples appear to be more tolerant to damage induced by freezing and thawing compared with oligozoospermic or asthenozoospermic samples. Cryopreservation of surgically retrieved epididymal and testicular spermatozoa is challenging, but a valuable component in effective treatment and management of severe male factor infertility. Cryopreservation of low numbers or single spermatozoa has multiple biological and technical aspects yet to be worked out; therefore, further research is required to introduce this technique into clinical practice. During cryopreservation, cells and tissue undergo dramatic transformation in chemical and physical characteristics as temperature drops from +37 to -196°C, thus risking cryoinjury. Velocity of cooling and warming is crucial and inaccurate cooling or thawing rates negatively correlate with sperm survival.

Spermatozoa cannot survive slow freezing without CPAs; CPAs have to be used at low concentrations with minimum exposure as CPAs are toxic and can cause osmotic damage. Gradual, stepwise introduction before freezing and removal of CPAs after thawing is essential. Conventional slow freezing with CPAs can offer cooling rates of 1–10°C/min. Vitrification, currently only an experimental technique, allows for extremely rapid freezing at rates of up to a 1000°C/min. LN_2 can offer long-term survival of spermatozoa due to essentially absent metabolic activity and aging of cells and tissues in the frozen state. Rigorous standards of operation and quality control are essential for sperm banks. Social, psychological, legal and ethical issues surrounding sperm banking are very complex and must be considered in each case.

The vitrification method uses no specially developed cooling program; it does not need permeable cryoprotectants; it is much faster, simpler and cheaper; and it can also provide a high recovery of motile spermatozoa after warming as effective protection of spermatozoa against cryodamage and helps to avoid many problems relevant to slow freezing such as ice formation; shifts in pH, extensive rehydration and osmotic damage.

Successful vitrification of human spermatozoa without toxic CPAs has been reported now by two independent groups. Moreover, live births were reported after vitrification of semen utilized for intrauterine insemination and IVF with ICSI procedures, making this freezing technique even more attractive in clinical practice.

8. References

Abdel Hafez, F.; Bedaiwy, M.; El-Nashar, S.A.; Sabanegh, E. & Desai, N. (2009). Techniques for cryopreservation of individual or small numbers of human spermatozoa: a systematic review. *Human Reproduction Update*, 15(2), pp. 153-164, ISSN 1460-2369

Arakawa, T.; Carpenter, J.F.; Kita, Y.A. & Crowe, J.H. (1990). The basis for toxicity of certain cryoprotectants. *Cryobiology*, 27, pp. 401-415, ISSN 0011-2240

Bahadur, G.; Chatterjee, R. & Ralph, D. (2000). Testicular tissue cryopreservation in boys. Ethical and legal issues. *Human Reproduction*, 15(6), pp. 1416-1420, ISSN 0268-1161

Bahadur, G.; Whelan, J.; Ralph, D. & Hindmarsh P. (2001). Gaining consent to freeze spermatozoa from adolescents with cancer: legal, ethical and practical aspects. *Human Reproduction*, 16(1), pp. 188-193, ISSN 0268-1161

Bleyer, A.; Budd, T. & Montello, M. (2006). Adolescents and young adults with cancer: the scope of the problem and criticality of clinical trials. *Cancer*, 107 (Suppl. 7), pp. 1645-1655, ISSN 1097-0142

Borges, E. Jr.; Rossi, L.M.; Locambo de Freitas, C.V.; Guilherme, P.; Bonetti, T.C.; Iaconelli, A. & Pasqualotto, F.F. (2007). Fertilization and pregnancy outcome after intracytoplasmic injection with fresh or cryopreserved ejaculated spermatozoa. *Fertility and Sterility*, 87(2), pp. 316-320, ISSN 1556-5653

Botchan, A.; Hauser, R.; Gamzu, R.; Yogev, L.; Paz, G. & Yavetz, H. (2001). Results of 6139 artificial insemination cycles with donor spermatozoa. *Human Reproduction*, 16, pp. 2298–2304 ISSN 0268-1161

Brackett, N.L.; Ferrell, S.M.; Aballa, T.C.; Amador, M.J.; Padron, O.F.; Sonksen, J. & Lynne, C.M. (1998). An analysis of 653 trials of penile vibratory stimulation in men with spinal cord injury. *The Journal of Urology*, 159(6), pp. 1931-1934, ISSN 0022-5347

British Andrology Society. (1999). *British Andrology Society guidelines for the screening of semen donors for donor insemination*, ISSN 0268-1161

Clarke, G. (1999). Sperm cryopreservation: is there a significant risk of cross-contamination? *Human Reproduction*, 14, pp. 2941–2943, ISSN 0268-1161

Cohen, J.; Garrisi, G.J.; Congedo-Ferrara, T.A.; Kieck, K.A.; Schimmel, T.W. & Scott, R.T. (1997). Cryopreservation of single human spermatozoa. *Human Reproduction*, 12(5), pp. 994-1001, ISSN 0268-1161

Craft, I.; Tsirigotis, M.; Bennett, V.; Taranissi, M.; Khalifa, Y.; Hogewind, G. & Nicholson, N. (1995). Percutaneous epididymal sperm aspiration and intracytoplasmic sperm injection in the management of infertility due to obstructive azoospermia. *Fertility and Sterility*, 63(5), pp. 1038-1042, ISSN 0015-0282

Curry, M.R. & Watson, P.E. (1994). Osmotic effects on ram and human sperm membranes in relation to thawing injury. *Cryobiology*, 31, pp. 39–46, ISSN 0011-2240

de Paula, T.S.; Bertolla, R.P.; Spaine, D.M.; Cunha, M.A.; Schor, N. & Cedenho, A.P. (2006). Effect of cryopreservation on sperm apoptotic deoxyribonucleic acid fragmentation in patients with oligozoospermia *Fertility and Sterility,* 86(3), pp. 597-600, ISSN 1556-5653

Denil, J.; Kuczyk, M.A.; Schultheiss, D.; Jibril, S.; Küpker, W.; Fischer, R.; Jonas, U.; Schlösser H.W. & Diedrich, K. (1996). Use of assisted reproductive techniques for treatment of ejaculatory disorders. *Andrologia,* 28 (Suppl. 1), pp 43-51, ISSN 0303-4569

Desai, N.N.; Blackmon, H. & Goldfarb, J. (2004). Single sperm cryopreservation on cryoloops: an alternative to hamster zona for freezing individual spermatozoa. *Reproductive Biomedicine Online,* 9, pp. 47-53, ISSN 1472-6483

Devireddy, R.V.; Swanlund, D.J.; Roberts, K.P.; Pryor, J.L. & Bischof, J.C. (2000). The effect of extracellular ice and cryoprotective agents on the water permeability parameters of human sperm plasma membrane during freezing. *Human Reproduction* 15(5), pp. 1125-1135, ISSN 0268-1161

Donnelly, E.T.; Steele, E.K.; McClure, N. & Lewis, S.E. (2001). Assessment of DNA integrity and morphology of ejaculated spermatozoa from fertile and infertile men before and after cryopreservation. *Human Reproduction,* 16(6), pp. 1191-1199, ISSN 0268-1161

Duty, S.M.; Singh, N.P.; Ruan, L.; Chen, Z.; Lewis, C.; Huang, T. & Hauser, R. (2002). Reliability of the comet assay in cryopreserved human sperm. *Human Reproduction* 17(5), pp. 1274-1280, ISSN 0268-1161

Ernst, E.; Ingerslev, H.J.; Schou, O. & Stoltenberg, M. (2007). Attitudes among sperm donors in 1992 and 2002: a Danish questionnaire survey. *Acta Obstetricia et Gynecologica Scandinavica,* 86, pp. 327–333, ISSN 0001-6349

ESHRE Task Force on Ethics and Law 11: Posthumous assisted reproduction (2006). *Human Reproduction,* 21(12), pp. 3050-3053, ISSN 0268-1161

Fahy, G.M. (1986) The relevance of cryoprotectant "toxicity" to cryobiology. *Cryobiology,* 23, pp. 1-13, ISSN 011-2240

Foreman, D.M. (1999). The family rule: a framework for obtaining ethical consent for medical interventions from children. *Journal of Medical Ethics,* 25, pp. 491–496, ISSN 0306-6800

Fountain, D.; Ralston, M.; Gorlin, J.B.; Uhl, L.; Wheeler, C.; Antin, J.H.; Churchill, W.H. & Benjamin, R.J. (1997). Liquid nitrogen freezers: a potential source of microbial contamination of hematopoietic stem cell components. *Transfusion,* 37, pp. 585–591, ISSN 0041-1132

Frith, L.; Blyth, E. & Farrand, A. (2007). UK gamete donors' reflections on the removal of anonymity: implications for recruitment. *Human Reproduction,* 22(6), pp. 1675-1680, ISSN 0268-1161

Fujikawa, S. & Miura, K. (1986). Plasma membrane ultrastructural changes caused by mechanical stress in the formation of extracellular ice as a primary cause of slow freezing injury in fruit-bodies of basidomycetes (Lyophyllum ulmariuam). *Cryobiology,* 23, pp. 371-382, ISSN 0386-7196

Gil-Salom, M.; Romero, J.; Rubio, C.; Ruiz, A.; Remohi, J. & Pellicer, A. (2000). Intracytoplasmic sperm injection with cryopreserved testicular spermatozoa. *Molecular and Cellular Endocrinology,* 169, pp. 15-19, ISSN 0303-7207

Golombok, S. (2005). Unusual families. *Reproductive Biomedicine Online,* 10 (Suppl. 1), pp. 9-12, ISSN 1472-6483

Gong, D.; Liu, Y.L.; Zheng, Z.; Tian, Y.F. & Li, Z. (2009). An overview on ethical issues about sperm donation. *Asian Journal of Andrology,* 11(6), pp. 645-652, ISSN 1745-7262

Gosálvez, J.; Cortés-Gutierez, E.; López-Fernández, C.; Fernández, J.L.; Caballero, P. & Nuñez, R. (2009). Sperm deoxyribonucleic acid fragmentation dynamics in fertile donors. *Fertility and Sterility*, 92, pp. 170-173, ISSN 1556-5653

Gottlieb, C.; Lalos, O. & Lindblad, F. (2000). Disclosure of donor insemination to the child: the impact of Swedish legislation on couples' attitudes. *Human Reproduction*, 15(9), pp. 2052-2056, ISSN 0268-1161

Hammadeh, M.E.; Askari, A.S.; Georg, T.; Rosenbaum, P. & Schmidt, W. (1999). Effect of freeze-thawing procedure on chromatin stability, morphological alteration and membrane integrity of human spermatozoa in fertile and subfertile men. *International Journal of Andrology*, 22(3):155-162, ISSN 0105-6263

Hauser, R.; Yogev, L.; Paz, G.; Yavetz, H.; Azem, F.; Lessing, J.B. & Botchan, A. (2006). Comparison of efficacy of two techniques for testicular sperm retrieval in nonobstructive azoospermia: multifocal testicular sperm extraction versus multifocal testicular sperm aspiration. *Journal of Andrology*, 27, pp. 28-33, ISSN 0196-3635

Health Canada. (2000). *Technical Requirements for Therapeutic Donor Insemination. Therapeutic Products Programme*, Health Canada, Ottawa, Canada, http://www.hc-sc.gc.ca/dhp-mps/brgtherap/applic-demande/guides/semen-sperme-acces/semen-sperme_directive-eng.php

Health Canada, *Therapeutic Products Programme Guidance on the Donor Semen Special Access Programme (DSSAP)*, Health Canada, Ottawa, Canada. (2011). http://www.hc-sc.gc.ca/dhp-mps/brgtherap/applic-demande/guides/semen-sperme-acces/dssap-passd_gui_doc-ori-eng.php

Henry, M.A.; Noiles, E.E.; Gao, D.; Mazur, P. & Critser, J.K. (1993). Cryopreservation of human spermatozoa. The effects of cooling rate and warming rate on the maintenance of motility, plasma membrane integrity, and mitochondrial function. *Fertility and Sterility*, 60(5), pp. 911-918, ISSN 0015-0282

Herrler, A.; Eisner, S.; Bach, V.; Weissenborn, U. & Beier, H.M. (2006). Cryopreservation of spermatozoa in alginic acid. *Fertility and Sterility*, 85(1), pp. 208-213, ISSN 1556-5653

Holt, W.V. (2000). Fundamental aspects of sperm cryobiology: the importance of species and individual differences. *Theriogenology*, 53, pp. 47–58, ISSN 0093-691X

Hovatta, O. (2003). Cryobiology of ovarian and testicular tissue. *Best Practice & Research Clinical Obstetrics & Gynaecology*, 17(2), pp. 331-342, ISSN 521-6934

Hovav, Y.; Almagor, M. & Yaffe, H. (2002). Comparison of semen quality obtained by electroejaculation and spontaneous ejaculation in men suffering from ejaculation disorder. *Human Reproduction*, 17(12), pp. 3170-3172, ISSN 0268-1161

Hwang, K. & Lamb, D.J. (2010). New advances on the expansion and storage of human spermatogonial stem cells. *Current Opinion in Urology*, 20(6), pp. 510-514, ISSN 1473-6586

Isaev, D.A.; Zaletov, S.Y.; Zaeva, V.V.; Zakharova, E.E.; Shafei, R.A. & Krivokharchenko, I.S. (2007). Artificial microcontainers for cryopreservation of solitary spermatozoa. *Human Reproduction*, 22, pp. i154, (*Abstracts of the 23rd Annual Meeting of the ESHRE*, Lyon, France, 2007)

Isachenko, E.; Isachenko, V.; Katkov, I.I.; Dessole, S. & Nawroth, F. (2003). Vitrification of mammalian spermatozoa in the absence of cryoprotectants: from past practical difficulties to present success. *Reproductive Biomedicine Online*, 6(2), pp. 191-200, ISSN 1472-6483

Isachenko, E.; Isachenko, V.; Sanchez, R.; Katkov, I.I. & Kreienberg, R. (2011). Cryopreservation of spermatozoa. Old routine and new perspectives, In: *Principles and Practice of Fertility Preservation*, J. Donnez & S.S. Kim (Ed.), pp. 177-198, Cambridge University Press, ISBN 978-953-7619-81-7, Cambridge, UK

Isachenko, E.; Isachenko, V.; Weiss, J.M.; Kreienberg, R.; Katkov, I.I.; Schulz, M.; Lulat, A.G.; Risopatrón, M.J. & Sánchez, R. (2008). Acrosomal status and mitochondrial activity of human spermatozoa vitrified with sucrose. *Reproduction*, 136(2), pp. 167-73, ISSN 1741-7899

Isachenko, V.; Isachenko, E.; Katkov, I.I.; Montag, M.; Dessole, S.; Nawroth, F. & Van der Ven, H. (2004). Cryoprotectant-free cryopreservation of human spermatozoa by vitrification and freezing in vapor: effect on motility, DNA integrity, and fertilization ability. *Biology of Reproduction*, 71, pp. 1167-1173, ISSN 0006-3363

Isachenko, V.; Isachenko, E.; Montag, M.; Zaeva, V.; Krivokharchenko, I.; Nawroth, F.; Dessole, S.; Katkov I.I. & Van der Ven, H. (2005). Clean technique for cryoprotectant-free vitrification of human spermatozoa. *Reproductive Biomedicine Online*, 10(3), pp. 350-354, ISSN 472-6483

Just, A.; Gruber, I.; Wober, M.; Lahodny, J.; Obruca, A. & Strohmer, H. (2004). Novel method for the cryopreservation of testicular sperm and ejaculated spermatozoa from patients with severe oligospermia: a pilot study. *Fertility and Sterility*, 82, pp. 445-447, ISSN 0015-0282

Katkov, I.I. (2002). The point of maximum cell water volume excursion in case of presence of an impermeable solute. *Cryobiology* 44, pp. 193-203, ISSN 0011-2240

Katkov, I.I.; Isachenko, V.; Isachenko, E; Kim, M.S.; Lulat, A.G-M.I.; Mackay, A.M. & Levine, F. (2006). Low- and high-temperature vitrification as a new approach to biostabilization of reproductive and progenitor cells. *International Journal of Refrigeration*, 29, pp. 346–357 ISSN 0140-7007

Katkov, I.I.; Katkova, N.; Critser, J.K. & Mazur, P. (1998). Mouse spermatozoa in high concentrations of glycerol: chemical toxicity vs osmotic shock at normal and reduced oxygen concentration. *Cryobiology*, 37, pp. 235–338, ISSN 0011-2240

Katkov, I.I. & Lulat, A.G. (2000). Do conventional CASA-parameters reflect recovery of kinematics after freezing? CASA paradox in the analysis of recovery of spermatozoa after cryopreservation. *CryoLetters* 21(3), pp. 141-148, ISSN 0143-2044

Koshimoto, C.; Gamliel, E. & Mazur, P. (2000). Effect of osmolality and oxygen tension on the survival of mouse sperm frozen to various temperatures in various concentrations of glycerol and raffinose. *Cryobiology*, 41, pp. 204–231, ISSN 0011-2240

Lania, C.; Grasso, M.; Fortuna, F.; De Santis, L. & Fusi, F. (2006). Open epididymal sperm aspiration (OESA): minimally invasive surgical technique for sperm retrieval. *Archivos espanoles de urologia*, 59, pp. 313-316, ISSN 0004-0614

Larson, J.M.; McKinney, K.A.; Mixon, B.A.; Burry, K.A. & Wolf, D.P. (1997). An intrauterine insemination-ready cryopreservation method compared with sperm recovery after conventional freezing and post-thaw processing. *Fertility and Sterility*, 68, pp. 143–148, ISSN 0015-0282

Lasso, J.L.; Noiles, E.E.; Alvarez, J.G. & Storey, B.T. (1994). Mechanism of superoxide dismutase loss from human sperm cells during cryopreservation. *Journal of Andrology*, 15, pp. 255-265, ISSN 0196-3635

Magelssen, H.; Brydoy, M. & Fossa, S. D. (2006). The effects of cancer and cancer treatments on male reproductive function. *Nature Clinical Practice Urology*, 3(6), pp. 312-322, ISSN 1743-4270

Mazur, P. (1984). Freezing of living cells: mechanisms and implications. *American Journal of Physiology*, 247, pp. 125-142, ISSN 0002-9513

Mazur, P.; Leibo, S.P. & Chu, E.H. (1972). A two factor hypothesis of freezing injury - evidence from Chinese hamster tissue culture cells. *Experimental Cell Research*, 71, pp. 345-355, ISSN 0014-4827

McGann, L.E. & Farrant, J. (1976). Survival of tissue culture cells frozen by a two-step procedure to -196°C. Holding temperature and time. *Cryobiology*, 13, pp. 261-268, ISSN 0011-2240

Morris, G.J.; Acton, E. & Avery, S. (1999). Novel approach to sperm cryopreservation. *Human Reproduction*, 14(4), pp. 1013-1021, ISSN 0268-1161

Mortimer, D. (2004). Current and future concepts and practices in human sperm cryobanking. *Reproductive Biomedicine Online*, 9, pp. 134-151, ISSN 1472-6483

Moskovtsev, S.I.; Jarvi, K.; Mullen, J.B.; Cadesky, K.I; Hannam, T. & Lo, K.C. (2010). Testicular spermatozoa have significantly lower DNA damage compared to ejaculated spermatozoa in patients with unsuccessful oral antioxidant treatment. *Fertility and Sterility* 93(4), pp. 1142-1146, ISSN 1556-5653

Moskovtsev, S.I.; Kuznyetsov, V.; Spiridonov, S; Lulat, A.; Crowe, M. & Librach, C.L. (2011). Comparison of vitrification and slow vapor protocols for cryopreservation of human spermatozoa. *In: Materials of 57th Canadian Fertility and Andrology Society Annual Meeting*, pp. 119, Toronto, Ontario, Canada, September 21- 24 2011

Moskovtsev, S.I; Willis, J; White, J. & Mullen, J.B. (2009). Sperm DNA damage: correlation to severity of semen abnormalities. *Urology*, 74(4), pp. 789-793, ISSN 527-9995

Mossad, H.; Morshedi, M.; Torner, J.P. & Oehninger, S. (1994). Impact of cryopreservation on spermatozoa from infertile men – implication for artificial insemination. *Archives of Andrology*, 33, pp. 51–57, ISSN 0148-5016

Nawroth, F.; Isachenko, V.; Dessole, S.; Rahimi, G.; Farina, M.; Vargiu, N.; Mallmann, P.; Dattena, M.; Capobianco, G.; Peters, D.; Orth, I. & Isachenko, E. (2002). Vitrification of human spermatozoa without cryoprotectants. *CryoLetters*, 23, pp. 93-102, ISSN 0143-2044

Nicopoullos, J.D.; Ramsay, J.W.; Almeida, P.A. &, Gilling-Smith, C. (2004). Assisted reproduction in the azoospermic couple. *An International Journal of Obstetrics and Gynaecology*, 111, pp. 1190-1203, ISSN 1470-0328

O'Connell, M.; McClure, N. & Lewis, S.E.M. (2002). The effects of cryopreservation on sperm morphology, motility and mitochondrial function. *Human Reproduction*, 17(6), pp. 704-709, ISSN 0268-1161

Patrizio, P.; Mastroianni, A.C. & Mastroianni, L. (2001). Disclosure to children conceived with donor gametes should be optional. *Human Reproduction*, 16(10), pp. 2036-2038, ISSN 0268-1161

Rivkees, S.A. & Crawford, J.D. (1988). The relationship of gonadal activity and chemotherapy-induced gonadal damage. *The Journal of the American Medical Assossiation*, 259, pp. 2123-21255, ISSN 0098-7484

Rueffer, U.; Breuer, K.; Josting, A.; Lathan, B.; Sieber, M.; Manzke, O.; Grotenhermen, F.J.; Tesch, H.; Bredenfeld, H.; Koch, P.; Nisters-Backes, H.; Wolf, J.; Engert, A. & Diehl, V. (2001). Male gonadal dysfunction in patients with Hodgkin's disease prior to treatment. *Annals of Oncology*, 12(9), pp. 1307-1311, ISSN 0923-7534

Said, T.M.; Gaglani, A. & Agarwal A. (2010). Implication of apoptosis in sperm cryoinjury. *Reproductive Biomed Online* 21(4), pp. 456-462, ISSN 1472-6491

Sanchez, R.; Isachenko, V.; Petrunkina, A.M.; Risopatron, J.; Schulz, M. & Isachenko, E. (2011). Live Birth after Intrauterine Insemination with Spermatozoa from an Oligo-Astheno-Zoospermic Patient Vitrified Without Permeable Cryoprotectants. *Journal of Andrology (in press)*, ISSN 1939-4640

Schlegel, P.N. & Su, L.M. (1997). Physiological consequences of testicular sperm extraction. *Human Reproduction*, 12(8), pp. 1688-1692, ISSN 0268-1161

Schlegel, P.N.; Palermo, G.D.; Goldstein, M.; Menendez, S.; Zaninovic, N.; Veeck, L.L. & Rosenwaks, Z. (1997). Testicular sperm extraction with intracytoplasmic sperm injection for nonobstructive azoospermia. *Urology*, 49, pp. 435-440, ISSN 0090-4295

Schover, L.R.; Rybicki, L.A.; Martin, B.A. & Bringelsen, K.A. (1999). Having children after cancer. A pilot survey of survivors' attitudes and experiences. *Cancer*, 86(4), pp. 697-709, ISSN 0008-543X

Schuster, T.G.; Keller, L.M.; Dunn, R.L.; Ohl, D.A. & Smith, G.D. (2003). Ultra-rapid freezing of very low numbers of sperm using cryoloops. *Human Reproduction*, 18(14), pp. 788-795, ISSN 0268-1161

Sereni, E.; Bonu, M.A.; Fava, L.; Sciajno, R.; Serrao, L.; Preti, S.; Distratis, V. & Borini, A. (2008). Freezing spermatozoa obtained by testicular fine needle aspiration: a new technique. *Reproductive Biomedicine Online*, 16, pp. 89-95, ISSN 1472-6483

Shimada, K. & Asahina, E. (1975). Visualization of intracellular ice crystals formed in very rapidly frozen cells at -27 degree C. *Cryobiology*, 12(3), pp. 209-218, ISSN 0011-2240

Thaler-DeMers, D. (2001). Intimacy issues: sexuality, fertility, and relationships. *Seminars in Oncology Nursing*, 17(4), pp. 255-262, ISSN 0749-2081

The American Society for Reproductive Medicine. (2004). Guidelines for sperm donation. *Fertility and Sterility*, 82, pp. S9-12, ISSN 0015-0282

Tomlinson, M. & Sakkas, D. (2000). Is a review of standard procedures for cryopreservation needed? Safe and effective cryopreservation—should sperm banks and fertility centres move toward storage in nitrogen vapour? *Human Reproduction*, 15(12), pp. 2460-2463, ISSN 0268-1161

Tomlinson, M. (2005) Managing risk associated with cryopreservation. *Human Reproduction*, 20, pp. 1751-1756, ISSN 0268-1161

Thomson, L.K.; Fleming, S.D.; Aitken, R.J.; De Iuliis, G.N.; Zieschang, J.A. & Clark, A.M. (2009). Cryopreservation-induced human sperm DNA damage is predominantly mediated by oxidative stress rather than apoptosis. *Human Reproduction* 24(9), pp. 2061-2070, ISSN 1460-2350

Tournaye, H.; Goossens, E.; Verheyen, G.; Frederickx, V.; De Block, G.; Devroey, P. & Van Steirteghem, A. (2004). Preserving the reproductive potential of men and boys with cancer: current concepts and future prospects. *Human Reproduction Update*, 10(6), pp. 525-532, ISSN 1355-4786

Tyler, J.P.; Kime, L.; Cooke, S. & Driscoll, G.L. (1996). Temperature change in cryo-containers during short exposure to ambient temperatures. *Human Reproduction*, 11(7), pp. 1510-1512, ISSN 0268-1161

Verza, Jr. S.; Feijo, C.M. & Esteves, S.C. (2009). Resistance of human spermatozoa to cryoinjury in repeated cycles of thaw-refreezing. *International Brazilian Journal of Urology*, 35, pp. 581-590, ISSN 1677-6119

Walmsley, R.; Cohen, J.; Ferrara-Congedo, T.; Reing, A. & Garrisi J. (1998). The first births and ongoing pregnancies associated with sperm cryopreservation within evacuated egg zonae. *Human Reproduction*, 13 (Suppl. 4), pp. 61-70, ISSN 0268-1161

Woods, E.J.; Benson, J.D.; Agca, Y. & Critser, J.K. (2004). Fundamental cryobiology of reproductive cells and tissues. *Cryobiology*, 48, pp. 146-156, ISSN 0011-2240

World Health Organization (1999). *WHO laboratory manual for the examination of human semen and sperm-cervical mucus interaction*, pp. 169-179, 4th ed. Cambridge University Press, ISBN 978-924-154-778-9, Cambridge, UK

4

Prevention of Lethal Osmotic Injury to Cells During Addition and Removal of Cryoprotective Agents: Theory and Technology

Dayong Gao[1] and Xiaoming Zhou[2]
[1]University of Washington, Seattle, WA
[2]University of Electronic Science and Technology of China, Chengdu,
[1]USA
[2]China

1. Introduction

Significant survival of cryopreserved cells became a reality only after the discovery and the use of cell-membrane-permeating cryoprotective agents (CPAs) (e.g. glycerol, Polge et al, 1949). Before freezing, one or various CPAs should be added to cell suspensions to prevent the cells from the cryoinjury during the freezing and thawing processes. Unfortunately, the CPAs, themselves, may have chemical toxicity to cells after thawing at room temperatures (Katkov el al, 1998). Therefore, a post-thaw washing of CPAs is required to remove CPAs from cells prior to scientific or medical applications. However, the addition of CPAs to cells before freezing and the removal of CPAs from cells after thawing may cause serious cell loss and damage if the processes are not properly handled.

"One-step" methods were formerly used for addition/removal CPAs. During the "one-step" CPA addition process, cells are directly (one-step) placed in a solution that is hyperosmotic with respect to the permeating CPA but isosmotic with respect to the impermeable salts/electrolytes. Cells first shrink because of the osmotic efflux of intracellular water and then increase in volume as the CPA permeates and as water concomitantly reenters the cells (as shown in Figure 1a). During the "one-step" CPA removal process, cells with a high intracellular concentration of CPA are directly exposed to an isotonic salt solution without CPA. Cells will swell because of an osmotic influx of extracellular water and then decrease in volume as the CPA diffuses out of the cells and as water concomitantly moves out (as shown in Figure 1b). As a result of these two aspects (i.e. addition and removal of CPAs) of the cryopreservation procedures, the cells may experience severe osmotic volume excursion causing significant cell "osmotic" injury (Sherman, 1973; Mazur and Schneider, 1984, 1986; Penninckx et al, 1984; Leibo, 1986, Crister et al, 1988a, Meryman, 2007).

Several possible reasons for the osmotic injury have been proposed, including (i) rupture of the cell membrane in hypo-osmotic conditions (i.e. expansion lysis); (ii) the water flux hypothesis: frictional force between water and potential membrane 'pores' caused cell membrane damage (Muldrew and McGann, 1994); (iii) the minimum volume hypothesis:

cell shrinkage in hyper-osmotic condition is resisted by cytoskeleton components, and the resultant interaction between shrunken cell membrane and the cytoskeleton damages the cells (Meryman, 1970); (iv) the maximum cell surface hypothesis: the cell shrinkage induces irreversible membrane fusion/change, and hence the effective area of cell membrane is reduced; when returned to isotonic condition, the cells lyse before their normal volume is recovered (Steponkus and Wiest, 1979); and (v) the solute loading hypothesis: hyperosmotic stress causes a net leak/influx of non-permeating solutes; when cells are returned to iso-osmotic conditions, they swell beyond their normal isotonic volume and lyse (Mazur et al., 1972).

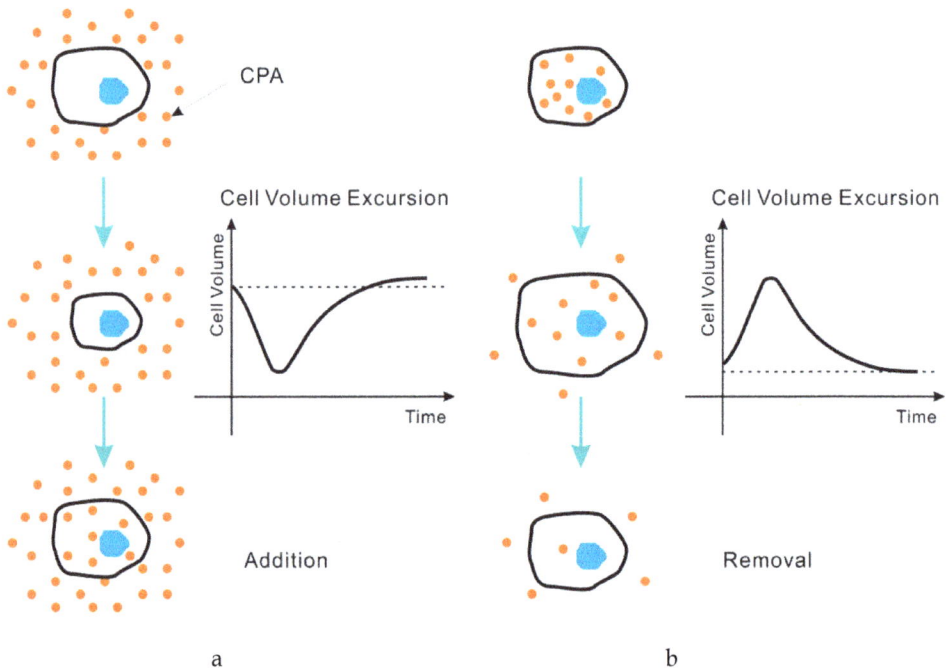

Fig. 1. Cell volume excursion during addition and removal of CPAs

In order to minimize osmotic injury, many efforts have been made and several techniques have been proposed. Basically, people utilize so-called "multi-step methods" instead of "one-step method" for addition and removal of CPAs, and the resulting cell recovery rate can be significantly improved. During the multi-step CPA addition process, solution with high CPA concentration is added into a cell suspension step by step and the CPA concentration in the cell suspension increases slowly and gradually. During the multi-step CPA removal process, an isotonic salt solution is added into the cell suspension step by step, and then by means of centrifugation CPAs in the cell suspension are removed (Figure 2). Although to some extend multi-step method reduces osmotic damage of cells, it is complex to operate, requires more laboratory staffs, and costs more time, which makes the addition and removal procedures more expensive and difficult practically.

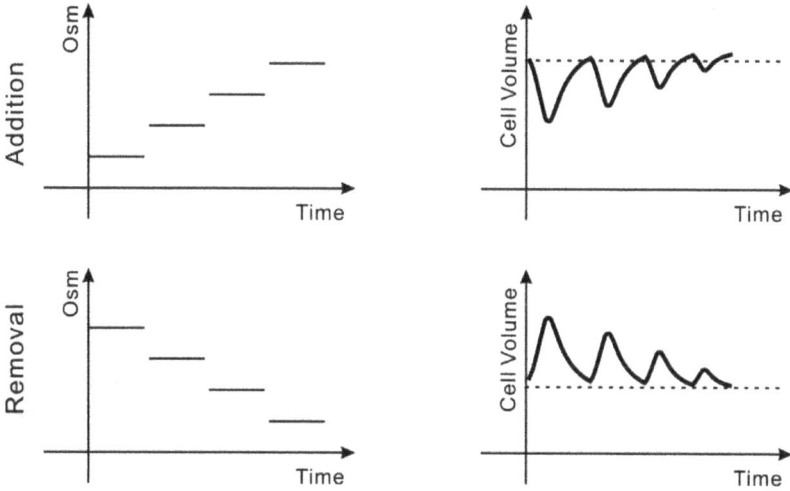

Fig. 2. Multi-step method for addition and removal of a CPA

In the past, attempts to develop procedures for the addition and removal of CPAs have been made based primarily on empirical approaches, i.e. for a given cell type, various temperatures, CPA types and concentrations, and number of procedures or steps for CPA addition and removal were empirically tested to find an acceptable procedure. Typical techniques includes (i) a multi-step addition and multi-step removal of permeating CPAs (Watson, 1979) and (ii) a multi-step addition and two-step removal (using a non-permeating solute as osmotic buffer) of CPAs (Rowe et al., 1968; Mazur and Leibo, 1977; Leibo 1981). New CPA addition-removal methods and automated devices have recently been developed based on fundamental cell membrane transport theory and engineering approaches (Gao, et al, 1995; Gilmore et al, 1997; Katkov, 1998; Myrthe, et al ,2004, Zhou, et al, 2011), which are introduced and discussed in this chapter.

2. Cell membrane transport models and mathematical formulatins

To date, a number of formalisms exist for describing the cell membrane transport process. These include a one-parameter model, a two-parameter model, and a three-parameter model, considering solute-solvent interactions.

i.   one-parameter model (Mazur et al, 1974, 1976),

The one-parameter model utilizes the hydraulic permeability (L_p) of cell membrane as the only parameter to describe the water transport across cell membrane. The model can be formulized as follows.

$$\frac{dV_w^i}{dt} = -L_p A_c \left(\Pi_e - \Pi_i \right) \qquad (1)$$

where, V_w^i is the volume of intracellular water, A_c is the area of cell membrane surface, Π_e and Π_i are the extracellular and intracellular osmotic pressures.

ii.   Two-parameter model

The two-parameter model was firstly presented by Jacob (1932-1933), and further developed by Kleinhans (1998), Katkov (2000) recently. The model utilizes the parameters L_p and P_s (CPA solute permeability) to characterize membrane permeability when water, a permeable solute and a nonpermeable solute are present:

$$\frac{dV_w^i}{dt} = -L_p A_c RT\left(M^e - M^i\right) \qquad (2)$$

$$\frac{dN_s^i}{dt} = P_s A_c\left(M_s^e - M_s^i\right) \qquad (3)$$

where N_s is the number of osmoles of solute inside cell, R is the universal gas constant, T is the absolute temperature, M_i and M_e are the intracellular and extracellular osmolality, respectively. The subscript 's' refers to permeable solute, and remaining symbols are as previously defined.

iii.   Three-parameter model

The classical formulation of coupled, passive membrane transport was developed by Kedem and Katchalshy (1958) using the theory of linear irreversible thermodynamics. The formulation includes two coupled first-order non-linear ordinary equations which describe the total transmembrane volume flux and the transmembrane permeable solute flux respectively.

In the model (so called Kedem-Katchalssky transport formalism or KK formalism), a reflection coefficient (σ) was introduced with Lp and Ps to describe water and solute (CPA) transport across the plasma membrane:

$$\frac{dV_c}{dt} = -L_p A_c RT\left[\left(M_n^e - M_n^i\right) + \sigma\left(M_s^e - M_s^i\right)\right] \qquad (4)$$

$$\frac{dN_s^i}{dt} = (1-\sigma)\bar{M}_s\frac{dV_c}{dt} + P_s A_c\left(M_s^e - M_s^i\right) \qquad (5)$$

Where V_c is cell volume, $\bar{M}_s$ is the average osmolality of intracellular and extracellular solution, and the subscript 'n' refers to nonpermeable solute, respectively.

The KK formalism used to be the most general of the three mentioned. However, more recent literature suggests that aquaporins in cell membrane are highly selective, with nonionic solute transport occurring mainly through the lipid bilayer or through other channels that are distinct from the aquaporins (Gilmore et al, 1995; Preston et al, 1992). In this case, the estimation of σ as independent parameter may be inappropriate and may not be relevant from a biological stand point (Kleinhans, 1998). By assuming that there is no interaction between water and solute during their transport through the membrane, the value of σ can be determined as $1-\left(P_s\bar{V}_s\right)/\left(RTL_p\right)$, where $\bar{V}_s$ = partial molar volume of permeating solute. In this manner, the KK formalism can still get correct result as two parameter model.

In the following context, two examples are demonstrated to show how to use cell membrane transport models and mathematical formulations to develop optimal conditions and technology/instrument for the addition and/or removal of the permeating CPAs in cells. **An important hypothesis** is that the degree of cell volume excursion can be used as an independent indicator to evaluate and predict the possible osmotic injury of the cells during addition and removal of CPAs.

Example 1: Development of optimal "multi-step methods" for addition and dilution of glycerol in human sperm

Glycerol is the most commonly used CPA in the cryopreservation of spermatozoa (Polge et al, 1949; Watson, 1979; Critser etl al., 1988a). Glycerol permeability characteristics for human spermatozoa have been very well studied and reported (Du et al, 1994; Gao et al., 1992). The hypothesis above was tested first using the following procedures: (i) to determine sperm osmotic injury as a function of its volume excursion limits (swelling/shrinking) in anisosmotic solutions containing only non-permeating solutes without glycerol; (ii) to simulate, by computer, the kinetics of water-glycerol transport through the sperm plasma membrane and to calculate the sperm volume excursion during different glycerol addition and removal processes using membrane transport equations and previously determined sperm membrane permeability coefficients for glycerol and water; (ii) combining information obtained from procedures (i) and (ii), to predict sperm osmotic injury caused by different procedures of glycerol addition and removal; and (iv) to perform experiments to test the predictions. If the hypothesis is confirmed, the above procedures also provide a methodology for predicting optimal protocols to reduce the osmotic injury associated with the addition and removal of high concentrations of glycerol in human spermatozoa.

2.1 Materials and methods

Preparation of sperm suspension

Human semen samples were obtained by masturbation from healthy donors after at least 2 days of sexual abstinence. Samples were allowed to liquefy in an incubator (5% CO2, 95% air, 37C, and high humidity) for ~1h. A total of 5 ul of the liquefied semen were used for a computer-assisted semen analysis (CASA) using CellSoft (Version 3.2/C, CRYO Resources, Ltd, Montgomery, NY, USA) (Jequier and Crich, 1986; Crister et al., 1988b). A swim-up procedure was performed to separate motile form immotile cells [layering 500ul of modified Tyrode's medium (TALP: Bavister et al., 1983) over 250 ul of semen, incubating for ~1 h in the incubator and carefully aspirating 400 ul of the supernatant in which >95% of spermatozoa were motile]. The motile cell suspensions were centrifuged at 400g for 7min and resuspended in the TALP medium (286~290 Osmol) supplemented with pyruvate (0.01 mg/ml) and bovine serum albumin (4 mg/ml), at a cell concentration of 1×10^9 cell/ml.

Assessment of human sperm membrane integrity

A methodology for the assessment of sperm membrane integrity, using dual florescent staining and flow cytometric analysis, has been developed by Garner et al. (1986) and previously validated in our laboratory (Gao et al., 1992, 1993; Noilles et al., 1993). Propidium iodide (catalogue no. P4170; Sigma Chemical Co., St Louis MO, USA) is a bright red, nucleic acid-specific fluorophore which permeates poorly into spermatozoa with intact plasma membrane, but is able to diffuse readily in to spermatozoa with a damaged membrane. 6-Carboxyfluoroscein diacetate (CFDA; Sigma, Catalog #C5041) is a membrane- permeable compound. After penetrating into cells, it is hydrolysed by intracellular esterase to 6-carboxyfluoroscein which is a bright green, membrane-impermeable fluorophore (Garner et al., 1986). When CFDA is added into the cell suspension with membrane-intact spermatozoa, the cells fluoresce bright green (Garner et al., 1986). Thus 5 ul CFDA (0.25 mg/ml DMSO) and 5 up propidium iodide (1 mg/ml water) stock solutions were added to each o.5ml of the treated sperm suspensions. A total of 1×10^5 spermatozoa per treatment were analyzed using a FACStar Plus Flow cytometer (Becton Dickinson, Rutherford, NJ, USA). The cells with CFDA staining and without propidium iodide staining were considered as intact cells. The percentage of intact cells was determined for each treatment.

The flow cytometer settings used for the experiments were (i) the gates were set using forward and 90° light scatter signals at acquisition to exclude debris and aggregates; (ii) instrument alignment was performed daily with fluorescent microbead standards to standardize sensitivity and setup; (iii) photomultiplier settings were adjusted with unstained overlap with individually stained cells; (iv) excitation was at 488 nm from a 4 W argon laser operating at 200 mW. Fluorescein emission intensity was measured using a 530/30 nm bandpass filter, and propidium iodide intensity using a 630/22 m bandpass filter.

Determination of osmotic injury as a function of sperm volume excursion in anisosmotic solutions of nonpermeating solutes

The anisosmotic solutions, ranging from 40 to 1500 mOsmol, were prepared as follows: hypo-osmotic solutions were made osmotic solutions were made by adding sucrose to TALP medium (sucrose and the solutes in TALP medium are essentially membrane-impermeable compounds). The final osmolality of each solution was measured and checked using a freezing-point depression osmometer (Adanced DigiMate Osmometer, Model 3D2; Advanced Instrument, Inc., Needham Heights, MA, USA). The osmotic tolerance of human spermatozoa was evaluated by exposing the cells to the anisosmotic solutions. A 10ul volume of isotonic cell suspension (286 mOsmol, 1×10^9 cells/ml) was mixed with 150µl of each anisosmotic solution. After 1 s to 30 min, spermatozoa in each anisosmotic solution were returned to near isotonic conditions (272-343 mOsmol) by adding 1500 µl isotonic TALP medium to 100 µl of each anisosmotic cell suspension. Sperm motility and plasma membrane integrity were measured by CASA and CFDA-propidium iodide dual fluorescent staining techniques respectively before and after the anisosmotic exposure. The centrifugal force used in sample preparation was 400 g for 7 min. All experiments were conducted at 22°C.

Thermodynamic modeling and mathematical formulation for glycerol and water permeating across the human sperm membrane

The next step was to compute the osmotic cell volume excursions associated with the addition and removal of hyperosmotic solutions of the permeating cryoprotectant glycerol to suspensions of human spermatozoa in isotonic saline. The classical KK formalism (shown as equations (4) and (5)) is used here and for the case of a solution consisting of a single permeable solute (e.g. glycerol) the average of extracellular and intracellular cryoprotective agent concentrations (osmolality) can be given as

$$\overline{M}_s = \left(M_s^e - M_s^i\right)\Big/\left[\ln\left(M_s^e/M_s^i\right)\right]$$

Since human spermatozoa behave as ideal osmometer (Du et al., 1993), intracellular concentrations of impermeable solute (salt) and permeable solute (cryoprotective agent) can be calculated as previously described (Mazur and Schneider, 1984).

$$M_n^i(t) = M_n^i(0)\left(\frac{V(0) - V_b - \overline{V}_s N_s^i(0)}{V(t) - V_b - \overline{V}_s N_s^i(t)}\right) \tag{6}$$

$$M_s^i(t) = \left(\frac{N_s^i(t)}{V(t) - V_b - \overline{V}_s N_s^i(t)}\right) \tag{7}$$

Where V_b= osmotically inactive cell volume (um3), and 0=initial condition (t=0). Initial conditions for V(0), $M_n^i(0)$, $M_s^i(0)$, $N_s^i(0)$ are known based on each experimental condition or protocol. In the computer simulations, it was assumed that extracellular concentrations of permeating or non-permeating solutes were constant, and that the mixture of solutions during the glycerol addition and removal was instantaneous, i.e. the mixing time =0.

Human sperm volume, surface area, v_b, water and glycerol permeability coefficients have been determined and previously published (Gao et al., 1992; Kleinhans et al., 1992; Noiles et al., 1993; Du et al., 1994). The values of these parameters are shown in Table 1. Assuming that there is no interaction between water and glycerol during their transport through the sperm membrane (or in other words, water and glycerol penetrate the cell membrane independently), the value of $\sigma = 1 - \left(P_s\overline{V}_s\right)\Big/\left(RTL_p\right)$ (Kedem and Katchalsky, 1958), can be calculated. From this equation and the data in Table 1, σ was calculated to be 0.99. This value was used in the present example.

Surface area	(A)	120μm2	Kleinhans et al (1992)
Volume	(V)	34μm3	(Kleinhans et al. (1992)
Osmotically inactive volume	(Vb)	16.6μm3	Kleinhans et al. (1992) Du et al. (1993)
Water permeability coefficient	(Lp)	2.4μm/min/atm	Noiles et al. (1993)
Glycerol permeability coefficient	(Ps)	1.68×10-3cm/min	Gao et al. (1993)

Table 1. Characteristic of human spermatozoa at 22°C

Using equations [4-7] kinetics of glycerol/water transport across the sperm plasma membrane as well as the cell volume excursion during different glycerol addition and removal procedures were calculated using a commercial differential equation solver, SLAB (Civilized Software, Inc., Bethesda, MD, USA). The sperm volume excursion and water transport through the membrane of cells in anisosmotic solution without glycerol were calculated using equation [4] and [5] with M_s=0 and N_s=0.

Addition of glycerol

A final 1.00 M glycerol in sperm suspension was achieved by 1:1 (v/v) mixing of the original, isotonic sperm suspension with 2.0M glycerol solution which contains an isotonic (non –permeating solute) salt concentration. Two approaches for mixing the 2.0 M glycerol solution with the sperm suspension were used, i.e. a fixed-volume-step (FVS) approach and a fixed –molarity-step (FMS) approach:

Approach 1: fixed-volume-step addition

A 2.0 M glycerol solution was added stepwise to the sperm suspension, and the volume of the 2.0 M glycerol solution added in each step was constant. For example, to make a four step addition of 1ml of 2.0 M glycerol solution to a 1 ml isotonic sperm sample, 0.25 ml of 2.0 M glycerol solution would be added four times to the isotonic sperm suspension. The time interval between any two steps was 0.5-1 min.

In the general case, the volume of cryoprotective agent stock medium added to cell suspension in each step can be calculated by the following equation:

$$V_i = \frac{M_f \times V_o}{M_o - M_f} \times \frac{1}{n} \tag{8}$$

Where M_f= the final CPA concentration (molarity) in the cell suspension, M_o = cryoprotective agent concentration (molarity) in the original stock cryoprotective agent medium, n= total number of steps, i=ith step addition, V_o= the original volume of isotonic cell suspension, and V_i= the volume of CPA stock medium added into cell suspension at the ith step.

Approach 2: fixed-molarity-step addition

Glycerol-containing medium was added stepwise into the cell suspension in such a way that the glycerol molar concentration in the cell suspension was increased by a fixed amount after each step of addition. For example, to increase the molarity by 0.25 M in each of four steps, 0.14, 0.19, 0.27 and 0.4 ml of 2.0 M glycerol stock solution should be added (step by step, four steps in total) to 1 ml of the sperm suspension. The time interval between any two steps was 0.5-1min.

In the general case, the volume of cryoprotective agent stock medium added to cell suspension at the ith step can be calculated by the following equation:

$$V_i = \frac{M_f \times V_o \times n \times M_o}{\left(nM_o - iM_f\right)\left[nM_o - \left(i - 1\right)M_f\right]} \quad \text{where i=1, ..., n} \tag{9}$$

$$V_i = \frac{1}{\lambda n\left(V_o/V_{i-1}^*\right) - 1} \times V_{i-1}^* \tag{10}$$

$$V_{i-1}^* = V_o + \sum V_k \quad \text{where k=1, ..., i-1} \tag{11}$$

$$\lambda = \frac{M_o}{M_f} \tag{12}$$

$$\Delta M = \frac{M_f}{n} \qquad (13)$$

Where Mf = the final cryoprotective agent concentration in the cell suspension (molarity), M_o = cryoprotective agent concentration in the original stock cryoprotective agent medium (molarity), n= total number of steps, $i =i$th-step addition, V_o= the original volume of isotonic cell suspension (ml), ΔM= increment of glycerol molarity in cell suspension after each step of glycerol addition, V_{i-1}^* = the total volume of cell suspension before the ith-step addition, V_i= volume of cryoprotective agent stock medium added to cell suspension at the ith step.

Removal of glycerol

To dilute the concentrated glycerol in the sperm suspension and remove glycerol from the cells, an isotonic without glycerol was added stepwise to the suspension. The FVS approach, FMS approach, and a two-step osmotic buffer approach were used for the dilution.

Approach 1: FVS dilution

Given the volume of the sperm suspension (V_o) with an initial cryoprotective agent concentration (M_o), the total volume of isotonic solution required to dilute the cryoprotective agent concentration from M_o to M_s can be calculated by the following equation:

$$V = \left[\frac{M_o}{M_s} - 1 \right] \qquad (14)$$

Using the FVS approach, the volume of isotonic solution which needs to be added to be cell suspension at the ith-step during the first n-1 steps (n steps in total) can be calculated as follows:

$$V_i = \frac{V}{n-1} = \frac{V_o}{n-1} \left[\frac{M_o}{M_s} - 1 \right] \qquad (15)$$

where M_s = cryoprotective agent concentration in the cell suspension (molarity) after n-1 step dilutions, M_o =cryoprotective agent concentration initial sperm suspension (molarity), n= total number of steps, i=the ith-step addition, V_o= original volume of cell suspension (ml) and V_i= volume of isotonic solution added into cell suspension at the ith step. After n-1 steps of addition of isotonic solution into the cell suspension, the diluted sperm suspension was centrifuged (400 g for 5-7 min), and then the sperm pellet was resuspended in an isotonic solution, which results in the last (nth) step removal of glycerol from the cells.

Approach 2: FMS dilution

Concentrated glycerol in the sperm suspension was diluted stepwise by addition of an isotonic solution. The decrement in the molarity of glycerol after each step dilution was fixed. In the general case, the following equation can be used to calculate the volume of isotonic solution added to cell suspension at the ith step during the first n-1 step (n steps in total):

$$\Delta M = \frac{M_o}{n} \qquad (16)$$

$$V_i = \frac{1}{n\left(V_o/V_{i-1}^*\right)-1} \times V_{i-1}^* \quad \text{where } i=1, \ldots, \text{n-1} \tag{17}$$

$$V_{i-1}^* = V_o + \sum V_k \quad \text{where } k=1, \ldots, \text{i-1} \tag{18}$$

where ΔM= the decrement in the glycerol molarity in the spermatozoa after each stepwise addition of the isotonic solution, M_o = cryoprotective agent concentration (molarity) in the initial sperm suspension, n= total number of steps, $i=i^{th}$-step addition, V_o=original volume of cell suspension, V_{i-1}^* = the total volume of cell suspension before the ith-step addition and V_i= volume of isotonic solution added into cell suspension at i^{th} step. After n-1 step of addition , the cryoprotective agent concentration in the cell was diluted to ΔM. Then spermatozoa were transferred to isotonic conditions, which is the last (the nth) step removal of glycerol, see Table 2 fore examples.

Approach 3: Two-step dilution with an osmotic buffer

Eight-step dilution	
Fixed-volume-step method	Fixed-molarity-step method
Add 100 µl of isotonic TALP seven times to 100 µl of sperm suspension to achieve a final glycerol concentration of 0.125 M. After centrifugation, 710 µl of supernatant is taken off. The remaining cell suspension is 90 µl	(1) Stepwise add 14.3, 19, 26.6 and 40 µl of isotonic TALP medium to 100 µl of sperm suspension with 1.0 M glycerol; (2) centrifuge the supernatant; stepwise volume add 10, 20 and 60 µl of isotonic solution to the remaining 30 µl of sperm suspension. After the seven dilution steps, the glycerol concentration in the sperm suspension is 0.125 M. The final suspension volume is 120 µl.
The final sperm suspensions (90 or 120 µl) were further diluted by adding 180 µl of TALP solution. The time interval between any two steps was ~0.5-1 min. The volume of diluent added in each step was calculated using equation [8] or [9]	
One-step dilution	
Add 2000 µl of isotonic solution directly to 100 µl of cell suspension with 1.0 M glycerol	

Table 2. Procedures used in one-step and eight-step removal of 1.0 M glycerol from human spermatozoa

1.	Add 2000 µl of sucrose buffer medium (TALP + sucrose, 600 mOsm to 100 µl of sperm suspension with 1.0 M glycerol. (The total length of time spermatozoa were in contract with sucrose was 0.5 min before centrifugation.
	Centrifuge the suspension (400 g for 7 min) and aspirate the supernatant.
	Resuspend the cell pellet with 500 µl of isotonic TALP medium

Table 3. Procedures used in the two-step removal of 1.0 M glycerol from spermatozoa using sucrose as an osmotic buffer

In the first step, glycerol was directly removed by transferring cells to a hyperosmotic medium (osmotic buffer, TALP with sucrose) containing no glycerol but only non-permeating solutes (salts and sucrose), and in the second step spermatozoa in the osmotic buffer were directly transferred to an isotonic solution (TALP), (Table 3) (Rowe et al, 1968; Mazur and Leibo, 1977; Leibo, 1981).

Experimental examination of the predicted osmotic injury during addition/removal of glycerol

Medium (TALP) with 2.0M glycerol was added either in one step or stepwise (using FVS or FMS approaches) to an equal volume of the isotonic sperm suspension to achieve a final 1.0 M glycerol concentration at 22°C. The glycerol in the spermatozoa was removed/diluted by a one-step or stepwise addition (using FVS or FMS approaches) of TALP medium, with or without an osmotic buffer (sucrose), to the cell suspension. Some detailed procedures for the removal of glycerol are described in Table 2 and 3. Sperm motility and plasma membrane integrity were measured before and after the different glycerol addition and removal procedures by CASA and the dual staining technique and flow cytometry respectively.

Statistical analysis

Data were analyzed using standard analysis of variance approaches with the General Linear Models procedure of the Statistical Analysis System (Spector et al., 1985). Comparisons were conducted using a protected LSD (least significant difference) approach (Zar, 1984).

2.2 Result

The percentage of spermatozoa which maintained motility or plasma membrane integrity after each treatment was normalized to that of untreated, isotonic control samples and the data are so presented.

Determination of osmotic injury as a function of sperm volume excursion

Human spermatozoa were exposed for 5min to hyper- or hypo-osmotic solutions of sucrose and TALP salts ranging in concentration from 60 to 1200 mOsmol, and their motilities were then determined by CASA while still in those solutions. Figure 3 shows that sperm motilities dropped significantly when the osmolality was >50 mOsmol above or below isotonic (286 mOsmol). Motilities approached zero when the osmolalities were <200 or >600 mOsmol.

The next step was to compare these motilities with those observed after spermatozoa were transferred from these anisosmotic solutions back to near isotonic solutions. Figures 4 and 5 show the motilities as a function of time after transfer from hyperosmotic or from hypo-osmotic exposures respectively. In both cases, the more the initial exposure departed from isotonicity, the greater the damage upon return to isotonicty. Most, or all, of the damage was evident in the first 30 s after the return, although in the case of transfer from hypertonic solutions to near isotonic, there was a further slight and gradual decline over the ensuing 30 min.

Figure 6 compares sperm motilities after a 5 min exposure to the various anisosmotic solutions before and after the return to near isotonic conditions. The reduction in the motilities of spermatozoa exposed to hypo-osmotic media was not affected by the return to isotonic media, but most of the apparent loss of motility of spermatozoa in hyperosmotic media of between 286 and 600 mOsmol was reversed when spermatozoa were returned to near isotonic. For example, although only 10% of spermatozoa were motile in 600 mOsmol

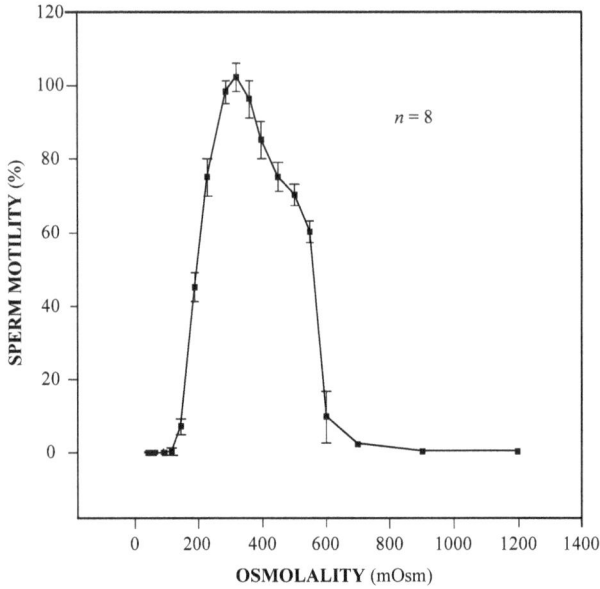

Fig. 3. Percent motility (mean±SEM, n=8) of human spermatozoa which were abruptly (one-step) exposed to different osmotic conditions for 5 min at 22°C.

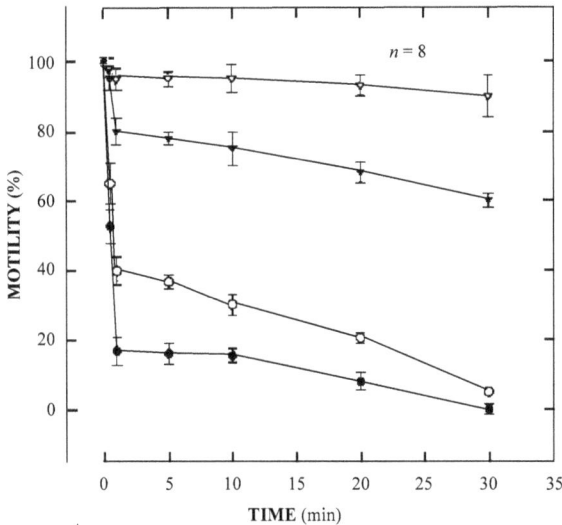

Fig. 4. Percent motility (mean±SEM, n=8) of human spermatozoa which were abruptly (one-step) returned to near isotonic conditions (305-343 mOsmol) after they had been exposed to different hyperosmotic conditions (TALP +sucrose) for different periods of time. ▽, 600 mOsmol; ▼, 700 mOsmol;○, 900 mOsmol; ●, 1200 mOsmol.

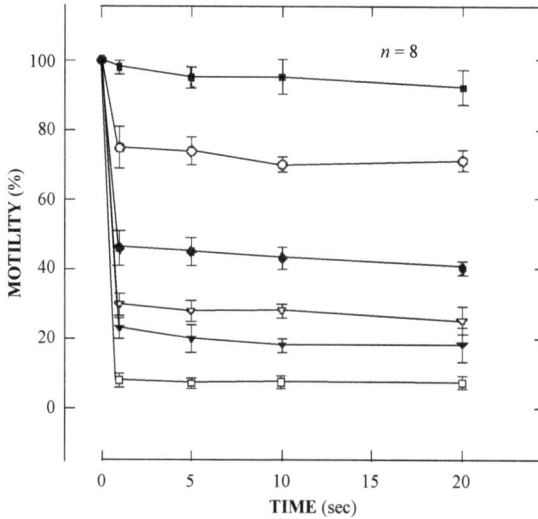

Fig. 5. Percent motility (mean±SEM, n=8) of human spermatozoa which were abruptly (one-step) returned to near isotonic conditions (273-284 mOsmol) after they had been exposed to different hyperosmotic conditions (TALP +water) for different periods of time. ■, 240 mOsmol; ○, 215 mOsmol; ●, 190 mOsmol; ▽, 143 mOsmol; ▼, 114 mOsmol; □, 90 mOsmol.

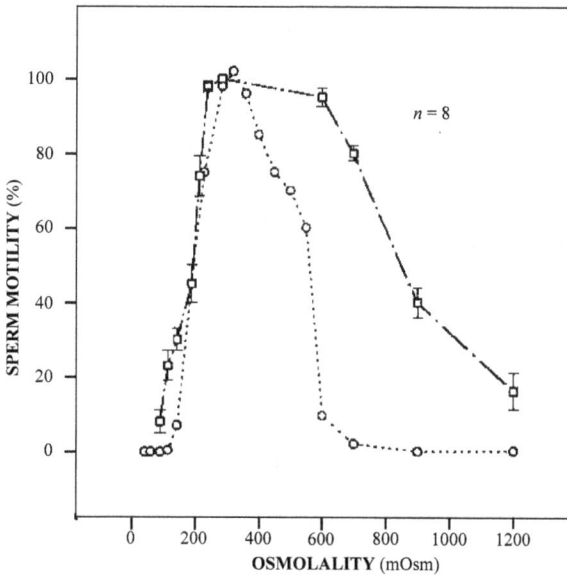

Fig. 6. A comparison of human sperm motility (% mean±SEM, n=8) after a 5 min exposure to the various hypo- and hyperosmotic solutions of non-permeating solutions before (○) and after (□) the return to near isotonic conditions (273-343 mOsmol).

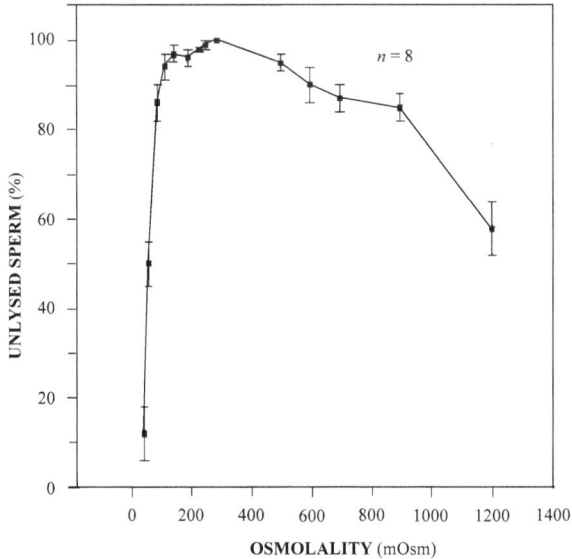

Fig. 7. Membrane integrity (CFDA and propidium iodide stain) (% mean±SEM, n=8) of human spermatozoa which were abruptly (one-step) returned to near isotonic conditions (273-343 mOsmol) after they had been exposed to different anisosmotic conditions for 5 min.

solutions, 95% of spermatozoa were motile after return to isotonic media. The return to near isotonic became especially damaging, however, when the initial hyperosmotic concentration was >600 mOsmol.

Figure 7 shows that integrity of the plasma membrane of spermatozoa (as assessed by CFDA /propidium iodide) was substantially more resistant to wide excursions from isotonicity than was motility. Thus, >90% of those spermatozoa exposed to a 90 mOsmol salt solution retained intact plasma membrane after return to near isotonic, whereas <10% remained motile both before and after return to isotonic. Loss of plasma membrane integrity in 50% of the spermatozoa occurred only when spermatozoa were exposed to a 60 mOsmol solution, a figure that agrees with a previous report (Noiles et al, 1993); that loss occurs whether or not spermatozoa are returned to isotonic. This has been interpreted to represent lysis from the attainment of a cell volume in excess of that tolerated by the surface area of the plasma membrane.

Using light microscopy, morphological changes in sperm cells were observed during the exposure to anisosmotic solutions. In a portion of the spermatozoa, the tail region became configured as a 'zigzag' pattern when exposed to a hyper-osmotic solution. The pattern of sperm tail curling in hypo-osmotic solutions was osmolality dependent, which is consistent with a previous report (Jeyendran et al., 1984). In addition, the curling of sperm tails occurred not only when the isotonic spermatozoa were exposed to a relative hypo-osmotic condition. (For example, the shrunken spermatozoa in hyperosmotic solutions were returned to iso-osmotic conditions. Iso-osmolality was 'hypo' relative to a given hyper osmolality.) The tail curling was irreversible. The mechanism(s) behind the morphological change is not clearly understood.

Calculated volume excursions associated with exposures to anisosmotic solutions

Since it has been shown that human spermatozoa behave as ideal osmometer over most of the range of osmolalities studied here (Du et al., 1993), a direct physical consequence of the exposures to anisosmotic conditions is major excursion in cell volume. The kinetics of volume excursion of spermatozoa in these hypo- and hyperosmotic solutions (containing only non-permeating solutes) were calculated and are plotted in Figure 8A and B

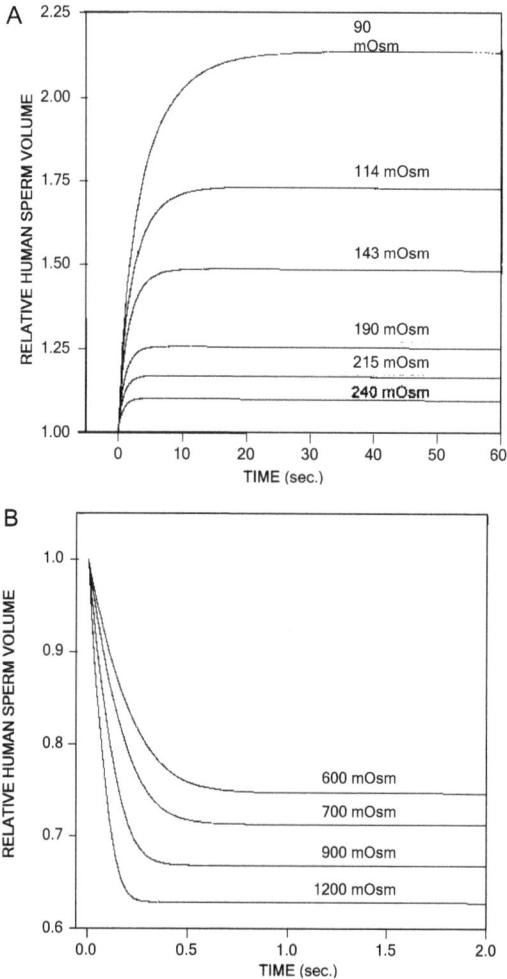

Fig. 8. (A) Calculated relative sperm volume (normalized to an isotonic sperm volume of 1) as a function of time after spermatozoa were one-step exposed to different hypo-osmotic solution containing non-permeating solutes. (B) Calculated relative sperm volume (normalized to the isotonic sperm volume of 1) as a function of time after the isotonic spermatozoa were one-step exposed to different hyperosmotic solutions containing non-permeating solutes.

respectively, indicating that only a short time was required for human spermatozoa to achieve osmotic equilibration (<1 s for shrinking, and ≤30 s for swelling). Figure 8A and B also show the maximum or minimum volume of spermatozoa when they were osmotically equilibrated with each anisosmotic solution. Sperm equilibration volume as a function of extracellular osmolality is shown in Figure 9, which can be calculated using equation (6) (no cryoprotective agent) or obtained directly from Figure 8A and B. To obtain a high (>95%) motility recovery, the lowest and highest osmolalities which human spermatozoa can tolerate (Figures 4 and 5) were found to be close to 240 and 600 mOsmol respectively. At these two osmolalities, the corresponding cell volume at osmotic equilibrium were directly estimated (Figure 9) to be ~1.1 (for 240 mOsmol) and 0.75 (for 600 mOsmol) times the isotonic sperm volume, indicating that spermatozoa can only swell or shrink in a relatively narrow range to maintain high post-anisosmotic motility recovery. Based on Figure 4, 5 and 9, Figure 10 was plotted, which clearly shows the post-anisosmotic injury (motility loss) as a function of osmotic equilibrium volume of spermatozoa in anisosmotic solutions. Defining lower volume limit (LVL) and upper volume limit (UVL) as cell volumes at which 5% of motile spermatozoa may irreversibly lose their motility, or, reciprocally, 95% of spermatozoa maintain their motility, one can obtain the LVL and UVL values for human spermatozoa from Figure 10 as follows: LVL =0.75×isotonic sperm volume, UVL=1.10× isotonic sperm volume.

Prediction of optimal protocols for glycerol addition/removal

The kinetics of human sperm volume excursion during one-step addition and removal of 0.5-2.0 M glycerol were calculated using equations (6-9) and are shown in Figure 11A and B respectively. The higher the glycerol concentration, the longer the time period taken for sperm volume recovery and the greater the volume excursion.

Two different approaches, i.e. fixed-volume-step (FVS) and fixed-molarity-step (FMS), for the addition/removal of glycerol in spermatozoa were considered and used in the present example. Based equations (6-9), the kinetics of water and glycerol transport through the

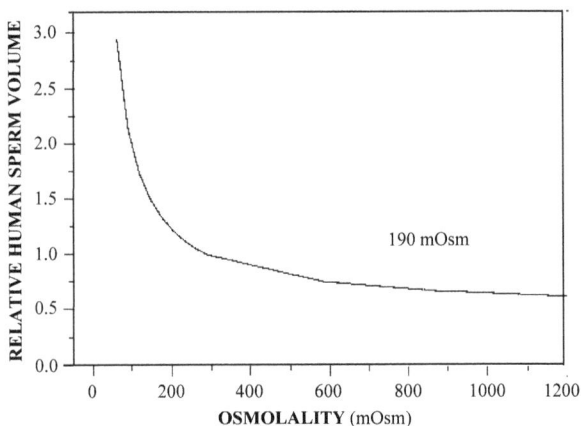

Fig. 9. Calculated relative sperm volume (normalized to the isotonic sperm volume of 1) after spermatozoa were osmotically equilibrated to different anisosmotic conditions.

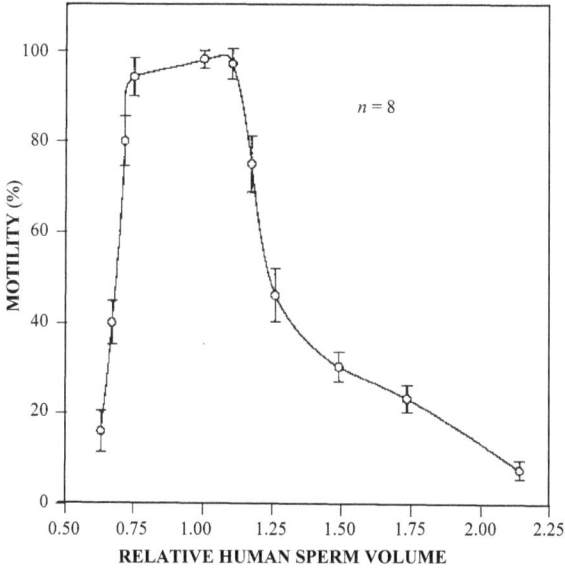

Fig. 10. Post-anisosmotic sperm motility recovery as a function of relative sperm volume (normalized to the isotonic sperm volume of 1) in different anisosmotic equilibrium states. Human spermatozoa were abruptly (one-step) returned to near isotonic conditions after exposure to anisosmotic conditions for 1 min.

sperm membrane were simulated by computer. Figure 12 shows the calculated sperm volume excursion during a one−step or four-step addition of glycerol achieve a final 1.0 M glycerol concentration at 22 C using the FMS and FVS approaches respectively. From Figure 12, a one-step addition of glycerol to spermatozoa was predicted to cause ~20% sperm motility loss because the minimum volume which the cells would achieve during this glycerol addition was ~72% of the cells would achieve during this glycerol addition was ~72% of the original cell volume, i.e. below the LVL (75% or 0.75 ×isotonic sperm volume). In contrast, a four-step FMS glycerol addition was predicted to be able to prevent sperm loss (<5% loss). Figure 12 also shows a comparison between a four-step FVS and FMS approach. A four-step FVS method was predicted to cause a lower minimum volume than a four-step FMS method. From Figure 13, a one-step removal of 1.0 M glycerol was predicted to cause >70% motility loss, because the maximum cell volume during the glycerol removal was calculated to be in excess of 1.6 times the isotonic cell volume, which is much higher than the UVL (1.1×isotonic sperm volume). Figure 14 shows that a four- or six-step FMS removal procedure was predicted to reduce sperm motility loss significantly, but these still may cause >*5 % motility loss, while an eight-step FMS removal was predicted to able to prevent sperm motility loss (<5% loss). Figure 13 also shows a comparison between an eight-step FMS and an eight-step FVS removal procedure. An eight-step FVS removal was predicted to cause a maximum cell swelling >1.2* isotonic cell volume (>UVL), while the maximum cell volume during an eight-step FMS removal was predicted to be much lower than the UVL, indicating the eight-step FVS removal is not as good as an eight-step FMS. Based on the data presented in Figures 11-14, it was also found, from calculations, that human spermatozoa will

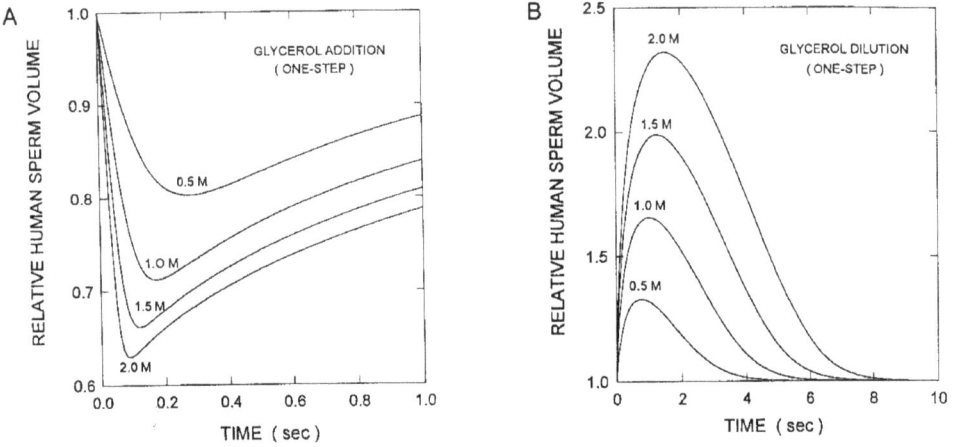

Fig. 11. (A) Calculated relative sperm volume (normalized to the isotonic sperm volume of 1) as a function of time after the isotonic sperm were exposed to different hyperosmotic glycerol solution isotonic with respect to non-permeating solutes (salt). (B) Calculated relative sperm volume (normalized to the isotonic sperm volume of 1) as a function of time after spermatozoa, which had been pre-equilibrated with different hyperosmotic glycerol solutions isotonic with respect to non-permeating solutes (salt), were one-step exposed to isotonic (286 mOsmol) saline solution without glycerol.

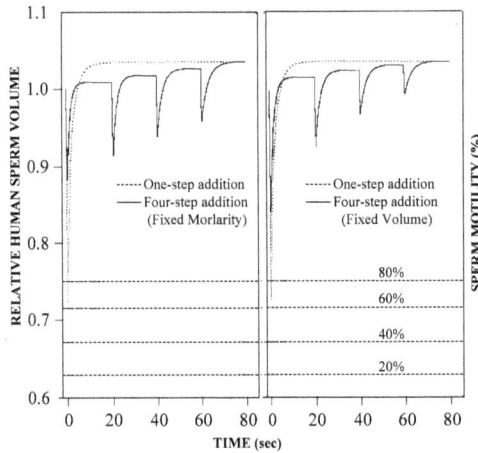

Fig. 12. (left) Calculated relative sperm volume (normalized to the isotonic sperm volume of 1) as a function of time after 1M glycerol was added to spermatozoa by either one-step or four fixed molarity steps. (right) Calculated relative sperm volume (normalized to the isotonic sperm volume of 1) as a function of time 1M glycerol was added to spermatozoa by either one step or four fixed-volume steps. The estimates of percent motility recovery as a function of sperm relative volume were obtained from Figure 8 and are indicated in the diagrams.

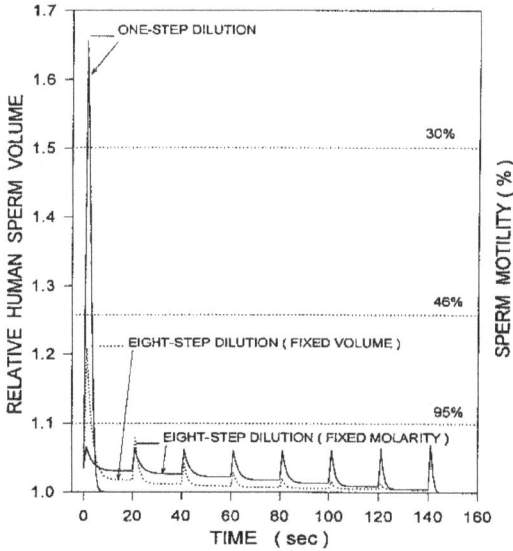

Fig. 13. Calculated relative sperm volume (normalized to the isotonic sperm volume of 1) as a function of time after 1 M glycerol was removed from spermatozoa by one-step, eight fixed-molarity steps or eight fixed-volume steps. The estimates of percent motility recovery as a function of sperm relative volume were obtained from Figure 10 and are indicated in the diagrams.

rapidly achieve an osmotic equilibrium (within 15 s) during any stepwise addition or removal of glycerol. For example, from the calculations, human spermatozoa achieve osmotic equilibrium within 15 s after each step addition of glycerol by either one-step of four-step addition (Figure 12). This indicates that only a short time interval between steps of glycerol addition/removal is required for cells to achieve corresponding osmotic equilibration volume after each step of glycerol addition and removal.

In the analysis above, sperm osmotic injury (motility loss) caused by different glycerol addition/removal procedures has been predicted and a four-step FMS addition and an eight-step FMS removal of 1.0 M glycerol were found to be acceptable protocols to prevent sperm motility loss (<5%).

Theoretical evaluation of two-step glycerol removal using an osmotic buffer

A two-step removal of cryoprotective agent from human spermatozoa using a non-permeating solute as an osmotic buffer has been previously used to avoid osmotic injury in other cell types (Rowe et al., 1968; Leibo and Mazur, 1978; Watson, 1979). The steps involved in this approach are (i) the cryoprotective agent is directly removed and cell swelling is reduced by transferring cells with the cryoprotective agent to a hyperosmotic medium (osmotic buffer) of non-permeating solutes; and (ii) the cells in the osmotic buffer are rehydrated by directly transferring them to isotonic solution. Since current results showed that 600 mOsmol was the hyperosmotic upper tolerance limit for human spermatozoa to maintain 95% motility, the osmolality of the osmotic buffer medium should not exceed 600 mOsmol. Using this liming criterion, a hyperosmolality of 600 mOsmol would be expected to provide the maximum 'buffer effect' to reduce sperm volume swelling during the first step of glycerol removal. Sperm volume excursion during this two-step glycerol removal process was calculated and is shown in Figure 15. It was predicted that the maximum volume spermatozoa would achieve is 1.25 times (15%) the isotonic cell volume, which is higher than the UVL of human spermatozoa, and could be expected to cause >40% sperm motility loss, as predicted from Figure 10.

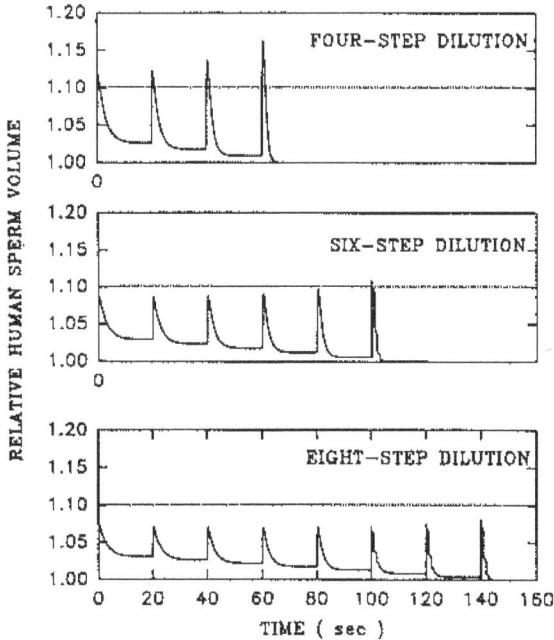

Fig. 14. Calculated relative sperm volume (normalized to the isotonic sperm volume of 1) as a function of time after 1 M glycerol was removed from spermatozoa by four, six and eight fixed-molarity steps. The dotted lines in this figure indicate the upper volume limit, 1.1, below which >95% of spermatozoa can maintain the motility. The four- or six-step dilution results in a cell volume excursion causing >5% motility loss.

Results from experimental examination

Glycerol was added to or removed from human spermatozoa using stepwise procedure to test the theoretical predictions. A one-step addition resulted in ~19.2% sperm motility loss or 81.8±8.7% ($\overline{X} \pm SEM$, n=15) motility recovery, while the four-step FMS or FVS addition significantly (P<0.001) increased in the motility recovery to 93.5±5.6% ($\overline{X} \pm SEM$, n=15) or 91±4.8% ($\overline{X} \pm SEM$, n=15) respectively. During different glycerol removal procedures (c.f. Table 2), <30% (28.5±3.8%, n=15) of motile spermatozoa kept their motility after a one-step removal of 1.0 M glycerol, while the majority of spermatozoa (92±8.2%, n=15) maintained motility after the eight-step FMS removal. In comparison, only 62±5.8% of spermatozoa maintained motility after eight-step FVS removal. The motility recovery after a two-step removal of glycerol (Table 3) using sucrose as an osmotic buffer was 43±5.3% ($\overline{X} \pm SEM$, n=15). The experimental result agreed well with the predictions generated from the computer simulations. Data analyses indicated that the different glycerol removal procedures caused different motility losses (P<0.001 between any two procedures). Over 90% of spermatozoa maintained membrane integrity under all experimental conditions.

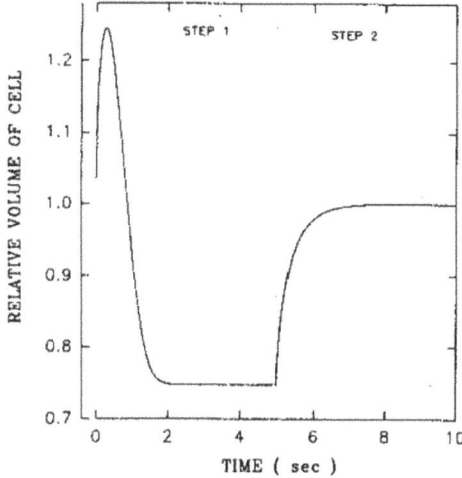

Fig. 15. Calculated relative sperm volume (normalized to the isotonic sperm volume of 1) as a function of time after 1 M glycerol was removed from spermatozoa by two steps using a 'hyperosmotic buffer' solution. Step 1: 1.0 M glycerol was removed from spermatozoa by one -step exposure of spermatozoa to 600 mOsmol hyperosmotic (salt+sucrose) solution without glycerol. Step 2: Spermatozoa in the 600 mOsmol solution were returned to isotonic condition (286 mOsmol) in one step.

Example 2: Development of a novel dilution-filtration method and instrument to remove glycerol from human red blood cells (RBCs)

Cryopreservation has been widely used today around the world for long term preservation of RBCs. In the USA, the FDA has approved the storage of frozen RBCs at -80°C for as long as 10 years (Meryman, 2007). However, the glycerol in RBCs must be reduced to final concentration below 1% before infusion to prevent hemolysis (Valeri et al, 2001). The step of removing CPAs may cause serious cell loss due to the cell volume excursion induced by osmotic disequilibria (Meryman, 2007). In the past decades, many efforts have been made to improve the process (Rowe et al, 1968; Meryman et al, 1972, 1977; Valeri et al, 1975, 2001; Castino et al,1996; Arnaud et al 2003).

Currently, multi-step centrifuging methods are most commonly used, and some of them can achieve favorable results (Rowe et al, 1968; Meryman et al, 1972, 1977; Valeri et al, 1975, 2001). However, the procedures are very difficult and time consuming for manual operation due to the large cell suspension volume or high CPA concentration. In addition, most of the systems are not closed and are thus open to contamination (Castino et al,1996; Valeri et al, 2006). Automatic centrifuging systems may significantly reduce human labor and contamination (Valeri et al, 2001), but the expensive cost limits their application in many areas. Recently, Dialysis was considered as an alternative method by some researchers (Castino et al,1996; Arnaud et al 2003; Ding et al 2007,2010). It can remove CPAs efficiently; however, due to the non-uniformity of distribution of hollow fibers, the mass transport in dialyzer is too complicated to be controlled, especially in the unsteady state. In addition, dialysis method is not efficient to remove large molecular substances (Daugirdas, et al, 2006), such as cell fragment and the released protein from broken cells. These factors limit the use of dialysis method in some applications.

In clinic, hemofiltration, which involves dilution and filtration to remove toxins from blood, has been proved to have better controllability as well as ability of removing large molecular substances than hemodialysis (Daugirdas, et al, 2006). By referencing to hemofiltration, a dilution-filtration system is developed recently for removing CPAs (Zhou et al, 2011). The closed system helps to avoid contamination to cells, and the continuous and automatic process could provide particular advantage in efficiency especially for large-scale samples. The related research work is introduced in the following.

2.3 Materials and methods

Technical Design

A dilution-filtration system is developed as shown in Fig.16 (Hemofilter: Plasmflo TM AP-05H/L, ASAHI; Pumps: 400F/M1, Watson-Marlow; silicone tubing: 985-75, Pall). For removing CPAs, thawed cell suspension is first transferred into the special blood bag (made by an infusion bag). Then, the suspension is driven by the blood pump to flow circularly among the bag, the mixer and the hemofilter. While going through the mixer, the suspension is quickly diluted by diluent, and the dilution ratio can be controlled to prevent lysis. In the hemofilter, extracellular solution containing CPA is partly ultrafiltrated while cells keep inside. Along with the circulation goes on, CPAs in cell suspension can be removed continuously. The whole process is conducted automatically in a closed system, and thus it is hopeful for this method to reduce human labor as well as the risk of contamination significantly.

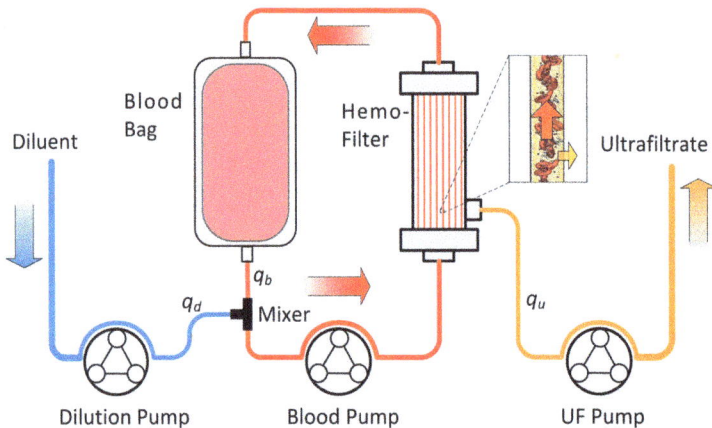

Fig. 16. Principle of the dilution-filtration system. Cell suspension is diluted and ultrafiltrated during circulating in the system, and then the CPAs inside can be continuously removed.

Theory of optimal operation protocol

Optimal operation protocol is defined here as the processes that minimize the operation time (to a final CPA concentration below 10g/L) as well as the osmotic cell volume excursion. A theoretical model was developed to predict the optimal operation protocols under the given experimental conditions (initial CPA concentration, cell density and total volume of cell suspension) and practical constraints. The detailed considerations for this procedure are described below.

Basic Assumptions and Formulation

The theoretical model of the dilution-filtration system is developed (as shown in Fig.17) under the following assumptions: (1) Both intra- and extra-cellular solutions in cell suspension consist of water, a permeable CPA (e.g. glycerol) and an impermeable salt (e.g. NaCl); (2) Blood bag, hollow fibers and their connecting tubing are filled with cell suspension, and cells are uniformly distributed in the suspension; (3) Extracellular solution is diluted/filtrated immediately and evenly at the diluting/filtrating point when cell suspension circulates in the system; (4) Suspension flow is one dimensional, and the convection factors can be neglected.

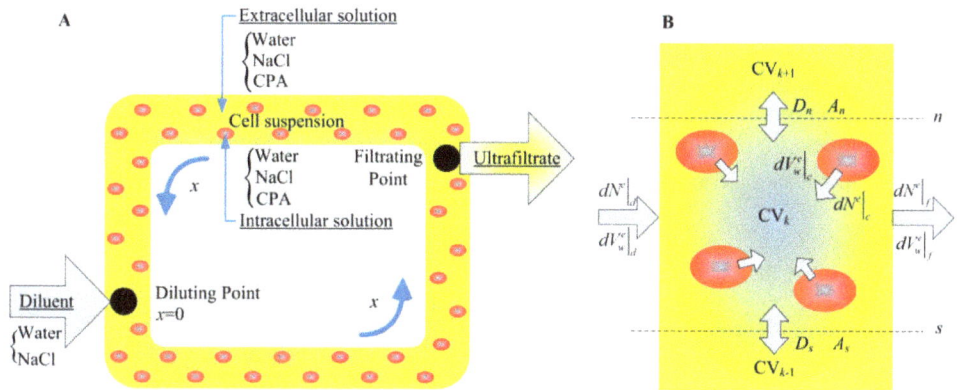

Fig. 17. Theoretical modeling of the system. A: the overall system, and B: a control volume.

Based on the assumptions, a governing equation about the mass transfer process can be derived by focusing on the extracellular solution:

$$\frac{\partial \phi^e}{\partial t} = \frac{1}{A}\frac{\partial}{\partial x}\left(DA \cdot \frac{\partial \phi^e}{\partial x}\right) + S \tag{19}$$

where, A refers to effective mass transfer area, D refers to diffusion coefficient, ϕ^e refers to extracellular solute concentration (in osmolality), and S is the mass source/sink term, respectively.

Source/Sink terms

The source/sink term can be derived by temporarily ignored the diffusion term:

$$S = \frac{\overline{d\phi^e}}{dt} = \frac{\overline{d\left(N^e/V_w^e\right)}}{dt} = \frac{1}{V_w^e}\frac{\overline{dN^e}}{dt} - \frac{\phi^e}{V_w^e}\frac{\overline{dV_w^e}}{dt} \qquad (20)$$

where N^e and V_w^e are the number of osmoles of solutes and water volume in extracellular solution, respectively. The overlines in the equation indicate the given deriving condition. The terms of dN^e and dV^e can be further specified as

$$\overline{dN^e} = dN^e\big|_d - dN^e\big|_f + dN^e\big|_c \qquad (21)$$

$$\overline{dV_w^e} = dV_w^e\big|_d - dV_w^e\big|_f + dV_w^e\big|_c \qquad (22)$$

where the subscripts "d", "f" and "c" refer to the effects of dilution, filtration and cell membrane transport, respectively. According to assumption (2), cell suspension inside the system can be equally divided into a finite number of control volumes (CVs), as shown in Fig.17B. For each CV, the values of the terms in the right hands of equation [21] and [22] can be determined as flows.

i.   Dilution/filtration

According to assumption (3), when a CV is going through the diluting point, the extracellular solution will be diluted immediately. Considering the pure filtration method used in the system, it is also assumed that ultrafiltration happens only at a certain location (the filtrating point, shown in Fig.17A), and the ultrafiltrate has the same composition as the extracellular solution. Thus the values of $dV_w^e\big|_d$, $dN^e\big|_d$, $dV_w^e\big|_f$ and $dN^e\big|_f$ of each CV can be

determined as

$$\overline{dV_w^e}\Big|_d = \begin{cases} \dfrac{Q_d V_{CV}}{Q_b} & \text{CV at the diluting point} \\ \\ 0 & \text{CVs at the other locations} \end{cases} \qquad (23)$$

$$\overline{dN^e}\Big|_d = \overline{dV_w^e}\Big|_d \cdot \phi^d \qquad (24)$$

$$\overline{dV_w^e}\Big|_f = \begin{cases} \dfrac{Q_f V_{CV}}{(Q_b + Q_d)(1 + \overline{V}_s \phi_s^e)} & \text{CV at the filtrating point} \\ \\ 0 & \text{CVs at the other locations} \end{cases} \qquad (25)$$

$$\overline{dN^e}\Big|_f = \overline{dV_w^e}\Big|_f \cdot \phi^e \qquad (26)$$

where Q_f, Q_b and Q_d are the flow rates of ultrafiltrate, cell suspension and diluent, ϕ^d is the solute concentration in diluent, ϕ_s^e is the extracellular CPA concentration, $\overline{V}_s$ is the partial molar volume of the CPA, and V_{CV} is the volume of a CV, respectively.

Prevention of Lethal Osmotic Injury to Cells During Addition and Removal of Cryoprotective Agents:
Theory and Technology

125

ii. Transportation across cell membrane

For the ternary system as considered in the present example, the mass transport across cell membrane can be described by the two-parameter formalism [2,3]. The total cell volume is the sum of the water, CPA and cell solid volumes:

$$V_c = V_w^i + V_s^i + V_{cb} \tag{27}$$

where the intracellular CPA volume can be determined as $V_s^i = \bar{V}_s N_s^i$. As soon as cell volume and intracellular solute concentrations are calculated the values of $dN^e\big|_c$ and $dV_w^e\big|_c$ can be further determined based on mass conservation:

$$dN_s^e\big|_c = -n_c dN_s^i, \quad dN_n^e\big|_c = 0 \tag{28}$$

$$dV_w^e\big|_c = -n_c dV_w^i = -n_c\left(dV_c - dV_s^i\right) \tag{29}$$

where n_c is the number of cells in a CV.

Numerical Simulation

With finite volume method, a fully implicit control volume integration of the governing equation will result in a finite difference scheme:

$$\left[a_{k-1} + a_{k+1} + a_k^{old} - \left(S_p V_{CV}\right)_k\right]\phi_k^e = a_{k-1}\phi_{k-1}^e + a_{k+1}\phi_{k+1}^e + a_k^{old}\phi_k^{e\,old} + \left(S_c V_{CV}\right)_k, k = 2,\cdots,K-1 \tag{30}$$

where a is the coefficient and K is the total number of CV in the system. The subscript 'k' refers to the kth CV in the system and the superscript 'old' refers to the previous time level. Sc and Sp are the constant portion and gradient of the linearized source term:

$$S_c = \frac{1}{V_w^e}\overline{\frac{dN^e}{dt}}, \quad S_p = -\frac{1}{V_w^e}\overline{\frac{dV_w^e}{dt}} \tag{31}$$

The subscript 'k-1' and 'k+1' in equation (30) refer to the previous and next CVs along the x direction, respectively. Noting that the cell suspension flows circularly in the closed system, the 1st CV is followed by the Kth one. Thus

$$\left[a_K + a_2 + a_1^{old} - \left(S_p V_{CV}\right)_1\right]\phi_1^e = a_K\phi_K^e + a_2\phi_2^e + a_1^{old}\phi_1^{e\,old} + \left(S_c V_{CV}\right)_1 \tag{32}$$

$$\left[a_{K-1} + a_1 + a_K^{old} - \left(S_p V_{CV}\right)_K\right]\phi_K^e = a_{K-1}\phi_{K-1}^e + a_1\phi_1^e + a_K^{old}\phi_K^{e\,old} + \left(S_c V_{CV}\right)_K \tag{33}$$

Here, the removal of glycerol from cryopreserved human red blood cells (RBCs) is discussed for an instance. For the ease of discussion, it is further restricted that blood volume keeps constant, i.e. ultrafiltrate flow rate keeps equal to diluent flow rate, although the presented system and model can adapt to more complicated situations. Besides, the concentration of NaCl in diluent and thawed blood are considered to be isotonic (0.29 Osmol/kg water). In this manner, the basic variables for a simulation consist of the experimental conditions (including the initial blood volume (V_b^0), hematocrit (h^0), and the concentrations of CPA (M_s^0) in extra/intracellular solution) as well as the operation parameters (including the flow rates of blood (Q_b) and diluent (Q_d)). The initial values of the other parameters in the model can be determined as

$$V_{CV}^0 = V_b^0 / K \tag{35}$$

$$V_c^0 = V_{iso} + V_s M_s^0 (V_{iso} - V_{cb}) \tag{36}$$

$$n_c^0 = V_{CV}^0 h^0 / V_c^0 \tag{37}$$

where V_{iso} is the isotonic volume of RBC. When terming the CV at the diluting point (x=0) when t=0 as the 1st CV (CV$_1$), the initial location of each CV can be allocated. Then the values of dN^e and dV^e for each CV can be calculated according to equations [21]-[31]. By alternatively calculating the source terms and solving the linearized governing equation, the concentration variation of extra-/intracellular solution as well as the responding cell volume excursion can be simulated. A typical process is shown in Fig.18, in which V_b^0 = 200ml, h^0 = 30%, M_s^0 = 6.28 Osmol/kg water (approximately 40% w/v), Q_b=200ml/min, and Q_d= 25ml/min.

To quantitatively evaluate the effect of an operation protocol, the maximum cell volume and the total time cost (to a final glycerol concentration below 10g/L (Brecher, 2002)) of the removing process can be taken as criteria for cell recovery rate and removing efficiency, respectively. Then the optimal protocol can be found out by applying different operation parameters to the given experimental conditions and comparing the simulated results. Hereinafter, the diffusion coefficients of glycerol and NaCl in water were set to be 5.43×10-10 m2/s and 14.41×10-10 m2/s, respectively (Ternstorm et al, 1996). The parameters about the dilution -filtration system and RBC membrane are also specified as listed in Table 4 and Table 5. These parameters may be different in various applications and systems.

Sections	Inner volume	Effective area
From the outlet of blood bag to the diluting point	5ml	1.25×10-5 m2
From the diluting point to the filtrating point	5ml	1.25×10-5 m2
From the filtration point to the outlet of hemofilter	85ml	5×10-4 m2
From the outlet of hemofilter to the inlet of blood bag	5ml	1.25×10-5 m2
Blood bag	Variable	5×10-3 m2

Table 4. Structural parameters of the dilution-filtration system used in the calculation

Surface area of RBC(A_c)	135 ×10-12 m2 [a]
Hydraulic permeability of cell membrane (L_p)	1.74 ×10-12 m/Pa/s [a]
Isotonic volume of RBC (V_{iso})	98.3 ×10-18 m3 [a]
Solid volume of RBC (V_{cb})	0.283 × V_{iso} [a]
Glycerol permeability to cell membrane (P_s)	6.61 ×10-8 m/s [a]

[a] From literature (Papanek, 1978);

Table 5. Membrane parameters of human RBC used in the calculation

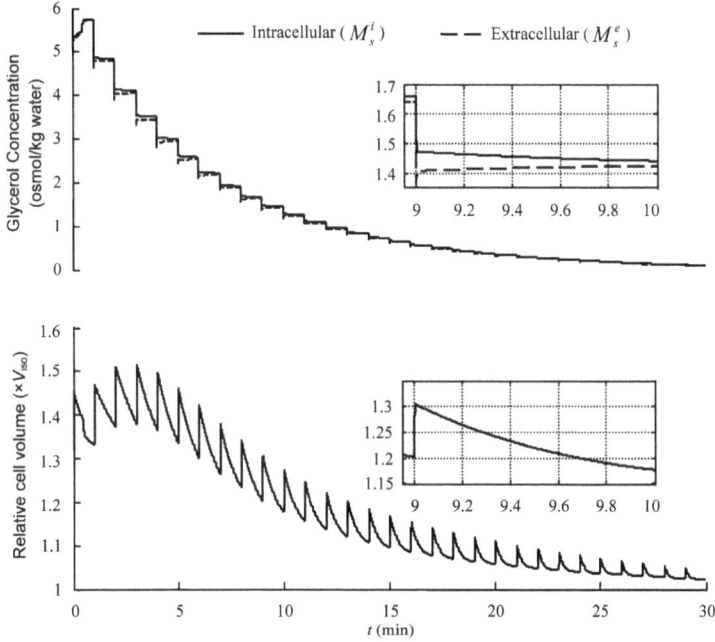

Fig. 18. Simulated glycerol concentration variation and cell volume excursion in CV_1 (initially at the diluting point) during a dilution-filtration process.

Experiments

Venous human blood was collected from healthy, adult blood donors in the Red Cross Transfusion Center of Heifei. For each donor, up to 200ml whole blood was collected into CPDA-1 anticoagulant solution in PVC plastic bag, and stored for up to 24 hours at 4°C. Then it was centrifuged at 1615×g for 4 minutes, and the platelets, leukocytes and plasma were removed to produce a hematocrit of 75±5 percent.

Each of the RBC suspensions was transferred into a 400-ml plastic bag, and then it was glycerolized by 57.1% w/v glycerol solution with a volume ratio of 2:1 (glycerol to blood) to achieve a final glycerol concentration about 40% (w/v) and a hematocrit of 25%-30%. Subsequently the blood bag was covered by PE foam sheet (thickness: 5mm) and then placed into a metal box (size: 200mm×150mm×20mm). After 30 minutes of equilibrium, the metal box was transferred to a -80°C freezer (MDF-U52V, SANYON, Japan) and the RBC suspension was frozen gradually. After cryopreservation in the freezer for 2~7 days, the RBC suspension was taken out and thawed in a 37°C water bath for about 10 minutes with gentle agitating.

Each unit of the thawed blood was deglycerolized with the dilution-filtration system as shown in Fig. 16, and the operation protocol was theoretically optimized. A typical experimental conditions (V_b^0 = 200ml, h^0 = 30%, M_s^0 = 6.28 Osmol/kg water) was studied first to reveal the general law. To evaluate the effect of each operation parameter, different protocols were applied respectively. Fig.19 shows that time cost is significantly reduced but maximum cell volume grows directly along with diluent flow rate increases, i.e. the washing efficiency can be improved by applying higher diluent flow rate but more hemolysis may be induced. Thus the diluent flow rate has to be carefully selected to achieve the optimal result. Comparatively, the effect of blood flow rate is not so complicated. Increasing of blood flow rate has little effect on glycerol clearance, but helps to reduce the maximum cell volume excursion.

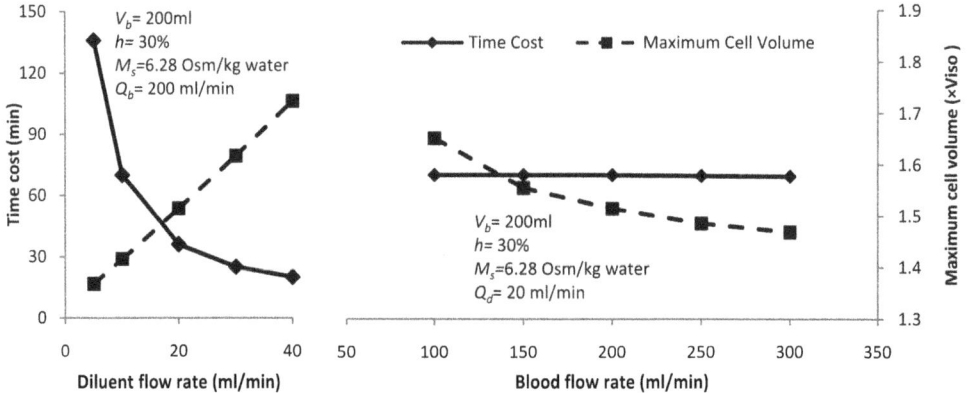

Fig. 19. Variations of time cost (real line and left Y-axis) and maximum cell volume (dash line and right Y-axis) with blood or diluent flow rates as parameters.

On the other hand, the effect of the operation parameters is also highly related to the blood conditions, especially the glycerol concentration. As shown in Fig.20, the same operation protocol (Q_b=200 ml/min and Q_d =20ml/min) is applied to several different conditions, in which V_b^0 =200ml, h^0 =30%, and M_s^0 varies from 0.56 Osmol/kg water (5% w/v) to 6.28 Osmol/kg water (40% w/v). When the glycerol concentration decreases, both the glycerol clearance and the maximum cell volume are reduced (glycerol clearance is defined here as the difference of initial and final numbers of osmoles of glycerol in blood over time cost). This phenomenon indicates us that along with the glycerol concentration drops during washing, diluent flow rate can be continuously increased to speed up the process without inducing extra cell volume excursion.

Based on the analysis above, it can be concluded that to achieve the optimal deglycerolization it is important to: a) use a low diluent flow rate at first, and stepwise increase it as CPA concentration drops; b) always use a high blood flow rate. The detailed operation parameters of the optimal protocol can be found out by the theoretical model with some practical constraints. During the in-vitro experiments, operation protocol for each unit was optimized theoretically according to the specific experimental conditions as well as the following constraints: maximum cell volume: 1.35 times of the isotonic volume (V_{iso}) of RBCs; maximum flow rate of pumps: 200 ml/min and maximum ultrafiltrate flow rate of hemofilter: 40 ml/min. The value of upper cell volume level was conservatively selected in order to achieve the best cell recovery rate, although the washing efficiency may be limited.

Samples were taken before and after deglycerolization. Cell count and hematocrit were measured by a hematology Analyzer (Ac·T diff II TM, Beckman COULTER®) The Freeze-Thaw-Wash (FTW) cell count recovery rates were calculated by comparing the total cell counts after thawing to that after washing (Valeri et al, 2001). Residual glycerol concentration in the washed blood was measured by a glycerol assay kit (K-GCROL, Megazyme®) and a spectrophotometer (756MC UV-VIS, Scientific Instrument®, Shanghai, China).

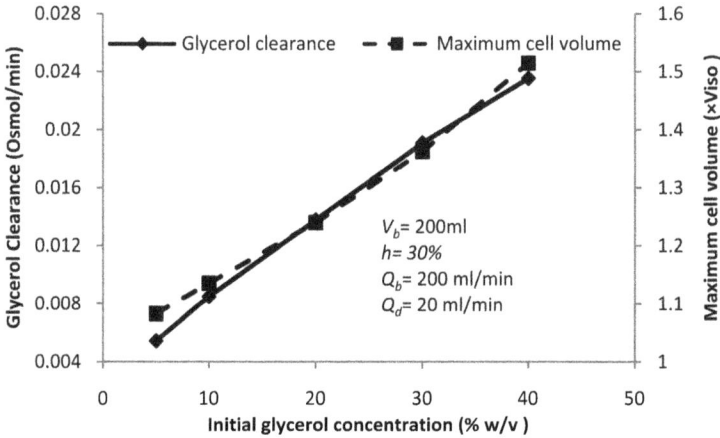

Fig. 20. Variations of glycerol clearance (real line and left Y-axis) and maximum cell volume (dash line and right Y-axis) with glycerol concentation as a paramter.

2.4 Results

A total number of ten units of blood were cryopreserved and deglycerolized by the dilution-filtration method, and the results are shown in Table 6. The residual glycerol concentration (5.57±2.81 g/L, n=10) is obviously lower than the standard value (10g/L) indicated by American association of blood banking (AABB). During the optimization of the operation procedures, the maximum cell volume constraints was critically applied (1.35×Viso) for the best of cell recovery, and thus the deglycerolizing efficiency is limited. However, each of the unit was processed within an hour, which is similar to the automatic centrifuging method (Valeri et al, 2001). The cell count recovery rate is 91.19±3.57% (n=10). Comparing to the reported methods (Diafiltration method: 70% (Castino et al, 1996), dialysis method, no in vitro data was presented (Ding et al, 2007, 2010), manual centrifuging method: >80% (Brecher, 2002), and automatic centrifuging method 89.4±3.0% (Valeri et al, 2001)), the recovery rate indicates an obviously advantage of our method in cell safety.

UNITS	Thawed Blood Volume (ml)	Thawed Blood Hct (%)	Cell Count Recovery (%)	Residual Glycerol (g/L)
1	221.8	30	93.64	2.60
2	204.1	27	85.92	3.39
3	229.8	29	91.64	12.26
4	219.0	23	90.91	4.90
5	217.6	25	91.38	3.39
6	210.0	28	90.76	6.56
7	216.7	24	81.60	3.80
8	200.5	29	94.18	7.34
9	204.5	21	92.24	5.35
10	205.0	30	93.80	6.08
Mean	212.9	27	91.19	5.57
S.D.	9.49	3	3.57	2.81

Table 6. *In-vitro* experiments of deglycerolization with dilution-filtration method

2.5 Discussion

An optimized method for addition and removal glycerol from cryopreserved human spermatozoa has been illustrated as an example. Although the mechanism(s) of the osmotic injury during cryopreservation is not clearly understood, the hypothesis has been tested and confirmed, i.e. human sperm volume excursion can be used as an indicator to predict possible osmotic injury to spermatozoa during glycerol addition and removal processes. Hence, the procedures used for testing the hypothesis provide a methodology to predict optimal protocols for cryoprotective agent addition/removal..

The FVS, multi-step procedure for the addition of glycerol to human spermatozoa before cryopreservation is a conventional, commonly used technique, i.e. 'drop by drop' (stepwise) addition of a solution with a relatively high glycerol concentration (the volume of each 'drop' is roughly constant) to the spermatozoa or sperm suspension in order to achieve a 0.6-1.0 M glycerol concentration in the final sperm suspension. In practice, the frozen-thawed sperm samples containing glycerol are either washed for intrauterine insemination or four in-vitro fertilization or directly transferred into the lower female reproductive tract for artificial insemination (e.g. intercervical insemination). In both cases, the glycerol is abruptly removed from spermatozoa by direct exposure to near isotonic conditions. In the example, it was predicted by computer simulation, and confirmed experimentally, that a one-step removal of glycerol would cause a high frequency of sperm motility loss even without freezing. Based on the results, the FMS removal (≥8 steps) of 1.0 M glycerol is recommended. Within the scope of the present investigation, a four-step FMS addition of glycerol to spermatozoa to achieve a final 1.0 M glycerol concentration and an eight-step

FMS removal of 1.0 M glycerol from spermatozoa were predicted and shown to be acceptable procedures which minimize osmotic injury. From calculations, the minimum or maximum cell volumes after each step of FVS addition or removal were shown to be unequal, some of which may exceed the lower or upper volume limits of the cells. In contrast, from calculations, the minimum or maximum cell volumes after each step of FMS addition or removal of glycerol were shown to be relatively even (Figures 12 and 13). For a fixed number of steps, the minimum or maximum of cell volume excursion during glycerol addition or removal using the FMS approach is much smaller than that using the FVS approach (see Figures 12 and 13).

In the example, it was postulated that the sperm osmotic injury as a function of cell volume excursion must be determined to predict the optimal glycerol addition and removal procedures. However, the definition and determination of 'sperm injury' is dependent upon the assays used. In the example, sperm motility was used as a standard of sperm viability because of its relatively high sensitivity to osmotic changes and the requirement of sperm motility for functional viability. If sperm membrane integrity was chosen as the endpoint to evaluate the sperm viability, as shown in Figure 7, different osmotic tolerance limits would be obtained. One can readily repeat the same procedures to predict the extent to which spermolysis is caused by the different glycerol addition/removal procedures used in the example, based on the information provided in Figure 5. For example, it was found (Figure 7) that >85% of spermatozoa maintained membrane integrity when they were returned to isotonic condition after having been exposed to anisosmotic conditions ranging from 90 and 700 mOsmol. The corresponding sperm volume excursion range was 0.7-2.1 times the isotonic sperm volume (Figure 9). From Figures 12 and 13, it can be seen that a one-step addition and one-step removal of 1.0 M glycerol would result in a minimum relative sperm volume of 0.72 and maximum volume of 1.68 respectively, which did not exceed the sperm volume excursion range 90.7-2.1 times relative volume) for maintaining >85% sperm membrane integrity. Based on this information, one can predict that the majority (>85%) of spermatozoa would maintain membrane integrity even using one-step addition and one-step removal of glycerol.

A dilution-filtration system for removing CPAs from cryopreserved cell suspension was also introduced here. The system realized continuous processing of cell suspension and the dilution & filtration were conducted simultaneously, thus it can achieve much better efficiency than traditional multi-step centrifuging methods. Moreover, dilution in the system is conducted to cell suspension flow in tubing but not whole suspension in container, thus the mixing process should be much rapider and then the osmotic disequilibrium during dilution can be significantly reduced.

A theoretical model was established to simulate the specific process. Based on the model, cell volume excursion and the variation of CPA concentration during the dilution-filtration process can be simulated. Theoretical analysis indicates the operation parameters, especially the flow rate of diluent, are critical for the dilution-filtration method. In the previous studies concerning removing CPAs with hollow fibers (Castino et al, 1996; Arnaud et al 2003; Ding et al, 2007, 2010), only the protocols with constant flow rates were discussed. However, it was found to be difficult to balance the requirements in removing efficiency and cell safety. This problem also exists in the presented dilution-filtration method. Removing efficiency can be improved by using higher diluent flow rate, but the cell recovery rate may be seriously reduced in the way. Besides, when using a constant diluent flow rate, the profile of glycerol concentration is nearly exponential, i.e., the removing efficiency starts at the highest value but gradually decreases as the process going on. However, when using a stepwise increased diluent flow rate, the removing efficiency can be maintained at a high level for a quite long period. Moreover, theoretical analysis also indicates stepwise increasing of the diluent flow rate may not cause any extra cell damage. Therefore, a stepwise increased diluent flow rate is necessary to achieve both high cell recovery rates and efficient glycerol clearance when using the dilution-filtration system. In addition, it was also deduced by the theoretical analysis that the removing effect of an operation protocol is highly related to the initial volumes and cell densities of cell suspensions. Therefore, the optimal operation protocols should be specialized and various from case to case. The theoretical model provides an effective tool to find out the optimal protocols for given applications.

The system was also investigated experimentally with deglycerolization from cryopreserved blood, and the operation procedures were optimized based on the theoretical model. It is clearly indicated by the results that the dilution-filtration method is safe and efficient for deglycerolization from cryopreserved RBCs. Comparing to the automatic centrifuging method, the cell recovery rate and removing efficiency are similar, but the equipment cost of the dilution-filtration system is much lower and thus it can be applied in more areas. We can also believe that with properly selected operation parameters, this system can also be applied to various CPA removal applications. In addition, all the media are processed in a closed system, and thus the system should have further advantages in avoiding contamination. It is hopeful for the cells to have a long shelf life after washing. These suppositions will be verified by further experiments.

3. References

Arnaud, F.G. and Pegg, D.E. (1990) Permeation of glycerol and propane-1,2-diol into human platelets. Cryobiology, 27, 107-118.

Arnaud, F., Kapnik, E., and Meryman, H. T., 2003, "Use of hollow fiber membrane filtration for the removal of DMSO from platelet concentrates," Platelets, 14(3), pp. 131-138.

Bavister, B.D., Leibfriend, M.L. and Lieberman, G. (1983) Development of preimplantation embryos of the golden hamster in a defined culture medium. Biol. Reprod., 28, 235-247

Brecher, M. E., 2002, Technical manual of the American Association of Blood Banks. 14th ed., American Association of Blood Banks, Bethesda.

Castino, F., and Wickramasinghe, S. R., 1996, "Washing frozen red blood cell concentrates using hollow fibers," Journal of membrane science, 110(2), pp. 169-180.

Crister, J.K., Huse-Benda, A.R., Aaker, D.V., Arneson, B.W. and Ball G.D. (1988a) Cryopreservation of human spermatozoa, 3. The effect of cryoprotectants on motility. Fertil. Steril., 50, 314-320.

Crister, J.K., Colvin, K.E. and Crister, E.S. (1988b) Effect of sperm concentration on computer assisted semen analysis results. Abstracts of the 1988 Annual Meeting of American Society of Andrology. J. Androl., 9 (Suppl.), Abstr. 105, p.45.

Curry, M.R. and Watson, P.F. (1994) Osmotic effects on ram and human sperm membranes in relation to thawing injury. Cryobiology, 31, 39-46.

Daugirdas, J. T., Blake, P., Ing, T. S. and Blagg, C., 2006, Handbook of Dialysis, Fourth Edition., Lippincott Willians & Wilkins, Philadelphia, pp.265-275.

Ding, W. P., Yu, J. P., Woods, E., Heimfeld, S., and Gao D. Y., 2007, "Simulation of removing permeable cryoprotective agents from cryopreserved blood with hollow fiber modules," Journal of membrane science, 288(1), pp. 85-93.

Ding, W. P., Zhou, X. M., Heimfeld, S., Reems, J., and Gao D. Y., 2010, "A Steady-State Mass Transfer Model of Removing CPAs from Cryopreserved Blood with Hollow Fiber Modules," Journal of Biomechanical Engineering, 132(1), pp. 011002

Du, Junying, Kleinhans, F.W., Mazur, P. and Crister, J.K. (1993) Osmotic behavior of human spermatozoa studied by EPR. Cryo-letters, 14, 285-294.

Du, Junying, Kleinhans, F.W., Mazur, P. and Crister, J.K. (1994) Human spermatozoa glycerol permeability and activation energy determined by electron paramagnetic resonance. Biochim. Biophys. Acta, 1994, 1-11.

Gao, D.Y., Mazur, P., Kleinhans, F.W., Watson, P.F., Noiles, E.E. and Crister, J.K. (1992) Glycerol permeability of human spermatozoa and its activation energy. Cryobiology, 29, 657-667.

Gao, D.Y., Ashworth, E., Watson, P.F., Kleinhans, F.W., Mazur, P. and Crister, J.K. (1993) Hyperosmotic tolerance of human spermatozoa: separate effect of glycerol, sodium chloride and sucrose on spermolysis. Biol. Reprod., 49, 112-123.

Gao, D.Y., Liu, J., Liu, C., McGann, L.E., Watson, P.F., Kleinhans, F.W., Mazur, P., Crister E.S. and Crister J.K. (1995). Prevention of osmotic injury to human spermatozoa during addition and removal of glycerol, Human Reproduction 10, 1109-1122.

Garner, D.L., Pinkel, D., Johnson, L.A. and Pace, M.M. (1986) Assessment of spermatozoal function using dual fluorescent staining and flow cytometric analyses. Biol. Reprod., 34, 127-138.

Gilmore, J.A., Liu, J., Gao, D.Y. and Crister, J.K. (1997) Determination of optimal cryoprotectant and procedures for their addition and removal from human spermatozoa. Human Reproduction 12, 112-118.

Gilmore, J.A., McGann, L.E., Liu, J., Gao, D.Y., Peter, A.T., Kleinhans, F.W., and Crister, J.K. (1995) Effect of cryoprotectant solutes on water permeability of human spermatozoa. Biol. Reprod. 53, 985-995.

Jequier, A. and Crich, J. (1986) Computer assisted semen analysis (CASA). In semen Analysis: A practical Guide. Blackwell Scientific, Boston, pp. 143-149.

Jeyendran, R.S., Van der Ven, H.H., Perez-Pelaez, M., Crabo, B.G. and Zaneveld, L.J.D. (1984) Development of an assay to assess the functional integrity of the human sperm membrane and its relationship to other semen characteristics. J. Reprod. Fertil., 70, 219-228.

Katkov, I. I., 2000, "A Two-Parameter Model of Cell Membrane Permeability for Multisolute Systems," Cryobiology, 40(1), pp. 64-83.

Katkov, I.I., Katkova, N., Crister, J.K., and Mazur, P. (1998a) Mouse spermatozoa in high concentrations of glycerol: Chemical toxicity vs osmotic shock at normal and reduced oxygen concentration. Cryobiology 37, 325-338.

Katkov, I.I. (1998b) Cell suspensions in high concentrations of a permeable cryoprotectant: Optimization of addition and dilution protocols. Cryobiology 37, 403-404

Kedem, O. and Katchalsky, A. (1958) Thermodynamic analysis of the permeability of biological membrane to nonelectrolytes. Biochim. Biophys. Acta, 27, 229-246.

Kleinhans, F. W., 1998, "Membrane Permeability Modeling: Kedem-Katchalsky vs a Two-Parameter Formalism," Cryobiology, 37(4), pp. 271-289.

Kleinhans, F.W., Travis, V.S., Du, Junying, Villines, P.M., Colvin, K.E. and Criter, J.K. (1992) Measurement of human sperm intracellular water volume by electron spin resonance. J. Androl. 13, 498-506.

Leibo, S.P. (1981) Preservation of ova and embryos by freezing. In Brackett, E.G. Seidel, G.E. and Seidel, S.M. (eds), New Technologies in Animal Breeding Academic Press, New York, pp. 127-139.

Leibo, S.P. (1986) Cryobiology: preservation of mammalian embryos. Basic Life Sci., 37, 251-272.

Leibo, S.P. and Mazur, P. (1978) Methods for the preservation of mammalian embryos by freezing. In Daniel, J.C., Jr (ed.), Methods in Mammalian Reproduction. Academic Press, New York, pp. 179-201.

Mazur, P. (1984) Freezing of living cells: mechanism and implications. Am. J. Physiol. 247, C125-C142.

Mazur, P. and Leibo, S.P. (1977) Mechanisms of freezing damage in bacteriophage T4 (Discussion). In Elliott, K. and Whelan, J. (eds), The Freezing of Mammalian Embryos. Ciba Foundation Symposium No. 52. Elsevier, Amsterdam, pp. 255-226.

Mazur, P. and Schneider, U. (1984) Osmotic consequences of cryoprotectant permeability and its relation to the survival of frozen-thawed embryos. Theriogenology, 21, 68-79.

Mazur, P. and Schneider, U. (1986) Osmotic responses of preimplantation mouse and boving embryos and their cryobiological implications. Cell Biophys., 8, 259-284.

Mazur, P., Leibo, S.P. and Chu, E.H.Y. (1972) A two-factor hypothesis of freezing injury. Exp. Cell Res. 71, 345-355.

McGann, L.E., Turner, A.R. and Turc, J.M. 91982) Microcomputer interface for rapid measurement of average volume using an electronic particle counter. Med. Biol. Eng. Comput., 20, 117-120.

Meryman, H.T. (1970) The exceeding of a minimum tolerable cell volume in hypertonic suspension as a cause of freezing injury. In Wolstenholme, G.E.W. and O'Connor, M. (eds), The Frozen Cell. CIBA Foundation Symposium. Churchill, London, pp. 51-67.

Meryman, H. T., 2007, "Cryopreservation of living cells: principles and practice," Transfusion, 47, pp. 935-945.

Meryman, H. T., and Hornblower M., 1972, "A method for freezing and washing red blood cells using a high glycerol concentration," Transfusion, 12(3), pp. 145-156.

Meryman, H. T., and Hornblower, M., 1977, "A simplified procedure for deglycerolizing red blood cells frozen in a high glycerol concentration," Transfusion, 17(5), pp. 438-442.

Muldrew, K. and McGann, L.E. (1994) The osmotic rupture hypothesis of intracellular freezing injury. Biophys. J., 66, 532-541.

Myrthe, T.W., and Barry A. B. (2004) Step-wise dilution for removal of glycerol from fresh and cryopreserved equine spermatozoa. Animal Reproduction Science 84, 147-156.

Noiles, E.E., Mazur, P., Watson, P.F., Kleinhans, F.W. and Crister, J.K. (1993) Determination of water permeability coefficient for human spermatozoa and its activation energy. Biol. Reprod., 48, 99-109.

Papanek, P. T., 1978, "The water permeability of the human erythrocyte in the temperature range +25 °C to -10°C," Ph.D. thesis, MIT.

Penninchx, P., Poelmans, S., Kerremans, R. and De Loecher, W. (1984) Erythrocyte swelling after rapid dilution of cryoprotectants and its prevention. Cryobiology. 21, 25-32.

Polge, C. Smith, A.U. and Parkes, A.S. (1949) Revival of spermatozoa after vitrification and dehydration at law temperatures, Nature, 164, 666-676.

Preston, G.M., Carroll, T.P., Guggion, W.B., and Agre, P.(1992) Appearance of water channels in Xenopus oocytes expressing red cell CHIP 28. Science 256, 385-387.

Rowe, A.W., Eyster, E. and Kellner, A. (1968) Liquid nitrogen preservation of red blood cells for transfusion: a low glycerol-rapid freeze procedure. Cryobiology, 5, 19-128.

Sherman, J.K. (1973) Synopsis of the use of frozen human sperm since 1964: state of the art of human semen banking. Fertil .Steril., 24, 397-412.

Spector, P.C., Goodnight, J.H., Sall, J.P., Sarle, S.W. and W.M. Stanish (1985) The GLM and the Catmod procedure. In SAS Institute, Inc., SAS User's Guide: Statistics, 5th end. SAS Institute. Inc., Cary, NC. Pp. 403-506.

Steponkus, P.A. and Wiest, S.C. (1979) Freeze-thaw induced lesions in the plasma membrane. In Lyons, J.M., Graham, D. and Raison, J.K. (eds), Low Temperature Stress in Crop Plants: The rolle of the Membrane. Academic Press, New York, pp. 231-350.

Ternstrom, G., Sjostrand, A., Aly, G., and Jernqvist, A. 1996, "Mutual Diffusion Coefficients of Water + Ethylene Glycol and Water + Glycerol Mixtures," Journal of chemical and engineering data, 41(4), pp. 876-979.

Valeri, C. R., 1975, "Simplification of the methods for adding and removing glycerol during freeze-preservation of human red blood cells with the high or low glycerol methods: biochemical modification prior to freezing," Transfusion, 15(3), pp. 195-218.

Valeri, C. R., and Ragno, G., 2006, "Cryopreservation of human blood products," Transfusion and apheresis science, 34(3), pp. 271-287.

Valeri, C. R., Ragno, G., Pivacek, L., and O'Neill E. M., 2001, "In vivo survival of apheresis RBCs, frozen with 40-percent (wt/vol) glycerol, deglycerolized in the ACP 215, and stored at 4 degrees C in AS-3 for up to 21 days," Transfusion, 41(7), pp. 928-932.

Watson, P. F. (1979) The preservation of semen in mammals, in Finn, C. A. (ed.) Oxford Reviews of Reproductive Biology, vol 1. Oxford University Press, Oxford, 283-350.

Zar, J.H. (1984) Biostatistical Aalysis. 2nd edn. Pretice-Hall, Inc., Englewood Cliffs, NJ, P. 718.

Zhou, X. M., Liu, Z., Shu, Z. Q., Ding, W. P., Du, P. A., Chung, J., Liu,C., Heimgeld,S.,and Gao,D. Y. (2011) A Dilution-filtration Method for Removing Cryoprotective Agents. J. Biomech. Eng-ASME Trans. 133, 021007.

Part 2

Stem Cells and Cryopreservation in Regenerative Medicine

5

Cryopreservation of Human Pluripotent Stem Cells: Are We Going in the Right Direction?

Raquel Martín-Ibáñez[1,2,3], Outi Hovatta[4] and Josep M. Canals[1,2,3]
[1]*Departament de Biologia Cellular, Immunologia i Neurociències, Programa de Terapia Cellular, Facultat de Medicina, Universitat de Barcelona, Barcelona*
[2]*Institut de Investigacions Biomèdiques August Pi i Sunyer (IDIBAPS), Barcelona*
[3]*Centro de Investigación Biomédica en Red Sobre Enfermedades Neurodegenerativas (CIBERNED)*
[4]*Division of Obstetrics and Gynecology, Department of Clinical Science, Intervention and Technology, Karolinska Institutet, K57, Karolinska University Hospital, Huddinge, Stockholm,*
[1,2,3]*Spain*
[4]*Sweden*

1. Introduction

The first derivation of human embryonic stem cells (hESCs) (Thomson et al., 1998) and the more recently development of human induced pluripotent stem cells (iPSCs) (Park et al., 2008; Takahashi et al., 2007; Takahashi & Yamanaka, 2006; Wernig et al., 2007; Yu et al., 2007) have marked the beginning of a new era in biomedical research. These two types of human pluripotent stem cells (hPSCs) are characterized by an unlimited capacity to self-renew while retaining their potential to differentiate into almost all cell types of the body (Odorico et al., 2001; Reubinoff et al., 2000; Silva & Smith, 2008). These remarkable properties turn hPSCs into one of the most interesting cell types for toxicology and drug discovery, tissue engineering and regenerative medicine (Battey, 2007; Mountford, 2008). In fact, work with hPSCs has already provided new and exciting developments that may eventually lead to the creation of novel cell-based therapies for the treatment of a wide range of human diseases including Parkinson's and other neurodegenerative diseases, diabetes, cardiac and vascular diseases (Kiskinis & Eggan, 2010; Ronaghi et al., 2010). However, a major challenge for the widespread application of hPSCs is the development of efficient protocols for cryopreservation.

To date, two techniques are mainly applied for the cryopreservation of hPSCs: conventional slow freezing and vitrification. The conventional slow-freezing and rapid-thawing procedure using dimethylsulfoxide (DMSO) as a cryoprotectant is the most commonly used method (Grout et al., 1990; Meryman, 2007). While this established technique is effective for somatic cell lines and even murine embryonic stem cells (mESCs), hematopoietic and mesenchymal human stem cells, this is not the case for hPSCs, due to low recovery rates and high levels of differentiation (Berz et al., 2007; Reubinoff et al., 2001; Richards et al., 2004; Thirumala et al., 2010). In contrast, vitrification of hPSCs by the "open pulled straw" method using high cryoprotectant concentrations together with flash-freezing in liquid nitrogen has reported higher cell survival rates (Li et al., 2010b; Reubinoff et al., 2001; Richards et al., 2004). However, there are several disadvantages preventing the widespread use of this technique. First, high concentrations of cryoprotectors, which are cytotoxic above 4°C, are needed. Second, these procedures are tedious to perform manually. Additionally, as vitrification is mostly performed in open pulled straws, contact between the liquid nitrogen and the cells is unavoidable, which carries the risk of contamination. Finally, and one of the most limiting disadvantages of this technique is that it is clearly unsuited for freezing bulk quantities of hPSCs.

During the last decade, several groups have been studying different approaches to improve the above described cryopreservation protocols. In the present work we will review the recent advances in the cryopreservation field trying to point out how a better understanding of the sensitivity of hPSCs to the cryopreservation process will help to develop more efficient protocols.

1.1 Human pluripotent stem cells

1.1.1 Human embryonic stem cells

The pioneering work on mESCs, and later advances in culturing techniques that were developed to culture nonhuman primate embryonic stem cell lines eventually led to the first successful generation of hESC lines by Thompson and coworkers and two years later by Reubinoff and coworkers (Evans & Kaufman, 1981; Martin, 1981; Reubinoff et al., 2000; Thomson et al., 1995; Thomson et al., 1996; Thomson et al., 1998). These hESCs were derived from human embryos that were produced by *in vitro* fertilization for clinical purposes. HESC lines were karyotypically normal and maintained the developmental potential to contribute to derivatives of all three germ layers, even after clonal derivation and prolonged undifferentiated proliferation (Amit et al., 2000). Since then, hundreds of stem cell lines have been derived world-wide from morula, later blastocyst stage embryos, fresh and cryopreserved supernumerary embryos, single blastomeres and parthenogenetic embryos (Klimanskaya et al., 2006; Lin et al., 2007; Mai et al., 2007; Revazova et al., 2007; Stojkovic et al., 2004; Strelchenko et al., 2004).

HESCs grow in tightly packed colonies and maintain defined borders at the periphery of colonies. High nucleus to cytoplasm ratio and prominent nucleoli are typical features of individual hESCs within colonies. HESCs are also characterized by high telomerase activity and expression of a number of cell surface markers and transcription factors including stage-specific embryonic antigen (SSEA)-4, SSEA-3, TRA antigens, Oct3/4, Nanog and absence of hESCs negative markers such as SSEA-1 (Carpenter et al., 2003; Chambers et al., 2003; Draper et al., 2004; Heins et al., 2004; Nichols et al., 1998). Functional confirmation of the multipotent nature of hESCs is generally achieved by examining their potential to differentiate into all three germ layers (ectoderm, mesoderm and endoderm) both *in vitro* and *in vivo*. *In vitro*, hESCs are allowed to randomly differentiate as embryoid bodies (EBs), which are aggregates of cells grown in suspension culture, followed by immunocytochemical analysis, or measurement of expression of genes associated with the three germ layers by RT-PCR (Reubinoff et al., 2000). The *in vivo* test for pluripotency of hESCs is normally teratoma formation in immunocompromised mice (Bosma et al., 1983).

1.1.2 Human induced pluripotent stem cells

An important revolution in the stem cell research field was accomplished when several groups in different studies demonstrated that using a cocktail of four factors, somatic cells could be reprogrammed into iPSCs (Maherali et al., 2007; Okita et al., 2007; Park et al., 2008; Takahashi et al., 2007; Takahashi & Yamanaka, 2006; Wernig et al., 2007; Yu et al., 2007). Consequently, it was shown that such cells could be generated from patient-specific cells for a wide variety of diseases (Kiskinis & Eggan, 2010; Raya et al., 2009; Raya et al., 2010) and from a wide variety of somatic cell types (Sun et al., 2010). These cells are morphologically similar to hESCs, express typical hESC-specific cell surface antigens and genes, differentiate into multiple lineages *in vitro*, and form teratomas containing differentiated derivatives of all three primary germ layers when injected into immunocompromised mice. Indeed, these new pluripotent cell lines satisfy all the original criteria proposed for hESCs (Thomson et al., 1998). Nevertheless, some differences have been observed between hESCs and iPSCs (Chin et al., 2009); but it remains unclear whether the small percentage of genes that are differentially expressed between these two types of hPSCs are shared among different lines and whether these differences are biologically significant.

Developing iPSCs into therapeutic reagents faces a number of practical hurdles, including risks associated with cell processing, the difficulty of ensuring the purity and characteristics of the reprogrammed population and the safety and efficacy of reprogrammed cells *in vivo* (Condic & Rao, 2008; Rao & Condic, 2008; Rao & Condic, 2009). Moreover, a case of rejection has been recently described after iPSCs autologous transplantation (Apostolou & Hochedlinger, 2011). Nonetheless, there is cause for considerable optimism that patient-specific iPSC lines will both enhance the study of human diseases and advance these studies toward clinical applications.

1.2 Cryopreservation of hPSCs

Cryopreservation is the process of cooling and storing cells, tissues or organs at sub-freezing temperatures, below −80°C and typically below −140°C, to maintain viability (Baust et al., 2009). The freezing process involves complex phenomena of water crystallization and changes in solute concentration both outside and inside the cell that can be detrimental to cell survival. In addition, exposure to low temperatures has been reported to induce a stress response resulting in biomolecular-based cell death for different cell types (Baust et al., 2001; Fu et al., 2001; Paasch et al., 2004; Xiao & Dooley, 2003).

In general, the major steps used in cryopreservation of most cell types can be summarized as follows (figure 1): i) harvesting the cells, ii) addition of cryoprotective agents within a carrier media to the cell suspension, iii) ice crystal induction in cell suspension following a determined cooling rate (ranging from -1 to -10°C/min), iv) long-term storage at cryogenic temperatures (normally in liquid nitrogen), v) rapid thawing by warming the cell suspension in a 37-40°C water bath, vi) removal of cryoprotective agent by centrifugation and vii) seeding down the cells to allow culture growth (Gao et al., 1998; Hubel, 1997).

Cryoinjury can be due to one or a combination of the following processes: 1) cytotoxicity of cryoprotective agents (Muldrew & McGann, 1994; Schneider & Maurer, 1983); 2) osmotic injury due to excursion of cryoprotective agents upon freeze-thawing (Gao et al., 1995; Mazur & Schneider, 1986); 3) intracellular ice formation in the cooling process (Fujikawa,

1980; Mazur et al., 1972) and 4) recrystallization of the intracellular ice during the warming process (Mazur & Cole, 1989; Trump et al., 1965). In addition, recent studies have linked numerous stress factors associated with cryopreservation to known initiators of molecular-based apoptotic cell death processes (Baust et al., 2009).

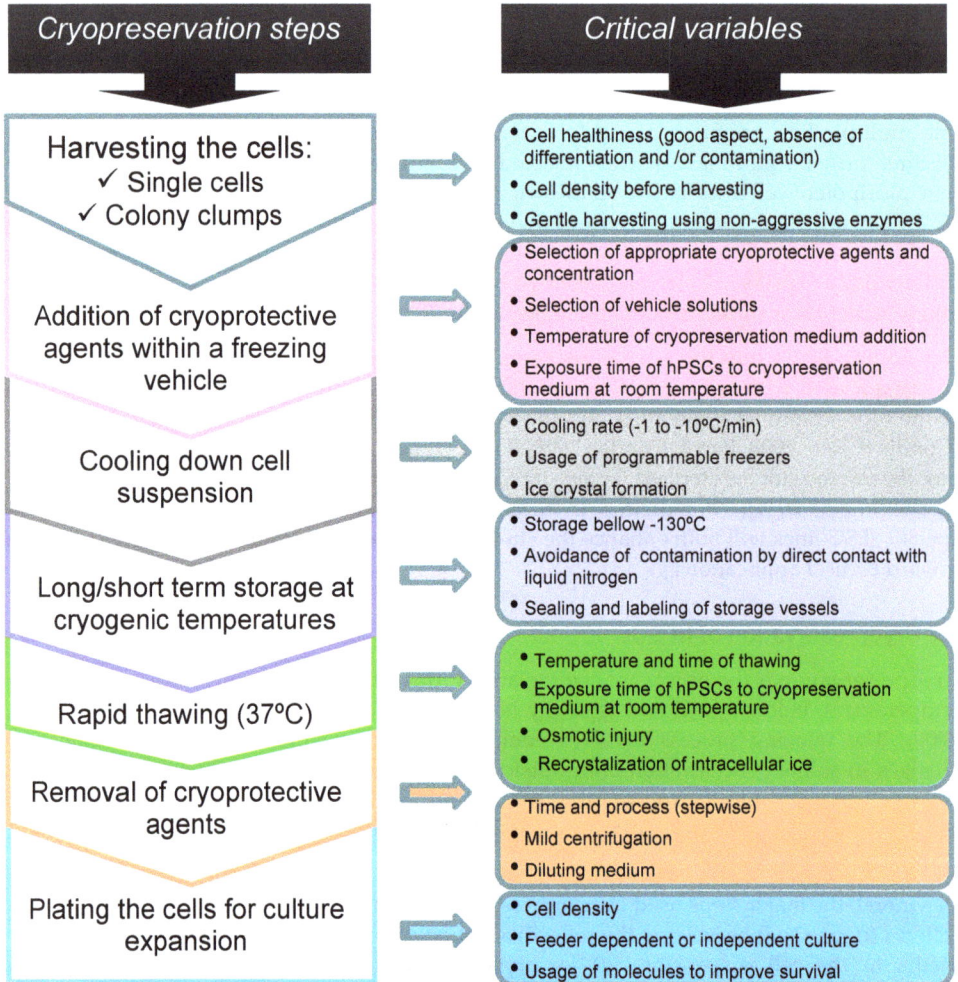

Fig. 1. Representative diagram of the main steps involved in a general cryopreservation process and the critical variables that should be considered in order to preserve cells with good recovery rates.

2. Sensitivity of hPSCs to cryopreservation

The techniques employed for the cryopreservation of hPSCs which include slow freezing-rapid thawing and vitrification, have been shown to be refractory for these cells that present very low survival rates (5-20% and 25-75% respectively) and many of the cells that do survive differentiate upon thawing and expansion (Reubinoff et al., 2001; Richards et al., 2004). The low efficiency of hPSCs cryopreservation has been attributed, in part, to the highly "cooperative" nature of these cells (as comparable with mESCs), which appear to require intimate physical contact between them within the colony (to permit cell-cell signaling) and an optimum clump size of about 100-500 cells during cryopreservation and serial passage (Amit et al., 2000; Reubinoff et al., 2000). All these statements mean that we are dealing with a cell type that presents extremely high sensitivity to cryopreservation. Therefore, the arising questions are: why are hPSCs so vulnerable to the cryopreservation process? And which are the processes involved in the low survival rates of hPSCs after cryopreservation?

Heng et al postulated for the first time that apoptosis instead of cellular necrosis, was the major mechanism inducing the loss of viability of cryopreserved hESCs during freeze-thawing with conventional slow-cooling protocols (Heng et al., 2006). They showed that most of the cells were viable (~98%) immediately after thawing (determined by the Trypan blue dye exclusion method) and that cell viability was gradually decreasing with time in culture at 37°C. Moreover, the kinetics of cell death could be reversibly slowed by a reduction in the temperature at which the cells were held post-thaw, indicating an apoptotic mechanism for cell death rather than an unregulated necrotic process. Based on these previous results, Xu et al investigated the apoptotic pathways activated during cryopreservation (Xu et al., 2010a). They described that the largest effect observed, mainly due to the freezing step, was an increase in the level of reactive oxygen species in hESCs. This presumably leads to the activation and translocation of p53 as strong expression of this protein was seen in the nucleus of thawed cells. Consequently, Caspase 9 was activated and a significant increase was also observed after thawing. In addition, Caspase 8 activity showed a similar increase post-thaw, indicating the possible activation of the extrinsic apoptotic pathway. They also stated that the elevated levels of F-actin observed during freezing could result in changes in apoptotic signals. These results led the authors to conclude that apoptosis in cryopreserved hESCs was induced through both, the intrinsic and extrinsic pathways (Xu et al., 2010a).

However, a remaining question is unanswered: why hPSCs are so sensitive to apoptosis compared to mESCs or other cell lines? In order to answer this issue, Wagh et al performed detailed microarray studies on hESCs at different time points after thawing and compared their transcriptomes with control cells that did not go through the cryopreservation process (Wagh et al., 2011). Viability, stemness, colony morphology and proliferation were also monitored at different times post-thawing. They observed a full recovery of the phenotypes of cryopreserved hESCs after 5 days of cultivation. However, the number of colonies was significantly smaller in the frozen hESCs compared to control groups. Furthermore, the colony growth rate was also reduced. Gene expression analysis showed very similar transcriptomes for the surviving fraction of 30 minutes frozen-thawed hESCs and the control unfrozen cells. Therefore, they concluded that the transcriptome of the surviving hESCs is preserved during cryopreservation. On the other hand, increases in the number of the up- and down-regulated genes occur continuously within 24 h after thawing and culturing, and those genes are declined or maintained within 48 h. This observation favored the hypothesis that physical cellular damage induced by freezing and/or thawing inhibits proper attachment during cultivation resulting in an induction of anoikis apoptotic cell death despite an almost stable transcriptome. Supporting this theory the analysis of the microarray showed differences in the expression of genes involved in cell communication, cell growth and maintenance, cell death, cell differentiation and cell proliferation (Wagh et al., 2011). In agreement, Li et al showed that increased cellular adhesion induced by the Rho associated kinase (ROCK) inhibitor Y-27632 enhances the survival of single hESCs after thawing (Li et al., 2009). To demonstrate this, they treated cryopreserved hESCs simultaneously with Y-27632 and EGTA, a calcium chelant that disrupts cadherin activity and therefore cell adhesion. This double treatment significantly reduced the capacity of hESCs to form colonies and cell viability after thawing (Li et al., 2009). These results point to a high sensitivity of cryopreserved hPSCs to the loss of adherence between cells and/or to the substrate, resulting in a detachment induced apoptosis or anoikis.

According to the high differentiation rates experienced by hPSCs after cryopreservation, Wagh et al observed a down-regulation of pluripotency markers such as nanog, sox2 or klf4 (Wagh et al., 2011). In agreement with these results, it has been shown that the pluripotency marker Oct-4 was significantly decreased after culturing cryopreserved hESCs for several days (Katkov et al., 2006). In addition, the freeze-thaw stress increases the expression of several genes involved in the differentiation processes such as embryonic morphogenesis, neurogenesis, ossification, tissue morphogenesis, regeneration and vasculature development (Wagh et al., 2011).

3. Improvement of existing cryopreservation protocols

Many laboratories have been working over the last 10 years in the development of new cryprerservation protocols for hPSCs. The main aim of the vast majority of these protocols has been the improvement of cell recovery including: an enhancement of cell survival and a reduction of cell differentiation. To this end different approaches have been adopted: development of new cryopreservation protocols such as vitrification, usage of different cryoprotective agents or molecules to improve survival, xeno-free cryopreservation media, cryopreservation of adherent hPSC colonies or single cells and/or utilization of devices to control changes in temperature. Although each of these works (summed up bellow), provide some improvements in hPSCs recovery after cryopreservation, not many of them have addressed the key question: how the changes introduced in the cryopreservation protocols contribute to the enhancement in cell recovery?

3.1 Vitrification and optimizations of the technique

One of the first attempts to overcome the low survival rates experienced by hPSCs after cryopreservation using the standard slow freezing-rapid thawing method was the adaptation of the vitrification protocol. This technique was developed for the cryopreservation of bovine ova and embryos (Vajta et al., 1998) and it was successfully adapted for hESCs freezing by Reubinoff and colleagues some years ago (Reubinoff et al., 2001). The protocol requires stepwise exposure of colony fragments to two vitrification solutions of increasing cryoprotectant concentrations. This exposure is sequential and brief (60 and 26 seconds respectively either at room temperature or at 37°C). The common components of the vitrification medium are DMSO and ethylene glycol. The composition of the vehicle solution varies, with differences in sucrose concentration, the presence or absence of serum and the buffer used. Using mixtures of cryoprotectants helps to reduce the intrinsic toxicity of each, and the method published by Reubinoff et al utilized 20% DMSO, 20% ethylene glycol and 0.5 mol/l sucrose (Reubinoff et al., 2001). However, no studies have been reported so far to determine the permeability of the cells (or colony fragments) to either cryoprotectant, or the intrinsic toxicity of these components to hESCs (Hunt, 2011).

Extremely rapid cooling rates are required to achieve vitrification using this two-component system. This is accomplished by direct immersion into liquid nitrogen of open-pulled straws containing small droplets (typically 1-20 µl) of vitrification solution within which the colony fragments (< 10) are held. Straws are then generally transferred to liquid nitrogen for long term storage (Vajta et al., 1997; Vajta et al., 1998).

The thawing process has to be as well, as rapid as possible to avoid ice crystallization. This is accomplished by direct immersion of the vitrified samples into pre-warmed culture medium containing sucrose, followed by stepwise elution of the cryopreotectants using sucrose as an osmotic buffer. An alternative method with direct exposure to growth medium without stepwise elution of the cryoprotectants has also been used with no noticeable deleterious effects (Hunt & Timmons, 2007; Reubinoff et al., 2001).

Vitrification has been adopted by many groups as the method of choice for hPSCs cryopreservation based on several comparative studies reporting recovery rates of undifferentiated colonies of more than 75% after vitrification compared to the 5-10% obtained after slow-cooling and rapid thawing (Li et al., 2010b; Reubinoff et al., 2001; Richards et al., 2004; Zhou et al., 2004). However, this technique presents some limitations: it is labor intensive and technically challenging, it is not suitable for large amounts of cells and the contact between the liquid nitrogen and the cells carries the risk of contaminations (Vajta & Nagy, 2006). Some attempts to improve these limitations have been done so far. One approximation was proposed by Heng et al for the cryopreservation of adherent hESCs colonies (Heng et al., 2005). They designed a culture plate made of detachable screw-cap culture wells resistant to storage at low temperatures in liquid nitrogen envisioned to develop automated systems for handling bulk quantities of cells (Heng et al., 2005). Alternatively, a method combining the large holding volume of slow-cooling rapid-thawing in cryotubs with the high efficiency of vitrification was described by Li et al (Li et al., 2008a). In this protocol hESCs clumps (>70um) were harvested after passage and transferred to a nylon cell strainer, exposed to vitrification solutions and vitrified by direct immersion in liquid nitrogen. Using this bulk vitrification method, 30 times more hESCs clumps (100-150) can be vitrified in a cell strainer compared to the open pulled straws. In addition, comparable results to those obtained for the classical vitrification method were reported for the recovery rate, the degree of differentiation and the maintenance of the pluripotency of the surviving cells (Li et al., 2008a). A refinement of this technique, using a cryovial fitted with stainless steel mesh, produced similar results (Li et al., 2010a). Although this method is easy and efficient to perform it still presents the limitation of direct contact of the cells with liquid nitrogen increasing the possibility of contamination and cell infection.

In order to avoid direct contact of the vitrification solution with the liquid nitrogen several methods have been developed. Usage of embryo straws sealed in both ends with a commercial plastic bag heat sealer was reported by Richards et al (Richards et al., 2004). This improvement of the cryopreservation technique presents a similar yield of hESCs recovery after thawing with low differentiation rates comparable with the results of Reubinoff et al (Reubinoff et al., 2001). The usage of cryovials for vitrification has also been explored showing interesting results. Nishigaki et al used a DMSO-based and serum-free vitrification medium to cryopreserve iPSCs in cryovials (Nishigaki et al., 2011). They compared various vitrification solutions containing different concentrations of DMSO, ethylene glycol and polyethylene glycol with knockout serum replacement (KSR) in both DMEM and Euro-collins vehicle solutions. Analysis of the thermal properties of the cryopreservation solutions during the cooling process by differential scanning calorimetry (DSC) indicated that they would vitrify at an optimal cooling rate of ~ -125 °C/min. Recovery rates between 20-30% are described one day after thawing using 40% ethylene glycol and 10% polyethylene glycol in Euro-Collins solution. Furthermore, cryopreserved cells express undifferentiation markers and keep pluripotency (Nishigaki et al., 2011). Therefore, using this protocol the vitrification of large amounts of cells is feasible and avoids the risk of contamination.

3.2 Usage of different cryoprotective agents and vehicles

During the cryopreservation course, as the cell suspension is cooled below the freezing point, ice crystals form and the concentration of the solutes in the suspension increases, being both processes damaging for the cells. Cryoprotective agents (CPAs) are necessary to minimize or prevent the damage associated with the freezing process. The mechanisms providing this protection in slow cooling-rapid thawing protocols, although not completely understood, appear to work primarily by altering the physical conditions of both the ice and the solutions immediately surrounding (external to) the cells. In contrast, vitrification overcomes the problems associated with ice crystallization in a different manner. Here cryoprotectants are used in high concentrations preventing ice formation entirely (Baust et al., 2009; Hunt, 2011).

Different CPAs have been identified so far to be used for the cryopreservation of mammalian cells (Klebe & Mancuso, 1983; Matsumura et al., 2010); however, the two most commonly used substances are glycerol and DMSO. Other substances used include sugars, polymers, alcohols and proteins. CPAs can be divided roughly into two different categories: (1) permeating CPAs: substances that permeate the cell membrane (e.g. DMSO and glycerol); and (2) nonpermeating CPAs: impermeable substances (e.g. polyethylenen glycol and trehalose); both types present a different impact on the freezing process. Permeating CPAs have a low molecular weight and thus can penetrate the cell membrane and gradually substitute the water present in the cells. The osmolality of the cells is thereby increased, and subsequently, the percentage of extracellular water that can form ice crystals before reaching the osmotic equilibrium is reduced and total dehydration of the cells is prevented. Nonpermeating CPAs cannot penetrate the cell membrane and stabilize the cell by forming a viscous glassy shell around its surface (Hubel, 1997; Karlsson, 2002; Karlsson & Toner, 1996; Meryman, 2007). Therefore, the selection of an appropriate CPA or a combination of them used in optimal concentrations within an effective vehicle solution will determine the efficiency of the cryopreservation process for a given cell type. In this sense, alternatives to DMSO as the cryoprotectant of choice for hPSCs have been tested due to the known effect of this solvent on inducing differentiation and cytotoxicity (Adler et al., 2006). See Table 1 for an overview of CPAs and freezing vehicles used in different protocols describe here.

Trehalose is a natural disaccharide that has been selected as an attractive CPA for several reasons. First of all, it has been shown to be effective in mammalian cell stabilization at low temperatures and water contents. Secondly, trehalose preserves cell viability by different mechanisms than DMSO (Crowe et al., 2001; Sum et al., 2003; Sum & de Pablo, 2003). Finally, trehalose addition to the cryopreservation medium containing DMSO and fetal bovine serum (FBS) has been proven to increase the viability of hematopoietic precursor cells from 7% to 20% and improved membrane integrity in cryopreserved fetal skin cells (Erdag et al., 2002; Zhang et al., 2003). Ji et al showed that trehalose loading into adherent colonies of hESCs prior to cryopreservation results in small, but significant improvements in cell viability when combined with DMSO treatment and high FBS concentrations (Ji et al., 2004). In the same line of results, it has been demonstrated that trehalose addition to the freezing and post-thawing medium of hESC colonies cryopreserved in suspension in freezing medium containing 10% DMSO, increased the recovery rate by ~3 folds (from 15 to 48%) (Wu et al., 2005). These results suggested that the protective mechanism of trehalose addition might be the reduction of osmotic changes during the freezing and thawing process, although this hypothesis has not been demonstrated. The addition of trehalose did not affect the normal karyotype of the cells neither their pluripotency capacity tested by teratoma formation (Wu et al., 2005).

A comparison between four different types of CPAs for iPSCs cryopreservation has recently been described: DMSO, ethylene glycol, propylene glycol and glycerol (Katkov et al., 2011). Interestingly, the toxicity of these four CPAs was analyzed after 30 minutes exposure of a 10% CPA solution at 37°C. The results showed that DMSO was the most toxic CPA for iPSCs while glycerol was the least harmful being the other two CPAs in between. Surprisingly, the protective effect exerted by the same CPAs after cryopreservation of small iPSC clumps by the slow cooling-rapid thawing protocol was the opposite, being DMSO the most protective CPA together with ethylene glycol while glycerol was the least protective one. The same result was obtained when iPSCs previously dissociated with Accutase™ were cryopreserved in the presence of a ROCK inhibitor in combination with the previous mentioned CPAs. Therefore, ethylene glycol was selected as the cryoprotectant of choice since it presents less toxicity than DMSO and exerts similar levels of protection (Katkov et al., 2011). In addition, these results give clear evidence that the low hPSCs recovery rate obtained after cryopreservation is mainly caused by the freezing-thawing procedure, rather than by the process of CPA addition/removal.

The combination of different CPAs has also been tested in comparison to the conventional freezing solution containing 10% DMSO in slow cooling-rapid freezing protocols. Ha et al performed a detailed study about the composition of the cryopreservation medium, initially analyzing the impact of both DMSO and FBS concentration in hESCs recovery (Ha et al., 2005). They reached the conclusion that a combination of 5% DMSO plus 50% FBS was the most effective one sustaining survival rates of 10%. Afterwards, they used this freezing medium composition as a starting point to test different concentrations of other CPAs such as ethylene glycol or glycerol. An increase of 3 fold in the survival rate (around 30%) was obtained when using a combination of 5% DMSO + 50% FBS +10% ethylene glycol that was selected as the most effective cryopreservation medium. Three passages after thawing cryopreserved hESCs retained the key properties and characteristics of hPSCs (Ha et al., 2005).

CPA composition	Freezing vehicle	Addition of other molecules	Cell type	Cell processing	Type of culture	Recovery rate	Reference
10 % EG	Not described	ROCK inhibitor	iPSCs	Colony clumps and single cells	MEF feeder layer	~20-50% recovery[1]	(Katkov et al., 2011)
		No		Adherent colonies		~60% recovery [1]	
10% DMSO	Growth medium	No	hESCs	Colony clumps	MEF feeder layer	60% (30 min) <10% (24 h) [2]	(Wagh et al., 2011)
						0-55% recovery [16]	(Li et al., 2010b)
		ROCK inhibitor	hESCs		Human foreskin feeder layer	50-60% survival[4]	(Martin-Ibanez et al., 2008)
			hESCs and iPSCs	Single cells	MEF feeder layer and feeder-free culture	7-8 fold increase in the number of recovered cells or colonies [8]	(Claassen et al., 2009)
		ROCK inhibitor + P53 inhibitor	hESCs	Single cells	MEF feeder layer and feeder-free culture	~80% survival[5]	(Xu et al., 2010a)
	90% FCS	ROCK inhibitor	hESCs and iPSCs	Single cells	Feeder-free culture	90% viable cells[4]	(Mollamohammadi et al., 2009)
	90% KSR	No	hESCs	Colony clumps	MEF feeder layer	8-53% survival[17]	(Lee et al., 2010)
		ROCK inhibitor	hESCs	Colony clumps	MEF feeder layer	85-95% survival[6]	(Li et al., 2008b)
				Single cells	Feeder-free culture	53-65% [7]	(Li et al., 2009)
	90% (DMEM/F12 + 20% FBS)	Z-VAD-FMK	hESCs	Adherent colonies	MEF feeder layer	18.7% [11]	(Heng et al., 2007)
		No				~98% (5 min) 20-30% (90 min) [12]	(Heng et al., 2005)
	60 % growth medium + 30% FBS	No	hESCs	Adherent colonies (microcarriers)	MEF feeder layer and feeder-free culture	1.5-1.9 fold increase in recovery rate[13]	(Nie et al., 2009)
5% DMSO 5% HES	80% (DMEM/F12 + 20% KSR)	No	hESCs	Colony clumps	MEF feeder layer	~80% recovery [9]	(T'joen et al., 2011)
10% DMSO + 0.2 mol/l Trehalose	90% KSR	No	hESCs	Colony clumps	MEF feeder layer	37-48% recovery[10]	(Zhang et al., 2003)

10% DMSO+ 35 mM Trehalose	40% Growth medium + 50% FBS	No	hESCs	Adherent colonies	MEF feeder layer and feeder-free culture	25% viability increase in respect to DMSO alone[14]	(Ji et al., 2004)
7.5% DMSO 2.5% PEG	Growth medium	ROCK inhibitor + P53 inhibitor	hESCs	Single Cells	MEF feeder layer and feeder-free culture	80-90% survival[5]	(Xu et al., 2010b)
10% DMSO + another undisclosed CPA	Phosphate buffer	No	hESCs and iPSCs	Colony clumps	Human foreskin feeder layer	90-96% viability[15]	(Holm et al., 2010)
5% DMSO 10% EG	50%FBS and DMEM/ F12	No	hESCs	Colony clumps	MEF feeder layer	30% colony recovery [3]	(Ha et al., 2005)

Table 1. Cryoprotectant agents and freezing vehicles used for the cryopreservation of hPSCs using the slow-freezing rapid-thawing protocol. The conditions and recovery rates showed in the table correspond to the best condition tested in the referenced works. (DMSO: dimethyl sulfoxide; EG: ethylene glycol; PEG: polyethylene glycol; HES: Hydrosyethylstarch; FBS: fetal bovine serum; FCS: fetal calf serum; KSR: Knockout serum replacement; MEF: mouse embryonic fibroblasts). The recovery rates were determined using different tests: (1) % cell recovery determined by QUANTA Coulter Counter measurement of Calcein-PM+/7AAD-. (2) % viability determined with FDA/EB staining at different time points after thawing. (3) Number of colonies 10 days after plating. (4) Cell viability was determined counting the number of cells by the Trypan blue method immediately after thawing. (5) Cell viability was determined by FACS using propidium iodide staining immediately after thawing. (6) Number of colonies at day 5/total colonies replated. (7) Flow cytometry analysis of apoptotis using Anexin V and propidium iodide immediately after thawing. (8) Fold increase in the number of recovered cells determined using a Z2 Coulter Counter and Size analyzer 4 days after thawing. Fold increase in the number of colonies determined by microscopy. (9) Recovery rate was calculated as follows: the amount of Grade A+B colonies at day 7 post-thawing versus the amount of frozen Grade A+B colonies. (10) Number of colonies 7 days after thawing. (11) MTT assay to measure % survival rate 24 h post-thawing. (12) Viability determined by Trypan blue exclusion method of adherent colonies at different time points after thawing. (13) Recovery was calculated as the number of cells one week after thawing divided by the number of cells at the time of freezing. The recoveries of hPSCs frozen using microcarriers are normalized to the recoveries of hPSCs frozen as free colonies. (14) Cell viability was measured by MTT assay or Alamar Blue assay several days after thawing. (15) Viability or percentage of surviving cells was calculated as a ratio between live hPSCs after thawing and total number of initially frozen cells. Cells were counted using the Trypan blue exclusion method. (16) Recovery rates were estimated as the % of attached and undifferentiated clumps counted 7-8 days after-thawing respect to the initially frozen. (17) Viability was assessed counting the number of colonies stained for alkaline phosphatase.

A new cryopreservation formula containing 7.5% DMSO plus 2.5% polyethylene glycol was analyzed in another work (Xu et al., 2010b). This study resulted in slight but significant increase in the hESCs recovery determined by counting the number of cells or colonies in feeder-independent or feeder-dependent culture respectively (Xu et al., 2010b). Recently, an alternative cryopreservation medium combining intracellular (5% DMSO) and extracellular (5% Hydrosyethylstarch) CPAs has been proven to be highly effective for the cryopreservation of small hESC clumps by the classical slow-freezing rapid-thawing method. These clumps are obtained by a combination of hESC colony detachment with Collagenase IV followed by 5 minutes dissociation using an undisclosed solution. This protocol is suitable for handling bulk amounts of hPSCs (T'joen et al., 2011).

Comparison of different freezing vehicles using DMSO as a cryoprotectant has also been studied for the cryopreservation of dissociated hESCs (Mollamohammadi et al., 2009). Three preservation media containing 10% DMSO plus: 90% fetal calf serum (FCS), 90% KSR or 90% hESCs medium containing 20% KSR and ROCK inhibitor were analyzed. The percentage of viable cells obtained by the Trypan blue exclusion method after thawing showed that cells were better preserved in the presence of 90% FCS as a vehicle (~90%). The other two freezing solutions caused lower survival rates (60-80%) (Mollamohammadi et al., 2009). Following a similar approach, Ha et al studied the impact of different FBS concentrations (5, 50 and 95%) in the vehicle freezing solution using a 5% DMSO as a CPA (Ha et al., 2005). A decrease in the survival rate is observed as the FBS concentration is reduced although no differences were found between 50 and 95%. Therefore, the authors established 50% of FBS as the optimal concentration to support hPSCs survival during the cryopreservation process (Ha et al., 2005).

3.3 Addition of molecules to enhance survival

One of the first molecules studied in the cryopreservation process to enhance cell survival was the caspase inhibitor Z-VAD-FMK (Heng et al., 2007). Results obtained in a previous work from the same group showing that apoptosis rather than necrosis was the responsible mechanism involved in the loss of viability during hESCs cryopreservation encouraged them to test a broad-spectrum irreversible inhibitor of caspase enzymes (Heng et al., 2006). Exposure to 100 mM Z-VAD-FMK in the freezing solution alone did not significantly enhace the post-thaw survival rate. However, when Z-VAD-FMK was added to the freezing solution as well as to the post-thawing solution a significant enhancement in the cell survival rate (~two fold) was observed. Nevertheless, the differentiation rates of cryopreserved hESCs were not reduced and therefore the culture recovery was not improved (Heng et al., 2007). Similarly, the addition of a specific Caspase-9 inhibitor to the post-thawing recovering medium failed to increase hESCs colony formation 4-5 days after thawing, although it did reduce Caspase 8 and 9 activity 2 h after cryopreservation (Xu et al., 2010a). These results suggested that Caspase activity was not the triggering mechanism contributing to the low hPSCs recovery after cryopreservation, but it could be a downstream effector.

A significant improvement in the cryopreservation field came up with the addition of a ROCK inhibitor. ROCK have been found to play a role in the regulation of multiple biological pathways such as apoptosis, cell cycle, differentiation, cell adhesion as well as gene expression (Amano et al., 1997; Hall, 1994; Ishizaki et al., 1997; Krawetz et al., 2009; Maekawa et al., 1999). Watanabe et al reported for the first time, that addition of the ROCK inhibitor Y-27632 improved the cloning efficiency of dissociated hESCs more than 25-fold when the cells were plated at low density (Watanabe et al., 2007). One year later, the same inhibitor was tested for the cryopreservation of hESCs. Li et al demonstrated that 10 µM Y-27632 added to the post-thaw medium during 1 day increased hESCs survival when cryopreserved as small clumps (Li et al., 2008b). In parallel our group reported that dissociated hESCs could be cryopreserved in the presence of ROCK inhibitor (Martin-Ibanez et al., 2008; Martin-Ibanez et al., 2009). The addition of Y-27632 to the freezing medium did not increase the formation of hESCs colonies compared to the control non treated cells although it increased cell survival. In contrast, the presence of ROCK inhibitor in the post-thawing recovery medium did increase the formation of hESCs colonies significantly (50-100 times). The addition of Y-27632 to both, the cryopreservation and the post-thawing medium was the condition tested contributing to the highest cell recovery after freezing.

Rock inhibitor addition	Type of cell and culture	Cell processing	Recovery rate (method of analysis)	Time of analysis	Reference
Freezing and post-thawing recovery media (1 day)	hESCs on human foresking feeders	Single cells	50-60% survival % survival after thawing analyzed by Trypan Blue exclusion method	Immediately after thawing	(Martin-Ibanez et al., 2008)
	hESCs and iPSCs in feeder-free culture	Single cells	60-80% viable cells % survival after thawing analyzed by Trypan blue exclusion method	Immediately after thawing	(Mollamoh ammadi et al., 2009)
	hESCs on MEF feeders and feeder-free culture	Single cells	80-90% survival % survival after thawing analyzed by propidium iodide staining	Immediately after thawing	(Xu et al., 2010a) (Xu et al., 2010b)
	iPSCs on MEF feeders	Single cells, clumps and adherent colonies	20-60% recovery depending on the type of culture % recovery after thawing by QUANTA Coulter Counter measurement of Calcein-PM+/7AAD-	Not stated	(Katkov et al., 2011)
Post-thawing recovery medium (1 day)	hESCs on MEF feeders	Colony clumps	85-95% survival Number of colonies at day 5/total colonies replated	5 days after thawing	(Li et al., 2008b)
	hESCs in feeder-free culture	Single cells	53-65% Flow citometry analysis of apoptosis using Anexin V and propidium iodide	Immediately after thawing and 24 h after thawing	(Li et al., 2009)
Post-thawing recovery medium (4 days)	hESCs on MEF feeders and feeder-free culture	Single cells	7-8 fold increase in the number of recovered cells 4 days after thawing (Z2 Coulter Counter and Size analyzer)	4 days after thawing	(Claassen et al., 2009)
Post-thawing recovery medium (2 days)	iPSCs on MEF feeders	Single cells	7 fold increase in the number of colonies 48 h post-thawing	2 days after thawing	(Claassen et al., 2009)

Table 2. Overview of the ROCK inhibitor treatments tested to improve the recovery rates after cryopreservation of hPSCs. Survival rates showed in the table are obtained using the best condition tested in the work referenced.

Moreover, we described a complete avoidance of hESCs differentiation just after cryopreservation showing that most of the colonies expressed the undifferentiation markers: Oct-4, nanog, SSEA-4, TRA-1-81 and TRA-1-60. The addition of Y-27632 increased the growth rates to control levels, did not affect hESCs normal karyotype and kept their pluripotency (Martin-Ibanez et al., 2008). Similar results have been shown not only for hESCs but also for iPSCs in both feeder-associated and feeder-free conditions (Claassen et al., 2009; Katkov et al., 2011; Mollamohammadi et al., 2009). See table 2 for a sum up of all the ROCK inhibitor treatments used for cryopreservation of hPSCs.

ROCK inhibitors have also been used in combination with other molecules such as Caspase inhibitors, p53 inhibitors or Bax inhibitors added always to the post-thawing culture medium. Xu et al showed that none of the three combinations pan-Caspase inhibitors + Y-27632, Caspase 9 inhibitor + Y-27632 and Bax inhibitor + Y-27632 enhanced the protective effect of ROCK inhibitor alone for cryopreserved hESCs (Xu et al., 2010a). Only the treatment with a p53 inhibitor + Y27632 induced a cell recovery similar to that of ROCK inhibitor. However, treatment with p53 alone did not account for an increase in cell survival (Xu et al., 2010a). Similar results were obtained by the same group in another report where they observed an enhancement of hESCs recovery when cryopreserved in 10% DMSO or 7.5% DMSO + 2.5% polyethylene glycol and treated with p53 inhibitor + Y-27632 in the post-thawing medium (Xu et al., 2010b).

Although most of the works studying the effect of ROCK inhibitor during cryopreservation did not address the mechanism of action of this molecule, at least two of them showed some interesting results (Li et al., 2009; Xu et al., 2010a). Both of them reported a reduction in hESCs apoptosis and/or Caspase activity one day after cryopreservation driven by Y-27632. This is in agreement with the previous report of Watanabe et al who pointed to an antiapoptotic role of this ROCK inhibitor (Watanabe et al., 2007). In addition, Li et al demonstrated that Y-27632 treatment increased the adherent properties of cryopreserved hESCs favoring cell aggregate formation and adhesion to the substrate. This effect, in turn, prevented anoikis and enhanced hESCs survival (Li et al., 2009; Mollamohammadi et al., 2009).

3.4 Cryopreservation of adherent versus suspension hPSC colonies

In view of the poor survival rates obtained after cryopreservation of hPSCs in suspension using the slow-cooling rapid thawing method, some authors decided to test cryopreservation of adherent cells. This decision was based on previous studies done with certain cell types difficult to preserve. For example, hepatocytes cryopreserved in alginate gels display a higher viability and lower apoptotic activity than hepatocytes cryopreserved in suspension (Mahler et al., 2003). Similarly, hepatocytes sandwiched between two layers of collagen provide enhanced viability and protein secretion compared with cells preserved in solution (Birraux et al., 2002; Koebe et al., 1990; Koebe et al., 1999). Taking these results as a proof of principle, hESCs were successfully cryopreserved as adherent colonies in 24 well plates in medium containing 10% DMSO + 30% FBS by Ji et al (Ji et al., 2004). This approach demonstrated that hESCs frozen as adherent colonies were five times more viable than clumps of colonies frozen in suspension. In addition, encapsulation of hESCs colonies inside Matrigel™ for 1 or 2 days increased viability significantly respect to unencapsulated adherent frozen colonies or colonies encapsulated for just 1 h. The percentage of adherent hESC colonies recovered 1 to 2 weeks after cryopreservation was about 80-90% and almost no differentiation was detected. In contrast, less than 2% of hESC colonies attached when frozen in suspension. A recent work by Katkov et al reported a refinement of the technique cryopreserving adherent iPSC colonies in the presence of ethylene glycol as a cryoprotectant and using a six-step programmed protocol (Katkov et al., 2011). Preservation of iPSCs under these conditions induced a six-fold increase in cell recovery after thawing respect the standard cryopreservation of cell clumps by the slow-freezing rapid thawing method (Katkov et al., 2011). Two mechanisms are postulated to explain the increased viability obtained preserving hESCs as adherent colonies. The first is that hESC colonies do not have to settle to the surface and attach. This is a decisive process for the survival of hPSC colonies frozen in suspension that is rarely achieved due to the massive cell death or cell damage experienced within the colony during cryopreservation. Second, the maintenance of a continuous extracellular matrix signaling may also play a role in the enhanced viability and reduced differentiation of hESCs cryopreserved in an adherent state (Ji et al., 2004). The disadvantage of this technique is that large scale storage is not feasible because hPSCs attached to plates cannot be stored at high density. In addition, culture plates are unable to be sealed like cryovials, increasing the risk of sample cross-contamination during storage in liquid nitrogen. However, methodologies such as preservation on microcarriers might provide the advantages of freezing adherent cells at higher densities that are not possible on flat surfaces. This is what has been described by Nie et al, who used Cytodex 3 microcarriers to cryopreserve adherent hESCs (Nie et al., 2009). These microcarriers consisted of a thin layer of denatured collagen covalently coupled to a matrix of cross-linked dextran. They were modified with Matrigel™ or irradiated MEF to enhance the adhesion of hESC colonies. In this work it was first demonstrated that hESCs colonies were effectively expanded in a pluripotent, undifferentiated state on both types of microcarriers (Matrigel™ and MEF coated). Then cryopreservation utilizing this system was compared to standard freezing of hESC colonies in suspension. hESCs-microcarriers were suspended in freezing medium consisting in 10%DMSO and 30%FBS at a cell density of 1×10^6 cells/ml on 10 cm^2 microcarriers. The suspension was transferred to cryovials, frozen inside a freezing container at a cooling rate of -1°C/min and moved into liquid nitrogen. Seven days after thawing viability was assessed by counting the number of cells. This number was compared to that of the conventional hESCs slow freezing method. Cryopreservation on microcarriers resulted in 1.7 times the recovery of hESCs frozen in free suspension (Nie et al., 2009). Although the enhancement of cell recovery is not very promising, further optimization of this methodology holds a great potential for future larger-scale cryopreservation.

3.5 Cryopreservation of dissociated single hPSCs versus clumps of colonies

hPSCs are colony-forming social cells that present a high vulnerability to apoptosis upon cellular detachment and dissociation (Amit et al., 2000; Watanabe et al., 2007). These characteristics could explain why most of the cryopreservation protocols rely on hPSCs small clumps to improve survival rates (Heng et al., 2006; Reubinoff et al., 2001; Richards et al., 2004; Zhou et al., 2004). However, the cryopreservation of clumps presents some associated problems such as limitations on cryoprotectant exposure inside the clump. In this sense, T'Joen et al demonstrated that the application of a cell dissociation solution before freezing, thereby creating a mixed population of very small hESC clumps and single cells, increased the recovery rate after cryopreservation (T'joen et al., 2011). In addition, the use of hPSC colony clumps also prevents a good estimation of freezing-thawing efficiency, as precise cell numbers cannot be estimated. Therefore, development of cryopreservation protocols for dissociated hPSCs is a pre-requisite for the widespread use of these cells in basic or clinical research.

Most of the studies carried out so far for the cryopreservation of dissociated hPSCs involved the usage of ROCK inhibitors (usually 10μM of Y-27632), since in its absence very few or none colonies are obtained. This inhibitor has been reported to significantly increase the survival rate of frozen/thawed single hESC as well as iPSCs (Claassen et al., 2009; Li et al., 2009; Martin-Ibanez et al., 2008; Mollamohammadi et al., 2009; Xu et al., 2010a; Xu et al., 2010b). Recent studies have demonstrated that Y-27632 increased not only the survival rate but also the adhesion of frozen–thawed dissociated single hPSCs in the presence and absence of feeder cells (Claassen et al., 2009; Katkov et al., 2011; Li et al., 2009; Martin-Ibanez et al., 2008; Mollamohammadi et al., 2009; Xu et al., 2010a; Xu et al., 2010b). In fact, Li et al. proposed that Y-27632 does not block apoptotic pathways, but rather prevents hPSCs from sensing their external environment, giving them time to make important cell-cell interactions and thus allowing them to escape anoikis (Krawetz et al., 2009; Li et al., 2009). Moreover, Mollamohammadi et al showed by RT-PCR analysis that the expression of integrin chains aV, a6 and b1 increased significantly in the presence of ROCK inhibitor (Mollamohammadi et al., 2009). They proposed that this increase in integrins expression may account for the maintenance of an undifferentiated state and an increase in cell adhesion of hESCs and iPSCs to the substrate allowing better cloning efficiency (Mollamohammadi et al., 2009).

The usage of ROCK inhibitors for continuous treatments has not induced any adverse effects on hPSCs pluripotency or chromosomal stability, even after substantial number of passages (Mollamohammadi et al., 2009; Watanabe et al., 2007). ROCK inhibitors such as Y-27632 or Fasudil are already used clinically in cardiovascular therapies (Hu & Lee, 2005), suggesting that they are safe for the treatment of hPSCs. Although the exact mechanism of action of ROCK inhibitors is, at the moment, unknown and a lot of cross talks between signaling pathways occur, the usage of this compound opens a new field of study to improve hPSCs cryopreservation and culture protocols.

3.6 Controlled-rate cryopreservation

The cooling rate is one of the cryobiological variables associated with damage during slow-cooling (Figure 2). When the cooling process is rapid, intracellular ice crystals form before complete cellular dehydration has occurred. These ice crystals disrupt cellular organelles and membranes and lead to cell death during the recovery (thawing) process. On the other hand, when the cooling process is slow, free intracellular water is osmotically pulled from the cells resulting in complete cellular dehydration and shrinkage. This can also cause cellular death but there is little agreement on the mechanisms involved. However, when the cooling rate is slow enough to prevent intracellular ice formation, but fast enough to avoid serious dehydration effects, cells may be able to survive the freezing and thawing process. This survival zone or window is readily observed in many bacteria and other prokaryotes, but for most eukaryotic cells it is nonexistent or very difficult to find without using cryoprotectants. These agents have little effect on the damage caused by fast freezing (intracellular ice crystal formation), but rather prevent or lessen the damage caused by slow freezing (dehydration and shrinkage) (Figure 2) (Mazur, 1984). Thus, a tight control of the cooling rate is crucial to reduce cellular damage during cryopreservation, even in the presence of CPAs. This is achieved by the usage of programmable freezers. These devices, although expensive and not always available, are technically more reliable and reproducible. Several works have studied the relevance of programmable freezers for the cryopreservation of hPSCs. Ware et al reported survival rates of 60-70% with no apparent increase in differentiation using DMSO as a cryoprotectant, a control rate freezing device and straws as containers (Ware et al., 2005). The results of this study identified three critical factors for successful hESCs freezing: ice crystal seed at some point above the temperature of spontaneous intracellular ice formation (between -7°C and -12°C), an appropriate freezing rate (between -0,3°C and -1,8°C/min) and rapid thawing (at 25-37°C) (Ware et al., 2005). Another study optimizing the same critical factors described an improved protocol consisting in: cooling the sample from 0 °C to -35 °C at a cooling rate of -0.5°C/min, seeding at -10 °C before being plunged immediately into the liquid nitrogen and rapid thawing. Under these conditions a survival rate of 80% was obtained (Yang et al., 2006). A successful usage of programmable freezing for the cryopreservation of adherent iPSCs has also been recently described (Katkov et al., 2011). The authors developed a six step programmed protocol including : 1) -1°C/min from 0°C (addition of CPA on ice) to -10°C; 2) hold for 30 min at -10°C; 3) -3°C/min to -40°C; 4) -1°C/min to -60°C; 5) -0.33°C/min to -80°C and 6) hold at -80°C for 5 min and then transfer to liquid nitrogen. Adherent iPSC colonies cryopreserved using ethylene glycol as a CPA under these conditions showed a 63% recovery, which represents a 6 fold increase respect the preservation without a programmable freezer using DMSO (Katkov et al., 2011).

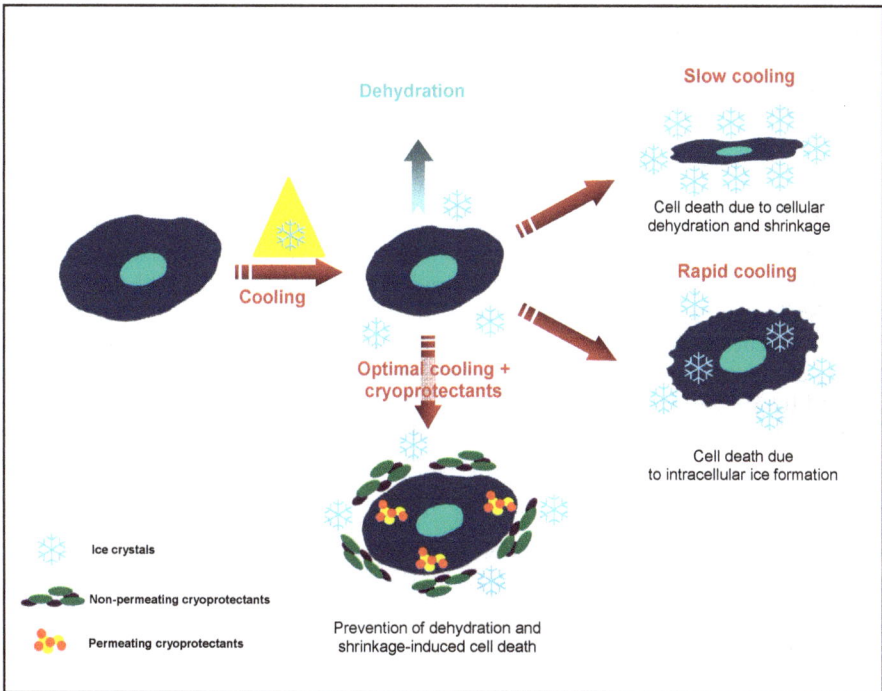

Fig. 2. Effects occurring during the cryopreservation of cells at different cooling rates. When the cooling process starts, ice crystals formation is induced and free intracellular water is osmotically pulled from the cells. If the cooling process is slow this effect lead to cellular cell death by dehydration and shrinkage. In contrast, if the cooling process is rapid, intracellular ice crystals form before complete cellular dehydration has occurred. These crystals induce cell death by cellular organelles and membrane disruption during the thawing process. An optimal cooling rate together with the usage of cryoprotectants in the freezing media avoids dehydration effects and intracellular ice formation allowing cell survival after thawing.

A recent report made an interesting comparison of three methods of cryopreservation of hESC clumps including: conventional slow freezing-rapid thawing using cryovials, vitrification and programmable cryopreservation in plastic straws (Li et al., 2010b). Assessing the efficiency of cryopreservation by counting the number of attached undifferentiated colonies 1-2 days and 7-8 days after thawing they reached the conclusion that conventional cryopreservation may not be appropriated for hESCs preservation since few colonies attached and most of them were differentiated. The usage of a programmable freezer increased significantly the cryopreservation efficiency (~50% colony recovery respect to ~5% of conventional freezing), although it was not better than the high efficiency obtained by vitrification (80-90% colony recovery). Both methodologies maintain unaffected the pluripotency and normal karyotype of the cells (Li et al., 2010b). Another comparative study published at the same time reported lower survival rates after programmable cryopreservation of hESC clumps (10-20% survival colonies), although they were significantly higher than the ones obtained after conventional slow-freezing (4-8%) (Lee et al., 2010). In this study the best cryopreservation condition was obtained using a stepwise transfer method for hESC clumps, which consisted in using a series of solutions with increasing serum replacement and DMSO concentrations to achieve a stepwise equilibration before freezing. The same inverse process was performed after thawing in order to gradually rehydrate the cells. The combination of stepwise methods with programmable freezers yielded survival rates of 30-50% with low numbers of differentiated cells (Lee et al., 2010).

3.7 Cryopreservation in xeno-free conditions

Clinical application of hPSCs would need hESC and iPSC lines derived, cultured, differentiated and cryopreserved in xeno-free conditions following good manufacturing practice (GMP) regulations. Several attempts to improve hPSCs culture conditions have been reported. These advances include: the derivation of clinical grade hESC and iPSC lines, the use of conditioned media together with Matrigel™ as an attachment substrate for hPSCs culture and the derivation and propagation of hESC lines on human feeder layers in xeno-free culture media (Amit et al., 2004; Hovatta et al., 2003; Rajala et al., 2007; Rajala et al., 2010; Richards et al., 2002; Richards et al., 2003; Skottman et al., 2006; Unger et al., 2008). Some approaches have also been done in the cryopreservation field towards the development of xeno-free effective cryopreservation protocols. The first one was an optimization of the established vitrification method previously described by Reubinoff et al (Reubinoff et al., 2001; Richards et al., 2004). In this new method they reported the successful vitrification of hESCs in sealed closed straws, their storage in the vapor phase of liquid nitrogen and the substitution of FCS with human serum albumin as the major protein source in the cryoprotectant solution. This refinement of the technique allowed the removal of animal components from the cryopreservation medium, therefore lowering the risk of cross-transfer of viruses and other pathogens to the hESCs. Moreover, sealing the straws the authors also prevented contact with potentially contaminated liquid nitrogen during cooling and storage. The efficiency of hESCs preservation was similar to the original vitrification protocol (Richards et al., 2004).

An effective serum and xeno-free chemically defined freezing procedure for hESCs and iPSCs has been recently developed (Holm et al., 2010). This protocol describes the usage of a commercially available freezing and post-thaw washing solution that presents the advantage of being chemically defined, sterile and batch tested. The cryopreservation solution named STEM-CELLBANKER™ contains 10% DMSO, glucose and a high molecular weight polymer (undisclosed) used as a second cryoprotectant, all dissolved in phosphate-buffered saline. hPSCs are preserved using this solution in cryovials and the slow-cooling rapid-thawing method, without any programmed freezer. After thawing, cells are recovered in the washing solution named CELLOTION™ containing NaCl, centrifuged to eliminate cryoprotectants and plated down on a feeder layer of human mitotically inactivated fibroblasts. Post-thaw recovery was substantially increased without any detrimental impact on proliferation or differentiation (Holm et al., 2010). Similar cryopreservation yields were obtained for both hESCs preserved as clumps and iPSCs preserved as single cells without ROCK inhibitor treatment. Therefore, this is a simple and efficient system that enables the cryopreservation of large quantities of hPSCs in a chemically defined medium that is clinical grade compatible (Holm et al., 2010). Employing a similar protocol but using a home-made cryopreservation solution containing 10% DMSO and 90% KSR, Li et al reported the preservation of single hESCs in serum and feeder-free conditions in the presence of ROCK inhibitor during the first day after thawing (Li et al., 2009).

4. Conclusion

Understanding the mechanisms involved in the high vulnerability of hPSCs to the cryopreservation process is essential to develop efficient protocols for cryopreservation. Most of the research being undertaken over the last years is still empirical and few advances have been achieved in the identification of the pathways involved in the enhancement of cell survival induced by different factors, cryoprotectants or preservation systems. However, from the results obtained in these studies it is becoming increasingly clear that cell-cell adhesion and/or paracrine signaling between hPSCs are essential for survival and control of their undifferentiated state (Amit et al., 2000; Reubinoff et al., 2000; Thomson et al., 1998). Gap junctions and cell adhesion molecules are highly expressed in hESCs and have been implicated in these processes (De et al., 2002; Richards et al., 2004; Sathananthan et al., 2002; Wong et al., 2004; Wong et al., 2008). Therefore, disruption of these structures during cryopreservation due to ice crystal formation outside the cells may induce anoikis contributing to the poor recovery of hPSCs after slow cooling. However, a better understanding of this process together with a systematic study of the critical cryobiological variables is still needed to improve the already existing cryopreservation protocols. Further advances in the field would also require the development of reliable and standardized assays to measure not only immediate post-thaw recovery but also the ability of single cells or clumps to re-attach, proliferate and maintain pluripotency. Moreover, it is necessary to establish the n-points at which these assays should be applied, in order to allow direct quantitative comparisons between different cryopreservation methods that are not feasible at the moment. Thus, all present and future investigations would likely provide a reproducible effective and efficient cryopreservation protocol for hPSCs large-scale storage that will fulfill GMP requirements, permitting the widespread use of hPSCs in basic and/or clinical research.

5. Acknowledgements

Our group is supported by grants from the Ministerio de Ciencia e Innovación (SAF2009-07774 and PLE2009-0089), Instituto de Salud Carlos III, Ministerio de Ciencia e Innovación

[CIBERNED and RETICS (RD06/0010/0006; Red de Terapia Celular)] and Generalitat de Catalunya (2009SGR-00326), Spain. The Cell Therapy Program is supported by the Centre of Regenerative Medicine in Barcelona (CMRB; Promt-0901), Generalitat de Catalunya, Spain.

6. References

Adler, S., Pellizzer, C., Paparella, M., Hartung, T., & Bremer, S. (2006). The effects of solvents on embryonic stem cell differentiation. *Toxicol. In Vitro*, Vol. 20, No. 3, pp. (265-271)

Amano, M., Chihara, K., Kimura, K., Fukata, Y., Nakamura, N., Matsuura, Y., & Kaibuchi, K. (1997). Formation of actin stress fibers and focal adhesions enhanced by Rho-kinase. *Science*, Vol. 275, No. 5304, pp. (1308-1311)

Amit, M., Carpenter, M.K., Inokuma, M.S., Chiu, C.P., Harris, C.P., Waknitz, M.A., Itskovitz-Eldor, J., & Thomson, J.A. (2000). Clonally derived human embryonic stem cell lines maintain pluripotency and proliferative potential for prolonged periods of culture. *Dev. Biol.*, Vol. 227, No. 2, pp. (271-278)

Amit, M., Shariki, C., Margulets, V., & Itskovitz-Eldor, J. (2004). Feeder layer- and serum-free culture of human embryonic stem cells. *Biol. Reprod.*, Vol. 70, No. 3, pp. (837-845)

Apostolou, E. & Hochedlinger, K. (2011). Stem cells: iPS cells under attack. *Nature*, Vol. 474, No. 7350, pp. (165-166)

Battey, J.F. (2007). Stem cells: current challenges and future promise. *Dev. Dyn.*, Vol. 236, No. 12, pp. (3193-3198)

Baust, J.G., Gao, D., & Baust, J.M. (2009). Cryopreservation: An emerging paradigm change. *Organogenesis.*, Vol. 5, No. 3, pp. (90-96)

Baust, J.M., Vogel, M.J., Van, B.R., & Baust, J.G. (2001). A molecular basis of cryopreservation failure and its modulation to improve cell survival. *Cell Transplant.*, Vol. 10, No. 7, pp. (561-571)

Berz, D., McCormack, E.M., Winer, E.S., Colvin, G.A., & Quesenberry, P.J. (2007). Cryopreservation of hematopoietic stem cells. *Am. J. Hematol.*, Vol. 82, No. 6, pp. (463-472)

Birraux, J., Genin, B., Matthey-Doret, D., Mage, R., Morel, P., & Le, C.C. (2002). Hepatocyte cryopreservation in a three-dimensional structure. *Transplant. Proc.*, Vol. 34, No. 3, pp. (764-767)

Bosma, G.C., Custer, R.P., & Bosma, M.J. (1983). A severe combined immunodeficiency mutation in the mouse. *Nature*, Vol. 301, No. 5900, pp. (527-530)

Carpenter, M.K., Rosler, E., & Rao, M.S. (2003). Characterization and differentiation of human embryonic stem cells. *Cloning Stem Cells*, Vol. 5, No. 1, pp. (79-88)

Chambers, I., Colby, D., Robertson, M., Nichols, J., Lee, S., Tweedie, S., & Smith, A. (2003). Functional expression cloning of Nanog, a pluripotency sustaining factor in embryonic stem cells. *Cell*, Vol. 113, No. 5, pp. (643-655)

Chin, M.H., Mason, M.J., Xie, W., Volinia, S., Singer, M., Peterson, C., Ambartsumyan, G., Aimiuwu, O., Richter, L., Zhang, J., Khvorostov, I., Ott, V., Grunstein, M., Lavon, N., Benvenisty, N., Croce, C.M., Clark, A.T., Baxter, T., Pyle, A.D., Teitell, M.A., Pelegrini, M., Plath, K., & Lowry, W.E. (2009). Induced pluripotent stem cells and embryonic stem cells are distinguished by gene expression signatures. *Cell Stem Cell*, Vol. 5, No. 1, pp. (111-123)

Claassen, D.A., Desler, M.M., & Rizzino, A. (2009). ROCK inhibition enhances the recovery and growth of cryopreserved human embryonic stem cells and human induced pluripotent stem cells. *Mol. Reprod. Dev.*, Vol. 76, No. 8, pp. (722-732)

Condic, M.L. & Rao, M. (2008). Regulatory issues for personalized pluripotent cells. *Stem Cells*, Vol. 26, No. 11, pp. (2753-2758)

Crowe, J.H., Crowe, L.M., Oliver, A.E., Tsvetkova, N., Wolkers, W., & Tablin, F. (2001). The trehalose myth revisited: introduction to a symposium on stabilization of cells in the dry state. *Cryobiology*, Vol. 43, No. 2, pp. (89-105)

De, M.A., Vega, V.L., & Contreras, J.E. (2002). Gap junctions, homeostasis, and injury. *J. Cell Physiol*, Vol. 191, No. 3, pp. (269-282)

Draper, J.S., Smith, K., Gokhale, P., Moore, H.D., Maltby, E., Johnson, J., Meisner, L., Zwaka, T.P., Thomson, J.A., & Andrews, P.W. (2004). Recurrent gain of chromosomes 17q and 12 in cultured human embryonic stem cells. *Nat. Biotechnol.*, Vol. 22, No. 1, pp. (53-54)

Erdag, G., Eroglu, A., Morgan, J., & Toner, M. (2002). Cryopreservation of fetal skin is improved by extracellular trehalose. *Cryobiology*, Vol. 44, No. 3, pp. (218-228)

Evans, M.J. & Kaufman, M.H. (1981). Establishment in culture of pluripotential cells from mouse embryos. *Nature*, Vol. 292, No. 5819, pp. (154-156)

Fu, T., Guo, D., Huang, X., O'Gorman, M.R., Huang, L., Crawford, S.E., & Soriano, H.E. (2001). Apoptosis occurs in isolated and banked primary mouse hepatocytes. *Cell Transplant.*, Vol. 10, No. 1, pp. (59-66)

Fujikawa, S. (1980). Freeze-fracture and etching studies on membrane damage on human erythrocytes caused by formation of intracellular ice. *Cryobiology*, Vol. 17, No. 4, pp. (351-362)

Gao, D.Y., Chang, Q., Liu, C., Farris, K., Harvey, K., McGann, L.E., English, D., Jansen, J., & Critser, J.K. (1998). Fundamental cryobiology of human hematopoietic progenitor cells. I: Osmotic characteristics and volume distribution. *Cryobiology*, Vol. 36, No. 1, pp. (40-48)

Gao, D.Y., Liu, J., Liu, C., McGann, L.E., Watson, P.F., Kleinhans, F.W., Mazur, P., Critser, E.S., & Critser, J.K. (1995). Prevention of osmotic injury to human spermatozoa during addition and removal of glycerol. *Hum. Reprod.*, Vol. 10, No. 5, pp. (1109-1122)

Grout, B., Morris, J., & McLellan, M. (1990). Cryopreservation and the maintenance of cell lines. *Trends Biotechnol.*, Vol. 8, No. 10, pp. (293-297)

Ha, S.Y., Jee, B.C., Suh, C.S., Kim, H.S., Oh, S.K., Kim, S.H., & Moon, S.Y. (2005). Cryopreservation of human embryonic stem cells without the use of a programmable freezer. *Hum. Reprod.*, Vol. 20, No. 7, pp. (1779-1785)

Hall, A. (1994). Small GTP-binding proteins and the regulation of the actin cytoskeleton. *Annu. Rev. Cell Biol.*, Vol. 10, pp. (31-54)

Heins, N., Englund, M.C., Sjoblom, C., Dahl, U., Tonning, A., Bergh, C., Lindahl, A., Hanson, C., & Semb, H. (2004). Derivation, characterization, and differentiation of human embryonic stem cells. *Stem Cells*, Vol. 22, No. 3, pp. (367-376)

Heng, B.C., Bested, S.M., Chan, S.H., & Cao, T. (2005). A proposed design for the cryopreservation of intact and adherent human embryonic stem cell colonies. *In Vitro Cell Dev. Biol. Anim*, Vol. 41, No. 3-4, pp. (77-79)

Heng, B.C., Clement, M.V., & Cao, T. (2007). Caspase inhibitor Z-VAD-FMK enhances the freeze-thaw survival rate of human embryonic stem cells. *Biosci. Rep.*, Vol. 27, No. 4-5, pp. (257-264)

Heng, B.C., Ye, C.P., Liu, H., Toh, W.S., Rufaihah, A.J., Yang, Z., Bay, B.H., Ge, Z., Ouyang, H.W., Lee, E.H., & Cao, T. (2006). Loss of viability during freeze-thaw of intact and adherent human embryonic stem cells with conventional slow-cooling protocols is predominantly due to apoptosis rather than cellular necrosis. *J. Biomed. Sci.*, Vol. 13, No. 3, pp. (433-445)

Holm, F., Strom, S., Inzunza, J., Baker, D., Stromberg, A.M., Rozell, B., Feki, A., Bergstrom, R., & Hovatta, O. (2010). An effective serum- and xeno-free chemically defined freezing procedure for human embryonic and induced pluripotent stem cells. *Hum. Reprod.*, Vol. 25, No. 5, pp. (1271-1279)

Hovatta, O., Mikkola, M., Gertow, K., Stromberg, A.M., Inzunza, J., Hreinsson, J., Rozell, B., Blennow, E., Andang, M., & Ahrlund-Richter, L. (2003). A culture system using human foreskin fibroblasts as feeder cells allows production of human embryonic stem cells. *Hum. Reprod.*, Vol. 18, No. 7, pp. (1404-1409)

Hu, E. & Lee, D. (2005). Rho kinase as potential therapeutic target for cardiovascular diseases: opportunities and challenges. *Expert. Opin. Ther. Targets.*, Vol. 9, No. 4, pp. (715-736)

Hubel, A. (1997). Parameters of cell freezing: implications for the cryopreservation of stem cells. *Transfus. Med. Rev.*, Vol. 11, No. 3, pp. (224-233)

Hunt, C.J. (2011). Cryopreservation of Human Stem Cells for Clinical Application: A Review. *Transfus. Med. Hemother.*, Vol. 38, No. 2, pp. (107-123)

Hunt, C.J. & Timmons, P.M. (2007). Cryopreservation of human embryonic stem cell lines. *Methods Mol. Biol.*, Vol. 368, pp. (261-270)

Ishizaki, T., Naito, M., Fujisawa, K., Maekawa, M., Watanabe, N., Saito, Y., & Narumiya, S. (1997). p160ROCK, a Rho-associated coiled-coil forming protein kinase, works downstream of Rho and induces focal adhesions. *FEBS Lett.*, Vol. 404, No. 2-3, pp. (118-124)

Ji, L., de Pablo, J.J., & Palecek, S.P. (2004). Cryopreservation of adherent human embryonic stem cells. *Biotechnol. Bioeng.*, Vol. 88, No. 3, pp. (299-312)

Karlsson, J.O. (2002). Cryopreservation: freezing and vitrification. *Science*, Vol. 296, No. 5568, pp. (655-656)

Karlsson, J.O. & Toner, M. (1996). Long-term storage of tissues by cryopreservation: critical issues. *Biomaterials*, Vol. 17, No. 3, pp. (243-256)

Katkov, I.I., Kan, N.G., Cimadamore, F., Nelson, B., Snyder, E.Y., & Terskikh, A.V. (2011). DMSO-Free Programmed Cryopreservation of Fully Dissociated and Adherent Human Induced Pluripotent Stem Cells. *Stem Cells Int.*, Vol. 2011, pp. (981606-

Katkov, I.I., Kim, M.S., Bajpai, R., Altman, Y.S., Mercola, M., Loring, J.F., Terskikh, A.V., Snyder, E.Y., & Levine, F. (2006). Cryopreservation by slow cooling with DMSO diminished production of Oct-4 pluripotency marker in human embryonic stem cells. *Cryobiology*, Vol. 53, No. 2, pp. (194-205)

Kiskinis, E. & Eggan, K. (2010). Progress toward the clinical application of patient-specific pluripotent stem cells. *J. Clin. Invest*, Vol. 120, No. 1, pp. (51-59)

Klebe, R.J. & Mancuso, M.G. (1983). Identification of new cryoprotective agents for cultured mammalian cells. *In Vitro*, Vol. 19, No. 3 Pt 1, pp. (167-170)

Klimanskaya, I., Chung, Y., Becker, S., Lu, S.J., & Lanza, R. (2006). Human embryonic stem cell lines derived from single blastomeres. *Nature*, Vol. 444, No. 7118, pp. (481-485)

Koebe, H.G., Dunn, J.C., Toner, M., Sterling, L.M., Hubel, A., Cravalho, E.G., Yarmush, M.L., & Tompkins, R.G. (1990). A new approach to the cryopreservation of hepatocytes in a sandwich culture configuration. *Cryobiology*, Vol. 27, No. 5, pp. (576-584)

Koebe, H.G., Muhling, B., Deglmann, C.J., & Schildberg, F.W. (1999). Cryopreserved porcine hepatocyte cultures. *Chem. Biol. Interact.*, Vol. 121, No. 1, pp. (99-115)

Krawetz, R.J., Li, X., & Rancourt, D.E. (2009). Human embryonic stem cells: caught between a ROCK inhibitor and a hard place. *Bioessays*, Vol. 31, No. 3, pp. (336-343)

Lee, J.Y., Lee, J.E., Kim, D.K., Yoon, T.K., Chung, H.M., & Lee, D.R. (2010). High concentration of synthetic serum, stepwise equilibration and slow cooling as an efficient technique for large-scale cryopreservation of human embryonic stem cells. *Fertil. Steril.*, Vol. 93, No. 3, pp. (976-985)

Li, T., Mai, Q., Gao, J., & Zhou, C. (2010a). Cryopreservation of human embryonic stem cells with a new bulk vitrification method. *Biol. Reprod.*, Vol. 82, No. 5, pp. (848-853)

Li, T., Zhou, C., Liu, C., Mai, Q., & Zhuang, G. (2008a). Bulk vitrification of human embryonic stem cells. *Hum. Reprod.*, Vol. 23, No. 2, pp. (358-364)

Li, X., Krawetz, R., Liu, S., Meng, G., & Rancourt, D.E. (2009). ROCK inhibitor improves survival of cryopreserved serum/feeder-free single human embryonic stem cells. *Hum. Reprod.*, Vol. 24, No. 3, pp. (580-589)

Li, X., Meng, G., Krawetz, R., Liu, S., & Rancourt, D.E. (2008b). The ROCK inhibitor Y-27632 enhances the survival rate of human embryonic stem cells following cryopreservation. *Stem Cells Dev.*, Vol. 17, No. 6, pp. (1079-1085)

Li, Y., Tan, J.C., & Li, L.S. (2010b). Comparison of three methods for cryopreservation of human embryonic stem cells. *Fertil. Steril.*, Vol. 93, No. 3, pp. (999-1005)

Lin, G., Ouyang, Q., Zhou, X., Gu, Y., Yuan, D., Li, W., Liu, G., Liu, T., & Lu, G. (2007). A highly homozygous and parthenogenetic human embryonic stem cell line derived from a one-pronuclear oocyte following in vitro fertilization procedure. *Cell Res.*, Vol. 17, No. 12, pp. (999-1007)

Maekawa, M., Ishizaki, T., Boku, S., Watanabe, N., Fujita, A., Iwamatsu, A., Obinata, T., Ohashi, K., Mizuno, K., & Narumiya, S. (1999). Signaling from Rho to the actin cytoskeleton through protein kinases ROCK and LIM-kinase. *Science*, Vol. 285, No. 5429, pp. (895-898)

Maherali, N., Sridharan, R., Xie, W., Utikal, J., Eminli, S., Arnold, K., Stadtfeld, M., Yachechko, R., Tchieu, J., Jaenisch, R., Plath, K., & Hochedlinger, K. (2007). Directly reprogrammed fibroblasts show global epigenetic remodeling and widespread tissue contribution. *Cell Stem Cell*, Vol. 1, No. 1, pp. (55-70)

Mahler, S., Desille, M., Fremond, B., Chesne, C., Guillouzo, A., Campion, J.P., & Clement, B. (2003). Hypothermic storage and cryopreservation of hepatocytes: the protective effect of alginate gel against cell damages. *Cell Transplant.*, Vol. 12, No. 6, pp. (579-592)

Mai, Q., Yu, Y., Li, T., Wang, L., Chen, M.J., Huang, S.Z., Zhou, C., & Zhou, Q. (2007). Derivation of human embryonic stem cell lines from parthenogenetic blastocysts. *Cell Res.*, Vol. 17, No. 12, pp. (1008-1019)

Martin, G.R. (1981). Isolation of a pluripotent cell line from early mouse embryos cultured in medium conditioned by teratocarcinoma stem cells. *Proc. Natl. Acad. Sci. U. S. A*, Vol. 78, No. 12, pp. (7634-7638)

Martin-Ibanez, R., Stromberg, A.M., Hovatta, O., & Canals, J.M. (2009). Cryopreservation of dissociated human embryonic stem cells in the presence of ROCK inhibitor. *Curr. Protoc. Stem Cell Biol.*, Vol. Chapter 1, pp. (Unit-

Martin-Ibanez, R., Unger, C., Stromberg, A., Baker, D., Canals, J.M., & Hovatta, O. (2008). Novel cryopreservation method for dissociated human embryonic stem cells in the presence of a ROCK inhibitor. *Hum. Reprod.*, Vol. 23, No. 12, pp. (2744-2754)

Matsumura, K., Bae, J.Y., & Hyon, S.H. (2010). Polyampholytes as cryoprotective agents for mammalian cell cryopreservation. *Cell Transplant.*, Vol. 19, No. 6, pp. (691-699)

Mazur, P. (1984). Freezing of living cells: mechanisms and implications. *Am. J. Physiol*, Vol. 247, No. 3 Pt 1, pp. (C125-C142)

Mazur, P. & Cole, K.W. (1989). Roles of unfrozen fraction, salt concentration, and changes in cell volume in the survival of frozen human erythrocytes. *Cryobiology*, Vol. 26, No. 1, pp. (1-29)

Mazur, P., Leibo, S.P., & Chu, E.H. (1972). A two-factor hypothesis of freezing injury. Evidence from Chinese hamster tissue-culture cells. *Exp. Cell Res.*, Vol. 71, No. 2, pp. (345-355)

Mazur, P. & Schneider, U. (1986). Osmotic responses of preimplantation mouse and bovine embryos and their cryobiological implications. *Cell Biophys.*, Vol. 8, No. 4, pp. (259-285)

Meryman, H.T. (2007). Cryopreservation of living cells: principles and practice. *Transfusion*, Vol. 47, No. 5, pp. (935-945)

Mollamohammadi, S., Taei, A., Pakzad, M., Totonchi, M., Seifinejad, A., Masoudi, N., & Baharvand, H. (2009). A simple and efficient cryopreservation method for feeder-free dissociated human induced pluripotent stem cells and human embryonic stem cells. *Hum. Reprod.*, Vol. 24, No. 10, pp. (2468-2476)

Mountford, J.C. (2008). Human embryonic stem cells: origins, characteristics and potential for regenerative therapy. *Transfus. Med.*, Vol. 18, No. 1, pp. (1-12)

Muldrew, K. & McGann, L.E. (1994). The osmotic rupture hypothesis of intracellular freezing injury. *Biophys. J.*, Vol. 66, No. 2 Pt 1, pp. (532-541)

Nichols, J., Zevnik, B., Anastassiadis, K., Niwa, H., Klewe-Nebenius, D., Chambers, I., Scholer, H., & Smith, A. (1998). Formation of pluripotent stem cells in the mammalian embryo depends on the POU transcription factor Oct4. *Cell*, Vol. 95, No. 3, pp. (379-391)

Nie, Y., Bergendahl, V., Hei, D.J., Jones, J.M., & Palecek, S.P. (2009). Scalable culture and cryopreservation of human embryonic stem cells on microcarriers. *Biotechnol. Prog.*, Vol. 25, No. 1, pp. (20-31)

Nishigaki, T., Teramura, Y., Nasu, A., Takada, K., Toguchida, J., & Iwata, H. (2011). Highly efficient cryopreservation of human induced pluripotent stem cells using a dimethyl sulfoxide-free solution. *Int. J. Dev. Biol.*, Vol. 55, No. 3, pp. (305-311)

Odorico, J.S., Kaufman, D.S., & Thomson, J.A. (2001). Multilineage differentiation from human embryonic stem cell lines. *Stem Cells*, Vol. 19, No. 3, pp. (193-204)

Okita, K., Ichisaka, T., & Yamanaka, S. (2007). Generation of germline-competent induced pluripotent stem cells. *Nature*, Vol. 448, No. 7151, pp. (313-317)

Paasch, U., Sharma, R.K., Gupta, A.K., Grunewald, S., Mascha, E.J., Thomas, A.J., Jr., Glander, H.J., & Agarwal, A. (2004). Cryopreservation and thawing is associated with varying extent of activation of apoptotic machinery in subsets of ejaculated human spermatozoa. *Biol. Reprod.*, Vol. 71, No. 6, pp. (1828-1837)

Park, I.H., Zhao, R., West, J.A., Yabuuchi, A., Huo, H., Ince, T.A., Lerou, P.H., Lensch, M.W., & Daley, G.Q. (2008). Reprogramming of human somatic cells to pluripotency with defined factors. *Nature*, Vol. 451, No. 7175, pp. (141-146)

Rajala, K., Hakala, H., Panula, S., Aivio, S., Pihlajamaki, H., Suuronen, R., Hovatta, O., & Skottman, H. (2007). Testing of nine different xeno-free culture media for human embryonic stem cell cultures. *Hum. Reprod.*, Vol. 22, No. 5, pp. (1231-1238)

Rajala, K., Lindroos, B., Hussein, S.M., Lappalainen, R.S., Pekkanen-Mattila, M., Inzunza, J., Rozell, B., Miettinen, S., Narkilahti, S., Kerkela, E., Aalto-Setala, K., Otonkoski, T., Suuronen, R., Hovatta, O., & Skottman, H. (2010). A defined and xeno-free culture method enabling the establishment of clinical-grade human embryonic, induced pluripotent and adipose stem cells. *PLoS. One.*, Vol. 5, No. 4, pp. (e10246-

Rao, M. & Condic, M.L. (2008). Alternative sources of pluripotent stem cells: scientific solutions to an ethical dilemma. *Stem Cells Dev.*, Vol. 17, No. 1, pp. (1-10)

Rao, M. & Condic, M.L. (2009). Musings on genome medicine: is there hope for ethical and safe stem cell therapeutics? *Genome Med.*, Vol. 1, No. 7, pp. (70-

Raya, A., Rodriguez-Piza, I., Guenechea, G., Vassena, R., Navarro, S., Barrero, M.J., Consiglio, A., Castella, M., Rio, P., Sleep, E., Gonzalez, F., Tiscornia, G., Garreta, E., Aasen, T., Veiga, A., Verma, I.M., Surralles, J., Bueren, J., & Izpisua Belmonte, J.C. (2009). Disease-corrected haematopoietic progenitors from Fanconi anaemia induced pluripotent stem cells. *Nature*, Vol. 460, No. 7251, pp. (53-59)

Raya, A., Rodriguez-Piza, I., Navarro, S., Richaud-Patin, Y., Guenechea, G., Sanchez-Danes, A., Consiglio, A., Bueren, J., & Izpisua Belmonte, J.C. (2010). A protocol describing the genetic correction of somatic human cells and subsequent generation of iPS cells. *Nat. Protoc.*, Vol. 5, No. 4, pp. (647-660)

Reubinoff, B.E., Pera, M.F., Fong, C.Y., Trounson, A., & Bongso, A. (2000). Embryonic stem cell lines from human blastocysts: somatic differentiation in vitro. *Nat. Biotechnol.*, Vol. 18, No. 4, pp. (399-404)

Reubinoff, B.E., Pera, M.F., Vajta, G., & Trounson, A.O. (2001). Effective cryopreservation of human embryonic stem cells by the open pulled straw vitrification method. *Hum. Reprod.*, Vol. 16, No. 10, pp. (2187-2194)

Revazova, E.S., Turovets, N.A., Kochetkova, O.D., Kindarova, L.B., Kuzmichev, L.N., Janus, J.D., & Pryzhkova, M.V. (2007). Patient-specific stem cell lines derived from human parthenogenetic blastocysts. *Cloning Stem Cells*, Vol. 9, No. 3, pp. (432-449)

Richards, M., Fong, C.Y., Chan, W.K., Wong, P.C., & Bongso, A. (2002). Human feeders support prolonged undifferentiated growth of human inner cell masses and embryonic stem cells. *Nat. Biotechnol.*, Vol. 20, No. 9, pp. (933-936)

Richards, M., Fong, C.Y., Tan, S., Chan, W.K., & Bongso, A. (2004). An efficient and safe xeno-free cryopreservation method for the storage of human embryonic stem cells. *Stem Cells*, Vol. 22, No. 5, pp. (779-789)

Richards, M., Tan, S., Fong, C.Y., Biswas, A., Chan, W.K., & Bongso, A. (2003). Comparative evaluation of various human feeders for prolonged undifferentiated growth of human embryonic stem cells. *Stem Cells*, Vol. 21, No. 5, pp. (546-556)

Ronaghi, M., Erceg, S., Moreno-Manzano, V., & Stojkovic, M. (2010). Challenges of stem cell therapy for spinal cord injury: human embryonic stem cells, endogenous neural stem cells, or induced pluripotent stem cells? *Stem Cells*, Vol. 28, No. 1, pp. (93-99)

Sathananthan, H., Pera, M., & Trounson, A. (2002). The fine structure of human embryonic stem cells. *Reprod. Biomed. Online.*, Vol. 4, No. 1, pp. (56-61)

Schneider, U. & Maurer, R.R. (1983). Factors affecting survival of frozen-thawed mouse embryos. *Biol. Reprod.*, Vol. 29, No. 1, pp. (121-128)

Silva, J. & Smith, A. (2008). Capturing pluripotency. *Cell*, Vol. 132, No. 4, pp. (532-536)

Skottman, H., Stromberg, A.M., Matilainen, E., Inzunza, J., Hovatta, O., & Lahesmaa, R. (2006). Unique gene expression signature by human embryonic stem cells cultured under serum-free conditions correlates with their enhanced and prolonged growth in an undifferentiated stage. *Stem Cells*, Vol. 24, No. 1, pp. (151-167)

Stojkovic, M., Lako, M., Stojkovic, P., Stewart, R., Przyborski, S., Armstrong, L., Evans, J., Herbert, M., Hyslop, L., Ahmad, S., Murdoch, A., & Strachan, T. (2004). Derivation of human embryonic stem cells from day-8 blastocysts recovered after three-step in vitro culture. *Stem Cells*, Vol. 22, No. 5, pp. (790-797)

Strelchenko, N., Verlinsky, O., Kukharenko, V., & Verlinsky, Y. (2004). Morula-derived human embryonic stem cells. *Reprod. Biomed. Online.*, Vol. 9, No. 6, pp. (623-629)

Sum, A.K. & de Pablo, J.J. (2003). Molecular simulation study on the influence of dimethylsulfoxide on the structure of phospholipid bilayers. *Biophys. J.*, Vol. 85, No. 6, pp. (3636-3645)

Sum, A.K., Faller, R., & de Pablo, J.J. (2003). Molecular simulation study of phospholipid bilayers and insights of the interactions with disaccharides. *Biophys. J.*, Vol. 85, No. 5, pp. (2830-2844)

Sun, N., Longaker, M.T., & Wu, J.C. (2010). Human iPS cell-based therapy: considerations before clinical applications. *Cell Cycle*, Vol. 9, No. 5, pp. (880-885)

T'joen, V., De, G.L., Declercq, H., & Cornelissen, M. (2011). An Efficient, Economical Slow-Freezing Method for Large-Scale Human Embryonic Stem Cell Banking. *Stem Cells Dev.*,

Takahashi, K., Tanabe, K., Ohnuki, M., Narita, M., Ichisaka, T., Tomoda, K., & Yamanaka, S. (2007). Induction of pluripotent stem cells from adult human fibroblasts by defined factors. *Cell*, Vol. 131, No. 5, pp. (861-872)

Takahashi, K. & Yamanaka, S. (2006). Induction of pluripotent stem cells from mouse embryonic and adult fibroblast cultures by defined factors. *Cell*, Vol. 126, No. 4, pp. (663-676)

Thirumala, S., Gimble, J.M., & Devireddy, R.V. (2010). Evaluation of methylcellulose and dimethyl sulfoxide as the cryoprotectants in a serum-free freezing media for cryopreservation of adipose-derived adult stem cells. *Stem Cells Dev.*, Vol. 19, No. 4, pp. (513-522)

Thomson, J.A., Itskovitz-Eldor, J., Shapiro, S.S., Waknitz, M.A., Swiergiel, J.J., Marshall, V.S., & Jones, J.M. (1998). Embryonic stem cell lines derived from human blastocysts. *Science*, Vol. 282, No. 5391, pp. (1145-1147)

Thomson, J.A., Kalishman, J., Golos, T.G., Durning, M., Harris, C.P., Becker, R.A., & Hearn, J.P. (1995). Isolation of a primate embryonic stem cell line. *Proc. Natl. Acad. Sci. U. S. A*, Vol. 92, No. 17, pp. (7844-7848)

Thomson, J.A., Kalishman, J., Golos, T.G., Durning, M., Harris, C.P., & Hearn, J.P. (1996). Pluripotent cell lines derived from common marmoset (Callithrix jacchus) blastocysts. *Biol. Reprod.*, Vol. 55, No. 2, pp. (254-259)

Trump, B.F., YOUNG, D.E., ARNOLD, E.A., & Stowell, R.E. (1965). EFFECTS OF FREEZING AND THAWING ON THE STRUCTURE, CHEMICAL CONSTITUTION, AND FUNCTION OF CYTOPLASMIC STRUCTURES. *Fed. Proc.*, Vol. 24, pp. (S144-S168)

Unger, C., Skottman, H., Blomberg, P., Dilber, M.S., & Hovatta, O. (2008). Good manufacturing practice and clinical-grade human embryonic stem cell lines. *Hum. Mol. Genet.*, Vol. 17, No. R1, pp. (R48-R53)

Vajta, G., Holm, P., Greve, T., & Callesen, H. (1997). Survival and development of bovine blastocysts produced in vitro after assisted hatching, vitrification and in-straw direct rehydration. *J. Reprod. Fertil.*, Vol. 111, No. 1, pp. (65-70)

Vajta, G., Holm, P., Kuwayama, M., Booth, P.J., Jacobsen, H., Greve, T., & Callesen, H. (1998). Open Pulled Straw (OPS) vitrification: a new way to reduce cryoinjuries of bovine ova and embryos. *Mol. Reprod. Dev.*, Vol. 51, No. 1, pp. (53-58)

Vajta, G. & Nagy, Z.P. (2006). Are programmable freezers still needed in the embryo laboratory? Review on vitrification. *Reprod. Biomed. Online.*, Vol. 12, No. 6, pp. (779-796)

Wagh, V., Meganathan, K., Jagtap, S., Gaspar, J.A., Winkler, J., Spitkovsky, D., HESCsheler, J., & Sachinidis, A. (2011). Effects of cryopreservation on the transcriptome of human embryonic stem cells after thawing and culturing. *Stem Cell Rev.*, Vol. 7, No. 3, pp. (506-517)

Watanabe, K., Ueno, M., Kamiya, D., Nishiyama, A., Matsumura, M., Wataya, T., Takahashi, J.B., Nishikawa, S., Nishikawa, S., Muguruma, K., & Sasai, Y. (2007). A ROCK inhibitor permits survival of dissociated human embryonic stem cells. *Nat. Biotechnol.*, Vol. 25, No. 6, pp. (681-686)

Wernig, M., Meissner, A., Foreman, R., Brambrink, T., Ku, M., Hochedlinger, K., Bernstein, B.E., & Jaenisch, R. (2007). In vitro reprogramming of fibroblasts into a pluripotent ES-cell-like state. *Nature*, Vol. 448, No. 7151, pp. (318-324)

Wong, R.C., Pebay, A., Nguyen, L.T., Koh, K.L., & Pera, M.F. (2004). Presence of functional gap junctions in human embryonic stem cells. *Stem Cells*, Vol. 22, No. 6, pp. (883-889)

Wong, R.C., Pera, M.F., & Pebay, A. (2008). Role of gap junctions in embryonic and somatic stem cells. *Stem Cell Rev.*, Vol. 4, No. 4, pp. (283-292)

Wu, C.F., Tsung, H.C., Zhang, W.J., Wang, Y., Lu, J.H., Tang, Z.Y., Kuang, Y.P., Jin, W., Cui, L., Liu, W., & Cao, Y.L. (2005). Improved cryopreservation of human embryonic stem cells with trehalose. *Reprod. Biomed. Online.*, Vol. 11, No. 6, pp. (733-739)

Xiao, M. & Dooley, D.C. (2003). Assessment of cell viability and apoptosis in human umbilical cord blood following storage. *J. Hematother. Stem Cell Res.*, Vol. 12, No. 1, pp. (115-122)

Xu, X., Cowley, S., Flaim, C.J., James, W., Seymour, L., & Cui, Z. (2010a). The roles of apoptotic pathways in the low recovery rate after cryopreservation of dissociated human embryonic stem cells. *Biotechnol. Prog.*, Vol. 26, No. 3, pp. (827-837)

Xu, X., Cowley, S., Flaim, C.J., James, W., Seymour, L.W., & Cui, Z. (2010b). Enhancement of cell recovery for dissociated human embryonic stem cells after cryopreservation. *Biotechnol. Prog.*, Vol. 26, No. 3, pp. (781-788)

Yu, J., Vodyanik, M.A., Smuga-Otto, K., ntosiewicz-Bourget, J., Frane, J.L., Tian, S., Nie, J., Jonsdottir, G.A., Ruotti, V., Stewart, R., Slukvin, I.I., & Thomson, J.A. (2007). Induced pluripotent stem cell lines derived from human somatic cells. *Science*, Vol. 318, No. 5858, pp. (1917-1920)

Zhang, X.B., Li, K., Yau, K.H., Tsang, K.S., Fok, T.F., Li, C.K., Lee, S.M., & Yuen, P.M. (2003). Trehalose ameliorates the cryopreservation of cord blood in a preclinical system and increases the recovery of CFUs, long-term culture-initiating cells, and nonobese diabetic-SCID repopulating cells. *Transfusion*, Vol. 43, No. 2, pp. (265-272)

Zhou, C.Q., Mai, Q.Y., Li, T., & Zhuang, G.L. (2004). Cryopreservation of human embryonic stem cells by vitrification. *Chin Med. J. (Engl.)*, Vol. 117, No. 7, pp. (1050-1055)

Part 3

Human Assisted
Reproduction Techniques (ART)

6

Cryopreservation of Testicular Tissue

Ali Honaramooz
University of Saskatchewan
Canada

1. Introduction

Immediate use of freshly collected testis tissue in diagnosis or in reproductive technologies is not always possible or desirable. Therefore, the ability to properly preserve the tissue for varying intervals is an essential step for maximizing the use of the source tissue. Preservation of gametes and gonads is a topic of interest in reproductive biomedicine. Other chapters in this book have elegantly covered current knowledge on the cryopreservation of sperm, oocytes, and early embryos as well as ovarian tissue, among other cells and tissues. However, the main objective of this chapter is to provide a focused discussion of the importance, methodology, potential applications, and limitations for applying cryopreservation to testicular tissue.

Cryopreservation of human testis tissue obtained by biopsy can be used as a potential future source of sperm. For adult cancer survivors whose only source of sperm is the testis parenchyma, cryopreservation of testis biopsies may be the only option remaining if they prefer to father their own biological progeny. This will require detection of sperm in frozen-thawed cell suspensions of testis tissues for use in intra-cytoplasmic sperm injection (ICSI). More importantly, cryopreservation of immature testis biopsies can offer a unique alternative for prepubertal boys undergoing gonadotoxic cancer treatments, whose only future source of spermatogenesis (*i.e.*, spermatogonial stem cells) is at risk. These strategies can also be applied to genetic preservation of endangered species/breeds through the cryopreservation of testis tissue from young animals that die prior to reaching maturity. Restoring the developmental potential of testis tissue after cryopreservation may also provide insight into proper banking of other immature tissues.

The effects of cryoprotectant concentration and cooling rate are not similar among tissues or species. Therefore, we will discuss the basis for a number of successfully applied strategies and workable protocols that have been used to effectively cryopreserve testis tissue in various species.

In summary, this chapter provides an overview of the current literature and contributions by the author and colleagues on cryopreservation of testicular tissue and its potential applications in experimental and clinical settings in reproduction medicine.

1.1 Developmental changes in the structure of testis tissue

In mammals at birth, all organs/tissues required for sustaining life display functional competence and histological similarity to those in mature individuals. Reproductive tissues, on the other hand, attain maturity much later and only when other bodily requirements of parenthood are also in place. Therefore, in discussion of testis tissue cryopreservation, the developmental stage of the tissue is an important factor to be considered. For instance, for cryopreservation of testis tissue from an immature individual, the differing tissue texture and need for maintaining its future developmental potential are to be taken into account.

Embryonic development of the testis begins when the SRY gene in a genetic male is expressed, driving the transformation of an indifferent early gonad to a testis. This in turn causes differentiation of Sertoli cells to enclose the fetal germ cells, to mark the differentiation of primordial germ cells into gonocytes, and results in the formation of seminiferous cords. In humans, this process begins at 7-9 wk gestation (Wilhelm et al., 2007) and is immediately followed by differentiation of fetal Leydig cells, located in the interstitial spaces between the seminiferous cords, to allow production of testosterone thus causing masculinization of the foetus (Scott et al., 2009).

In early postnatal humans and most domestic species, the testis still contains interstitial tissue and seminiferous cords, with gonocytes as the only type of germ cells present (Franca et al., 2000). Initially, gonocytes reside in the centre of the seminiferous cords (**Fig. 1A**), but they gradually migrate toward the periphery of the cords and remain in close contact with Sertoli cells and peritubular myoid cells at the basement membrane to form the stem cell niche (Pelliniemi, 1975; Van Straaten & Wensing, 1977). Gonocytes eventually give rise to spermatogonial stem cells (SSCs), which have the ability to both self-renew and give rise to differentiating germ cells. Postnatal development of the testis also involves proliferation and maturation of Sertoli cells to transform testicular cords into seminiferous tubules (containing a lumen), followed by sequential division and differentiation of germ cells to generate sperm (**Fig. 1B**) (Hughes & Varley, 1980; Ryu et al., 2004). Therefore, SSCs form the foundation of spermatogenesis and are responsible for a lifetime supply of sperm.

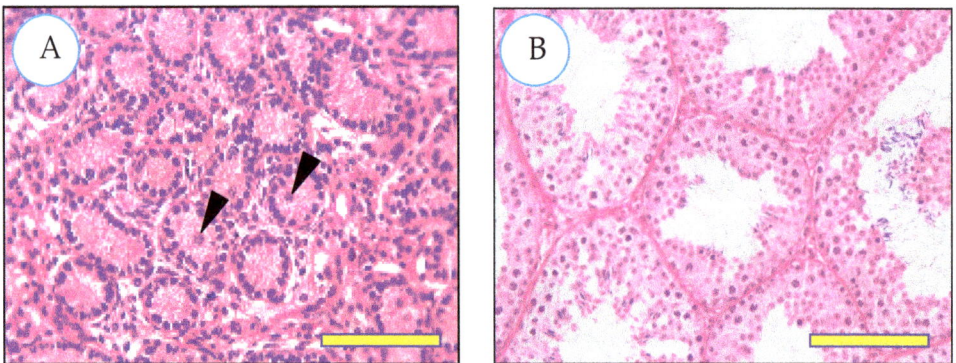

Fig. 1. Histological differences between an immature and a mature testis tissue. In the immature testis (**A**), seminiferous cords contain only one type of germ cells - gonocytes (arrow heads). In the mature testis (**B**), on the other hand, seminiferous tubules are much larger in diameter, contain a lumen, and a repertoire of germ cell types. The composition and extent of the interstitial tissue also changes over development. These differences may affect the response of the tissue to a given cryopreservation protocol even within the same donor species. Scale bar = 100 mg. Images modified from Abbasi & Honaramooz (2011).

As highlighted in **Figure 1**, the cellular composition of a typical mature testis is quite different from that of an immature testis; for instance, the latter hosts a considerably higher number of differentiating germ cells, known to be more sensitive to manipulations and temperature changes (Franca et al., 2000; Frankenhuis et al., 1981). Consequently, the tissue composition of the testis changes during development and proportionally larger volumes of the mature testis are occupied by the seminiferous tubules. Therefore, the developmental state of the testis affects the tissue composition and has important implications for its cryopreservation.

2. Rationale for preserving testis tissue from human and animal donors

Preservation of testicular tissue could be pursued for multiple reasons. An estimated 1 in 650 children will be diagnosed with malignancies by age 16, of which 80% will be cured (Stiller et al., 2006). However, irreversible gonadotoxic insult of chemo/radio-therapy remains a major concern in the use of these life-saving treatments, which render about 20% of boys sterile in the long term, likely as a result of the loss of spermatogonial stem cells (Apperley & Reddy, 1995; Naysmith et al., 1998). With improved treatments, the proportion of childhood cancer survivors is expected to increase, posing an even greater challenge for reproductive medicine and oncologist practitioners in the decades to come. A routine strategy to offer preservation of future fertility for adult men undergoing sterilizing cytotoxic treatments is to freeze semen samples; however, some men may be azoospermic at the time of cancer diagnosis. More critically, in pre-adolescent boys, collection of sperm is not possible because spermatogenesis has not yet started. In such cases, cryopreservation of testicular biopsies collected prior to the start of the treatment may provide a potential source for future use in emerging reproductive technologies.

In animal conservation, preventing the permanent loss of a male's potential contribution to the genetic variability of a rare or endangered species/breed is feasible through the collection of sperm before or even shortly after death by retrieval from the ejaculate, epididymis, or testes, which is then cryopreserved for future use in assisted reproduction (Gañán et al., 2009; Kishikawa et al., 1999; Martínez et al., 2008; Maksudov et al., 2009). Preservation of sperm, however, is not an option when young offspring die prior to reaching sexual maturity. Cloning has been used for a number of species and especially where the goal has been to produce a genetically exact replica of an individual animal. However, development of cloning for a new species is technically demanding and costly but, more importantly, does not immediately provide the genetic diversity that would otherwise be offered by gametes. In such cases, cryopreservation of testicular tissue can again provide an alternative strategy for *ex situ* generation of sperm from these neonatal/immature animals for use in reproductive technologies (Abbasi & Honaramooz, 2011).

3. Methodology for cryopreservation of testicular tissue

A number of cryogenic strategies have been developed to serve as a means to maintain functional properties of the preserved cells and tissues. Apparently, the first successful cryopreservation of cells was carried out by accidental freezing of fowl sperm in diluents containing glycerol (Polge et al., 1949). Later, cryopreservation of bull sperm using glycerol (Polge & Lovelock, 1952; Smith, 1961), set the stage for revolutionizing the bovine artificial insemination industry. At about the same time, cryopreservation of unfertilized oocytes was also studied following exposure to glycerol and low temperatures (Smith, 1952). After initial success with *in vitro* embryo manipulation in the 1950s (McLaren & Biggers, 1958), research involving embryo freezing intensified. Many methods have now been developed for embryo cryopreservation and, since the 1980s, some have become routine procedures (Whittingham et al., 1972; Whittingham, 1977; Wilmut, 1972). Cryopreservation of mature oocytes has also been achieved (Fabbri et al., 2001; Porcu, 2001; Porcu et al., 1997), with high survival rates and development of normal pregnancies after *in vitro* fertilization (IVF).

Cryopreservation of structurally intact tissues in certain situations is more desirable than cryopreservation of isolated cells. This is especially important for complex tissues in which preservation of the target cells' functionality depends on that of other cell types present within the tissue. In case of testicular tissue, not only germ cells but also the intra-tubular supporting - Sertoli - cells as well as androgen producing interstitial - Leydig - cells are of particular interest. However, this requires devising suitable freezing protocols to maintain the existing relationship among different compartments of the tissue.

The first gonadal tissue to be successfully cryopreserved was ovarian tissue, using exposure to glycerol, resulting in preservation of cell viability and normal function after being autografted back into the animals (Deanesly, 1954; Green et al., 1956; Parkes, 1958). Subsequent reports of live rat offspring, sheep ovarian cyclic function, and pregnancy after grafting cryopreserved ovaries represented important steps in demonstrating the feasibility of this approach (Gosden et al., 1994; Parrot, 1960). Restoration of spermatogenesis was then obtained after cryopreserved testis cells were transplanted into recipient testes (Avarbock et al., 1996; Brinster & Nagano, 1998; Ogawa et al., 1999).

Cryopreservation of testicular tissue to be used as tissue *per se*, however, was not widely considered, perhaps due to lack of its potential applications. This need changed when we and others were first to show that cryopreservation of immature testis tissue prior to its xenografting can be done so as to maintain its potential for development of complete spermatogenesis (Honaramooz et al., 2002a; Schlatt et al., 2002). In a short period of time since then, major advances in cryopreservation of testicular tissue have opened new possibilities for preservation of male fertility in animals and humans. More recently, induction of complete spermatogenesis *in vitro* has further highlighted the importance of applying cryopreservation to testicular tissue for future applications. Overall, major advances have been made in the cryopreservation of reproductive tissues. The following sections review the primary contributing factors to be considered for optimal cryopreservation.

3.1 Biophysics of cryopreservation

A clear understanding of biophysical behaviour of cells at the time of freezing and exposure to different cryoprotectants is critical in providing conditions to improve the cell structural and functional potential after freezing-thawing. During slow rate of cooling, extracellular ice crystal formation begins with the presence of a nucleation site in the extracellular medium. Because ice is pure crystalline water, the extracellular space becomes hypertonic due to the removal of water as ice crystals develop. Intracellular water, therefore, moves outward across the cell membrane due to the differential osmotic gradient, and cells dehydrate and shrink. This is the opportunity when certain cryoprotective compounds come into play, permeating the cells and protecting them against high solute concentration or ice crystal damage. Because various cryoprotectant agents (CPAs) permeate different cell types at varying rates, it is of benefit to understand the biophysics of cryopreservation to minimize damage (Fuller & Paynter, 2004; Pegg, 2007).

3.2 Freezing injuries

Two main rival theories have been proposed to explain cell damages due to freezing. One emphasizes the direct and primarily mechanical damage to live cells by ice crystals puncturing through the cell membranes, and the other highlights the secondary effects of ice formation via osmotic changes. Perhaps, both mechanisms are important and what is recently agreed upon is that for individual cells, for example those in suspensions, intracellular freezing is very hazardous, while the extracellular ice may not be as harmful (Pegg, 2007). Unlike cell suspensions, the cellular organization and structural composition of the tissue may be seriously affected by cryogenic damage through widespread extracellular ice formation (Hunt et al., 1982; Taylor & Pegg, 1983). Ice formation within a tissue, initiated in the extracellular space, leads to an osmotic gradient across the cell membranes, causing intracellular water to move toward the concentrated extracellular space surrounding the cells (Bagchi et al., 2008; Fuller, 2004). Due to the differential destructive effects of extracellular ice formation between cell suspensions and complex tissues, conventional approaches to cryopreservation of cells, even testis cells for instance, may not necessarily be suitable for multicellular tissues such as the testis tissue. Optimal cooling rates for various cell and tissue types have been shown to differ and be directly associated with the degree of water permeability of cell membranes at different temperatures during freezing (Leibo et al., 1970; Mazur, 1990; Pegg, 2007).

When extracellular ice formation causes elevated solvent concentrations, it leads to cell dehydration; prolonged exposure to which can permanently damage cell membranes and destabilize proteins (Fuller, 2004). However, short exposure of cells to optimized concentrations of hypertonic media before freezing might protect them from retention of supercooled water within cells and subsequent crystallization during freezing (Fuller, 2004). When cooling is faster than optimal, intracellular ice formation could occur due to inadequate time for water to follow the osmotic gradient across the cell membrane (Fuller, 2004; Fuller & Paynter, 2004; Pegg, 2007). The osmotic tolerance of cells is another critical factor to be considered during addition and removal of different cryoprotectants. Physical destruction, subsequent organelle disruption, and functional damage are some of the known consequences of ice crystal formation (Mazur, 2004).

3.3 Protection mechanism and toxicity of cryoprotectants

Sufficient concentration of cryoprotectants could minimize ice crystallization and/or promote amorphous solidification (vitrification). Glycerol was introduced as a CPA in 1949 (Polge et al., 1949) and, a decade later, cryoprotective properties of dimethyl sulfoxide (DMSO) were also reported (Lovelock & Bishop, 1959). These two cryoprotectants have mainly been used since then as classic cryoprotective additives, although many other CPAs have been introduced. Permeating CPAs, such as DMSO, glycerol, methanol, propanediol, ethylene glycol, and dimethyl acetaldehyde, as well as non-permeating CPAs, including sucrose, dextran, albumin, polyvinyl pyrollidone, and hydroxyethyl starch, have also been shown to afford effective cryoprotection (Bagchi et al., 2008; Fuller, 2004).

Cryoprotective agents are known to act through different pathways to protect cells against freezing injuries. This includes modulation of hydrogen bonding and interaction with water molecules, which give CPAs solubility and high permeability across cell membranes (Fuller, 2004). As a second mechanism, CPAs may provide a salt-buffering effect. During freezing, cells experience osmotic dehydration and shrinkage; therefore, the addition of CPAs into the cells maintains salt dilution. Basically, the CPA replaces water in cells, which dilutes the intracellular salts and prevents intracellular crystal formation. The amount of CPAs and water that permeates into the cells depends on the concentration of permeable solutes and the final cell volume. The properties of CPAs and those of cell membranes will influence the degree of cryoprotection for different cell types (Fuller, 2004; Fuller & Paynter, 2004). A third potential pathway is the stabilization of biomembrane critical macromolecules. Under normal conditions, water stabilizes the membrane bilayers. Loss of water during cryopreservation may disrupt normal membrane permeability and damage the membrane itself. The CPAs stabilize proteins as well as phospholipid bilayers of cell membranes and help to protect the membrane against freezing and dehydration stresses (Crowe, et al., 1990). Studies have collectively demonstrated that CPAs, including DMSO and disaccharide sugars such as sucrose and trehalose, may electrostatically interact with membrane phospholipids to provide stabilization (Anchordoguy et al., 1987; Rudolph & Crowe, 1985). The fourth mechanism by which CPAs protect the cells and tissue is through scavenging oxygen free radicals and preventing oxidative stress to the cells (Fuller, 2004). CPAs block the action of unstable intermediate products, such as oxygen free radicals, by binding their hydrogen atoms to them (Benson, 2004; Fleck et al., 2000). The fifth possible pathway for the protective effects of CPAs is the inhibition of nucleation, through which ice formation occurs in the media. During cooling, initial heterogeneous nucleation sites, such as small particles, change in shape and increase in size within media, eventually reaching a stage that forms ice crystals. Alternatively, induced nucleation could be beneficial to provide consistent extracellular crystallization. This phenomenon is the basis for "seeding", which induces nucleation onto supercooled media enabling proper cryopreservation (Fuller, 2004). Seeding can be achieved by clamping the side of vials or straws with a forceps cooled in liquid nitrogen to stimulate local ice growth in the solutions. Intracellular nucleation can also be lethal or damaging for cells and tissues. Some CPAs, such as DMSO or glycerol, inhibit nucleation by increasing the high viscosity of intracellular water (Fuller, 2004). Non-permeating CPAs, on the other hand, increase and promote cellular dehydration by increasing the extracellular solute concentration thereby reducing intracellular crystallization (Bagchi et al., 2008).

Despite the protective potential of CPAs, a side effect of their addition is cytotoxicity. Tissue tolerance to CPAs is limited and overexposure may cause damage (Pegg, 2002); however, measuring this toxicity is difficult to precisely assess (Fuller, 2004). Cytotoxicity is further exacerbated by increasing CPA concentrations during ice formation. Optimizing the freezing rate as well as the addition or removal of CPAs could reduce their toxicity (Pegg, 2002).

3.4 Choice of cryopreservation strategies

For cryopreservation of testicular tissue, two popular strategies are slow freezing and vitrification. These techniques differ mainly in the concentration of CPAs used.

Cryopreservation of cells within intact tissues is obviously more demanding than for cells within suspensions. Theoretical differences include heterogeneity of cells, slower rates of solute diffusion, and heat exchange through the mass of a complex tissue. However, judging from evidence from other tissue types, if a sufficient concentration of CPAs is provided, finding a proper cooling rate can yield high survival for different cell types within the tissue (Pegg, 2007). Critical factors for effective cryopreservation, such as cell permeability to water or CPA and subsequent osmotic changes, are directly affected by the rate of cooling (Mazur, 1990). Therefore, finding the optimal cryopreservation protocol for testicular tissue of a particular species/maturational state depends on the application of a proper concentration of the cryoprotectant with a suitable cooling rate.

Slow (controlled) freezing is considered the conventional method for cryopreservation of testicular tissue, in which the CPA is used at low concentrations (usually 0.5 to 2 M) to minimize both cell damage and CPA toxicity. During slow freezing (e.g., -1°C/min), the CPA is given a chance to slow down the formation of extracellular ice crystals (and prevent the intracellular ones) but especially to moderate the indirect solution effects as freezing proceeds. However, prolonged exposure to CPA before completion of cryopreservation can also cause cell toxicity (Fuller, 2004). On the other hand, if the cell is cooled more rapidly, then water will not leave the cells fast enough to avoid intracellular freezing, which is very damaging to the cells (Pegg, 2007). Using automated systems, freezing curves (**Fig. 2**) can be customized to maximize cell viability after cryopreservation of the tissue.

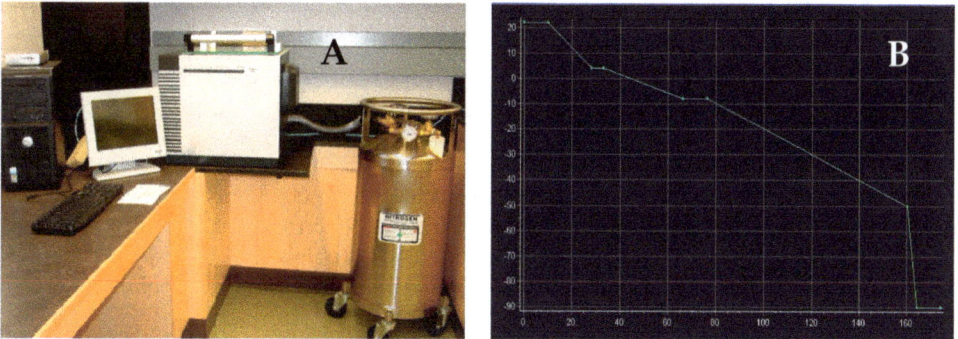

Fig. 2. A programmable automated freezing system. Although requiring larger capital investments, automated cell/tissue freezing systems (**A**), consisting of a freezing chamber attached to a computer and a liquid nitrogen tank, allow customization of the freezing curve (**B**) to achieve pre-defined temperatures (Y-axis) for desired lengths of time (X-axis), in an accurate and consistent manner.

As indicated earlier, the formation of extracellular ice, which may not pose a problem for freezing of cell suspensions, is likely the main problem for tissues. Therefore, an alternative route to avoid ice crystal formation and solute damage within the tissue is to avoid ice crystal formation altogether using transformation of aqueous milieu of the cell/tissue to the amorphous character of a glassy state, known as vitrification. Vitrification is a cryopreservation method in which ice crystal formation is prevented because the cells or tissues are exposed to very high concentrations of CPAs (e.g., 5 to 8 M) and undergo ultra rapid freezing rates (e.g., up to -2500°C/min) (Fuller, 2004; Pegg, 2002, 2007). However, this approach is compromised by the cytotoxic effects of CPAs at such high concentration, especially with increased exposure times (Fuller, 2004; Fuller & Paynter, 2004). For small volumes of cell suspension, CPA concentrations can be reduced somewhat by using very rapid cooling and warming rates. However, especially with increasing size and complexity of the tissue, the limits of temperature exchange rates are more restricted, hence the use of very high concentrations of CPAs are unavoidable (Pegg, 2007). To overcome this problem, the use of a combination of CPAs to improve vitrification while reducing toxicity has been suggested. Proper media may include disaccharides, such as sucrose or trehalose, and proteins or polymers (Kasai & Mukaida, 2004; Sutton, 1992). The optimal CPA concentrations and exposure times to prevent toxicity must be specifically considered for each tissue type. (Fuller & Paynter, 2004; Pegg, 2007). We have used a solid-surface vitrification method to minimize the volume surrounding the tissue pieces, while avoiding liquid nitrogen (LN_2) vapour formation and preventing direct contact with LN_2 to prevent potential contamination (**Fig. 3**, Abrishami et al., 2010a).

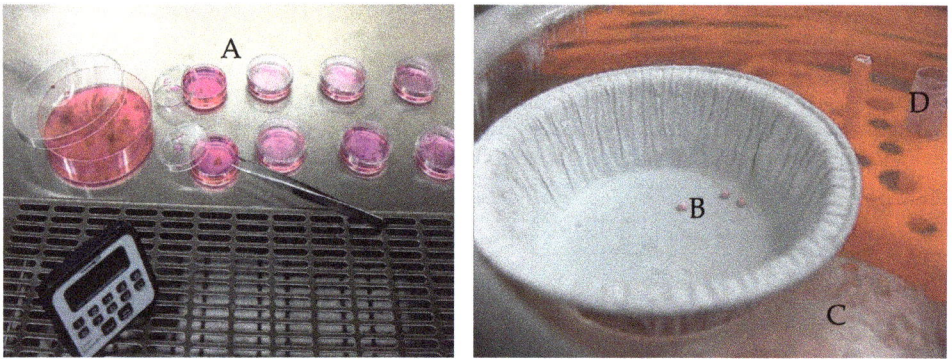

Fig. 3. Solid-surface vitrification procedure for testicular tissue fragments. After exposure of testis tissue fragments to differing concentrations of vitrification solutions for varying lengths of time (**A**), testis tissue fragments are placed on a sterile aluminum boat (**B**) floating on liquid nitrogen (**C**), then transferred into cooled cryovials (**D**) followed by plunging into liquid nitrogen (images modified from Abrishami, 2009).

3.5 Thawing methods

Whether freezing is permitted (conventional cryopreservation) or prevented (vitrification), the CPA that has reached the internal compartments of a multicellular system must diffuse back through numerous membranes in the tissue, with each acting as a barrier. Therefore, optimal thawing and CPA removal procedures are also critical factors for cell/tissue survival after freezing (Bagchi et al., 2008). Earlier studies pointed out that consistent cooling and thawing rates (slow-freezing followed by slow-thawing, or fast-freezing followed by fast-thawing) can improve cell/tissue survival after cryopreservation (Whittingham et al., 1972). Moreover, extreme osmotic changes during CPA removal might damage the cells by extensive cell shrinkage or swelling associated with the rapid movement of water into the cell as compared to the slower movement of the CPA out of the cell. However, a limited amount of water replacement is needed to restore osmotic equilibrium and physiologic cell volume (Pegg, 2007).

3.6 Post-thawing analysis

For successful cryopreservation of a complex vascularized tissue, such as testis tissue, the majority of essential cells need to be viable for the tissue to survive and retain its function. However, there is not yet a comprehensive and universally applied method for post-thawing analysis of cryopreserved testis tissue; subsequently, multiple approaches have been used to assess tissue/cell viability and extent of cryogenic injuries. These approaches commonly include histopathological examination of tissue sections for morphological changes. Using light microscopy, for instance, such objective criteria as seminiferous cord/tubular diameter or cell density within tubule cross sections can be measured, or semi-quantitative morphometric analyses applied to subjectively score such criteria as health or integrity of tissue compartments (Abrishami et al., 2010a; Curaba et al., 2011; Milazzo et al., 2008; Travers, et al., 2011). Transmitted electron microscopy, although not widely used, can be invaluable in the examination of subcellular components most likely to be affected by testis tissue cryopreservation, including cytoplasm integrity, nuclear membrane, and various organelles (Keros et al., 2007). Other valuable morphological analyses may include assessment of cell-specific changes, for example, using double-staining of proliferation markers (*e.g.*, Ki67) and MAGE-AH, vimentin, or CD34 for identification of spermatogonia, Sertoli cells, or peritubular cells, respectively (Keros et al., 2007; Wyns, et al., 2007).

A quantitative measure of tissue damage due to cytotoxicity after cryopreservation can be achieved through lactate dehydrogenase release assays (Curaba et al., 2011) or through viability assessment of dissociated cells after digestion of frozen-thawed tissues using Trypan blue exclusion assays or the various cell viability kits using a flow cytometer analyzer (Abrishami et al., 2010a; Gouk et al., 2011). Assessment of apoptosis, using for instance, caspase-3 (Wyns et al. 2008), or TUNEL assay for detection of DNA fragmentation provides insight into the extent of cell damage (Milazzo, et al., 2008). Detection of phophatidylserine translocation from the inner to the outer layer of the plasma membrane, using fluorescent-labelled Annexin V, also allows more targeted assessment of apoptotic-associated changes within the cryopreserved testis tissue (Milazzo et al., 2008).

Having merely high cell survival rates or lacking visible damage does not guarantee functional preservation of the tissue as a whole. A thorough post-thawing analysis should include a form of testing for the functionality of the cryopreserved tissue. Post-thawing *in vitro* organotypic culture of the cryopreserved testis tissue has allowed assessment of its survival in the short-term (Curaba et al 2011; Keros et al., 2007) and measurement of its hormone release into culture media (Gouk et al., 2011). Perhaps more robust examination is provided by grafting, where the survival and developmental competence (both in terms of germ cell differentiation and androgen release) of the cryopreserved tissue *in vivo* as grafts allows a longer-term functional assessment (Abrishami et al., 2010a; Jahnukainen et al., 2007; Wyns et al., 2007).

3.7 Effects of tissue size

To offer cryoprotection, the CPAs need to diffuse rapidly in and out of the tissue; therefore, the size of testis tissue samples undergoing cryopreservation can be an important intuitive consideration. The results of studies differ depending not only with respect to the donor species but also potentially on the protocols employed. For instance, while cryopreservation of immature rat testis using similar procedures demonstrated better results for 7.5 mg pieces than 15 mg pieces (Travers et al., 2011), cryopreservation of immature mouse testis using whole testes with punctured tunica albuginea was deemed more suitable than using whole testes with intact tunica, whole testes without tunica, or testis halves (Gouk et al., 2011). Mouse testes have considerably less connective tissue content than most other species; therefore, tissue fragment size is especially a concern for testis tissues from species with higher interstitial tissue density. For cryopreservation of (cryptorchid) testes from prepubertal boys, fragments sizes of 2-9 mm^3 were used successfully (Wyns et al., 2007). We also reported that immature porcine testis tissues undergoing the same cryopreservation treatments were not affected by the original size of the testis tissue fragment (5, 15, 20, or 30 mg) (Abrishami et al., 2010a). Although not used for cryopreservation, no effect of tissue sample size was observed for one-wk old piglet testes (as intact or fragments of 100 or 30 mg) when used for hypothermic preservation for 6 days (Yang et al., 2010). It remains to be seen if whole human testes can be cryopreserved as has been accomplished for whole ovaries (Courbiere et al., 2006; Jadoul et al., 2007; Martinez-Madrid et al., 2007).

4. Applications of testis cryopreservation for new reproductive technologies

Given that properly cryopreserved testis biopsies can last decades in liquid nitrogen and that most prepubertal cancer patient boys donating biopsies may not need to resort to assisted reproductive technologies for a couple of decades, it is advisable that cryopreservation of testicular biopsies be offered to such patients in a hope that our ability to use such tissues will be further improved and the options expanded in the coming years.

A number of potential applications already exist for the use of cryopreserved testicular tissue in experimental and clinical settings in reproduction medicine/science. Such technologies allow retrieval of existing sperm from mature donor samples and, more importantly, offer hope for production of sperm in samples of cryopreserved testis immature testis. If the preserved testis tissue contains endogenous spermatogenesis (*e.g.*, from obstructive azoospermic adult patients), it can be used to extract sperm, elongated spermatids, or even round spermatids to be used for fertilization of oocytes through ICSI (Rosenlund et al., 1998; Schrader et al., 2002; Gianaroli et al., 1999; Tesarik et al., 2000; Schoysman et al., 1999).

If preserved testis samples are obtained from neonatal/immature donors, they can still be used to induce spermatogenesis through the following approaches.

4.1 Germ cell transplantation

The technique for germ cell transplantation has allowed (re)establishment of spermatogenesis after introduction of donor testis cell suspensions into the seminiferous tubules of infertile recipient testes. Once deposited in the tubular lumen, donor SSCs are recognized by the host Sertoli cells and allowed passage to the stem cell niche, where new colonies of spermatogenesis can begin and expand. This approach has allowed production of sufficient numbers of sperm to allow infertile recipient mice to sire donor-derived progeny (Avarbock & Brinster, 1994; Brinster & Zimmermann, 1994). Later, the capability of cryopreserved mouse testis cells after transplantation into recipient testes to start spermatogenesis was also confirmed (Avarbock et al., 1996; Brinster & Nagano, 1998; Ogawa et al., 1999). While heterologous transplantation of human germ cells into recipient mice did not lead to completion of spermatogenesis (Nagano et al., 2002), the transfer technique has been tested using human testes (Schlatt et al., 1999; Brook et al., 2001). Although autologous/homologous transplantation of germ cells for humans is currently considered purely experimental, one possibility for prepubertal human testis samples taken and frozen prior to treatments is to isolate testis cells and transfer them back to the individual. As a major problem with this approach is the risk of reseeding a systemic cancer, solutions to this (e.g., soring out tumour cells) and other safety issues are under investigation.

We have expanded the technique for germ cell transplantation into farm animals (**Fig. 4**), showed the feasibility of SSC engraftment in unrelated recipient individuals (of the same species) without a need for immune-suppression, and further demonstrated the applicability of the approach through donor-derived sperm production by the recipients and birth of progeny carrying the donor characteristics (Honaramooz et al., 2002b 2003a, 2003b; Honaramooz & Yang, 2011). Therefore, although experimental at this stage, the approach may offer promise in salvaging genetic material from cryopreserved testicular tissue from immature endangered species.

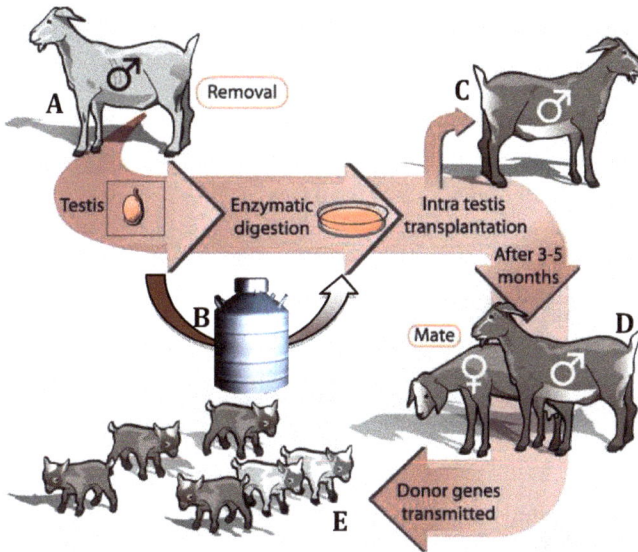

Fig. 4. Schematic overview of germ cell transplantation from a donor male into the testes of a recipient. The testes are collected from a donor animal (**A**), which could theoretically include post-mortem testis recovery from a recently deceased juvenile individual of an endangered species. The testis tissue could be cryopreserved (**B**) until conditions for its use are in place. At the time of transplantation, a single-cell suspension is prepared and the cells are infused into the seminiferous tubules of a recipient animal (**C**). Mating of the recipient (**D**) produces progeny (**E**), some of which will carry the donor genome (image modified from Honaramooz et al., 2003b).

4.2 Testis tissue (xeno)grafting

Another potential strategy for the use of cryopreserved testis tissue is represented by testis tissue xenografting. Grafting of both fresh and cryopreserved testis tissue fragments from donors of different species under the back skin of recipient mice results in the production of functional sperm (Honaramooz et al., 2002a). The approach has especially been successful using neonatal/immature donors (**Fig. 5**), from laboratory animals to domestic animals, primates, and even humans (Honaramooz et al., 2002a, 2004, 2008; Schlatt et al., 2002; Oatley et al., 2004; Snedaker et al., 2004; Rathi et al., 2005, 2006; Arregui et al., 2008; Abrishami et al., 2010b).

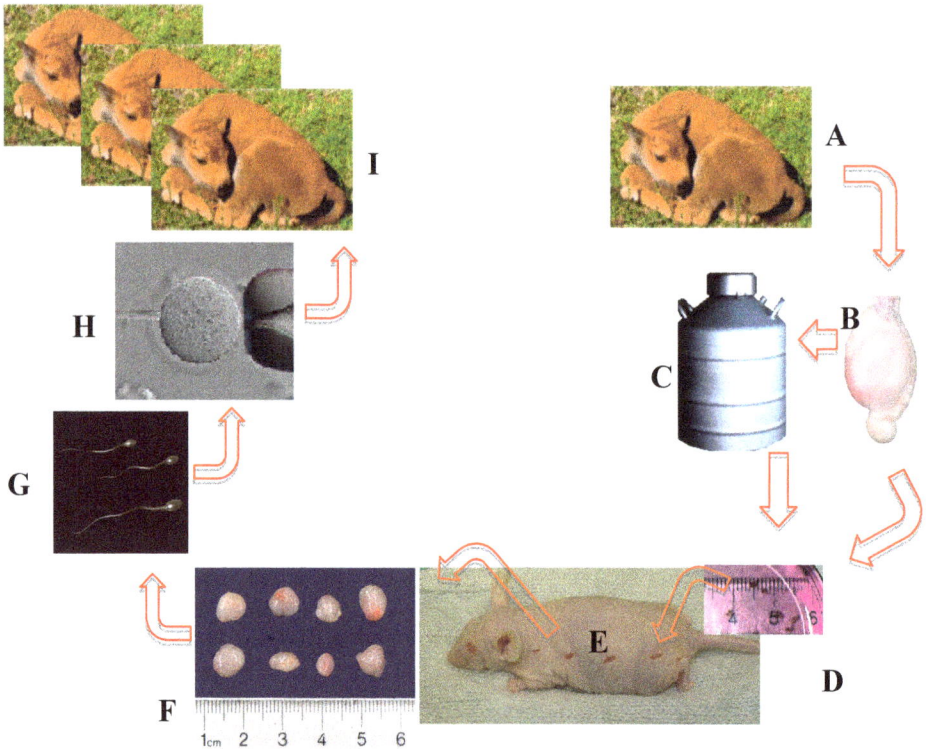

Fig. 5. Schematic representation of testis tissue (xeno)grafting from an immature donor individual into the back skin of a host mouse. The testes are collected from a donor animal (**A**), which could include post-mortem testis recovery from a recently deceased newborn animal of an endangered species. The testis tissue (**B**) could be cryopreserved (**C**) until grafting. At the time of grafting, tissue fragments of ~0.5 mm³ (**D**) are prepared and the fragments are grafted subcutaneously into an immunodeficient host mouse (**E**). When given enough time, the grafts can grow in size (**F**) and undergo development, leading to the production of complete spermatogenesis, including fertilization-competent sperm (**G**). The sperm can then be extracted from the grafts and used in intracytoplasmic sperm injection (ICSI) (**H**), which after embryo transfer can potentially lead to birth of progeny (**I**).

The sperm recovered from such grafts, including those from primates, have been shown to be fertilization competent after ICSI (Honaramooz et al., 2002a, 2004, 2008), leading to the birth of healthy progeny (Schlatt et al., 2003; Nakai et al., 2010). We recently showed that testes recovered post-mortem from newborn bison calves, as a model for closely-related rare or endangered ungulates can be used for this application, and when allowed to develop in the host mouse, lead to full spermatogenesis (Abbasi & Honaramooz, 2011). Therefore, testis tissue xenografting can be used as unique solution for genetic conservation of immature males by producing sperm from these otherwise resource-less donors in xenografts, followed by extraction and cryopreservation of sperm for future use in ICSI (**Fig. 5**).

However, xenografting of human gonadal tissues into animals to harvest the resultant gametes for use in IVF for humans is prohibited in Canada, and possibly in other countries, due to the potentially serious risk of animal viral transmission or contamination with animal genetics. Nevertheless, the promising results from animal research suggest a potential hope for future use of cryopreserved testis biopsies from pre-adolescent boys to be grafted back to the individual; whether this technique can be used to produce viable sperm for future use from prepubertal boys undergoing gonadotoxic treatments remains to be determined. However, the same safety risks as for autologous germ cell transplantation exist and require addressing before such an option can be offered clinically.

4.3 *In vitro* maturation of germ cells

In theory, cryopreserved testicular tissues can also be used for *in vitro* induction of differentiated germ cells and ideally production of sperm or spermatids to be used for ICSI. If successful, this approach can circumvent the potential risk of reintroducing cancer cells into post-recovery patients. Many labs have experimented with the idea, and some have had success with maturation of later stages of human spermatogenesis (but not from SSCs), including live births (Tesarki et al., 1999). Availability of a culture system to support complete *in vitro* spermatogenesis from the SSC stage was, however, elusive until very recently when it was reported that all spermatogenic lineage cells including fertilization-competent sperm could be produced from neonatal mouse testes maintained exclusively in a culture system (Shinohara et al., 2011). This is a very promising step, indicating that similar results may be achievable in future using immature human testis biopsies.

5. Current trends in testis tissue cryopreservation

Since the first reports of successful germ cell transplantation and xenografting of testis tissue raised new interest in this field, several promising cryopreservation protocols have been introduced. Perhaps not surprisingly, the results differed and at times conflicted depending on the tissue donor species/developmental stage. These first reports of cryopreservation of pig and mouse testis tissues were based on DMSO-based slow freezing protocols originally developed for isolated testis cells or for ovarian tissue, respectively (Honaramooz et al., 2002a; Schlatt et al., 2002). Later, other detailed studies comparing multiple protocols showed high cell viability with programmed slow-freezing of immature mouse testis tissue using 1.5M DMSO as a cryoprotectant (Milazzo et al., 2008; Traverse et al., 2011). DMSO has also been found to be a more suitable cryoprotective agent than ethylene glycol for immature mouse and rat testis tissue (Goossens et al., 2008; Jezek, 2001). Shinohara et al. (2002) reported the birth of mouse offspring from sperm retrieved from cryopreserved pre-pubertal testis tissue with DMSO after transplantation under tunica albuginea of the recipient testes (Shinohara et al., 2002). Similar results were obtained using primate testis tissue, where 1.4M (but not 0.7M) DMSO was able to protect some of the developmental potential of grafts from rhesus monkeys (Jahnukainen et al., 2007) but the 0.7M DMSO protocol was successful for cryopreservation of human testis tissue (Wyns et al., 2007) at one age/developmental stage but not others (Wyns et al., 2008; Keros et al., 2005, 2007). Somewhat different from reports in other species, and after an extensive study of several strategies for cryopreservation of immature testis tissue, we concluded that glycerol was a better cryoprotectant for pig tissues (Abrishami et al., 2010a). These results suggest that each species and donor developmental age may need a different cryopreservation protocol, with a concomitant need to adjust the concentration of cryoprotectant or even adopt different cryoprotectants. These differences may be related to testicular architecture, morphology, or lipid composition.

In a first report of immature testis tissue vitrification, we also showed maintenance of cell viability and developmental potential to actively (re)establish complete spermatogenesis after xenografting into immunodeficient mice (Abrishami et al., 2010a). Recently, similar or much higher cell viability results were obtained using immature mouse testis tissue with vitrification compared with conventional slow freezing (Gouk et al., 2011; Curaba et al., 2011). With proper tissue handling, and the use of an appropriate choice of final cryoprotectant exposure, vitrification can provide preferential conditions for tissue freezing with proven superior results in restoration of immature testis tissue. Vitrification also does not require the extensive laboratory equipment commonly used for programmed slow freezing; however, direct plunging of tissues into liquid nitrogen, a common procedure in routine vitrification, poses a greater risk of contamination. The solid-surface vitrification of testis tissue (**Fig. 3**) is an easy, safe, and applicable cryopreservation technique for the preservation of tissue structural integrity and developmental potential.

6. Conclusion

Although cryopreservation of isolated testis cells has been successfully achieved for animals and humans, only in the past 10 years has intense attention been paid to cryopreservation techniques aimed at maintaining the developmental potential of structurally intact testis tissue. Cryopreservation of testis tissue theoretically offers a practical method when other techniques such as cryopreservation of ejaculated sperm are not available or applicable. Preservation of testis tissue has many applications, including conservation of fertility for prepubertal boys undergoing gonadotoxic cancer therapies. Ovarian and testicular toxicity are the inevitable long-term consequences of certain therapeutic oncological regimens, leading to premature fertility failure or sterility in cancer patients. Cryopreservation of gonadal cells or tissue before high-dose gonadotoxic chemo- and radio-therapy may therefore be considered in a comprehensive treatment and recovery plan. This could provide an alternative method for preserving the fertility potential of prepubertal boys with cancer or azoospermic men, as spermatogenesis is not completed in these patients. Although successful gamete and gonadal tissue restoration could have major impact on the enhancement of fertility preservation, serious ethical implications associated with collection and preservation of human gametes and gonadal tissues have yet to be resolved. Salvaging the genetic potential of immature endangered and valuable animals through banking of gonadal tissue is also a subject of clinical significance in animal reproduction and conservation. Optimal cryoconservation methods could also be combined with transplantation, xenografting, or culturing techniques to overcome some of the complications in the biodiversity crisis of rare or endangered species. In fact, experimental methods for the generation of fertility-competent gametes from cryopreserved ovarian or testis tissues have paved the way for future clinical use in human patients. Therefore, experimental conservation of gonadal tissue and cells by cryopreservation can serve as a platform for further evaluation of the potential for long-term storage.

Many challenges are associated with the optimal maintenance of tissue structure and the subsequent functional restoration of cryopreserved samples. It is intuitively known that optimal cryopreservation requires refinement of freezing and thawing rates, osmotic conditions, choice and concentration of cryoprotectants, and equilibration times in cryoprotective solutions. Indeed, improvement of all aspects of freezing techniques will ensure survival rates of tissue structure and subsequent functional restoration of cryopreserved cells within those tissues. Several studies have examined cryopreservation of testis cell suspensions or tissue fragments using glycerol, ethylene glycol, DMSO, or propanediol. In most cases, analyses of the cryopreserved samples lacked functional assessments of the preserved testicular cells/tissues. We now know that even if many cells of a multicellular system survive freezing and thawing, preservation of all functional compartments of the tissue is not guaranteed. Merely maintaining the physical characteristics of the cryopreserved testis tissue is not adequate, and an efficient approach to overcome the deficiencies in developmental (re)establishment of spermatogenesis is also required.

7. Acknowledgement

The author would like to thank the Natural Sciences and Engineering Research Council (NSERC) of Canada, and the Saskatchewan Health Research Foundation for grants to support the work from the current laboratory summarized here.

8. References

Abbasi S, and Honaramooz A. (2011) Xenografting of testis tissue from bison calf donors into recipient mice as a strategy for salvaging genetic material. Theiogenology; 76 607-614.

Abrishami M, Anzar M, and Honaramooz A. (2010a) Cryopreservation of immature porcine testis tissue to maintain its developmental potential after xenografting into recipient mice. Theriogenology 73, 86–89.

Abrishami M, Abbasi S, and Honaramooz A. (2010b) The effect of donor age on progression of spermatogenesis in canine testicular tissue after xenografting into immunodeficient mice. Theriogenology 73, 512–522.

Abrishami M. (2009) Cryopreservation and xenografting of testis tissue. MSc Thesis. Cryopreservation and xenografting of testis tissue., University of Saskatchewan, Saskatoon.

adoul P, Donnez J, Dolmans MM, Squifflet J, Lengele B, and Martinez-Madrid B. (2007). Laparascopic ovariectomy for whole human ovary cryopreservation: technical aspects. Fertil Steril 87, 971-975.

Anchordoguy TJ, Rudolph AS, Carpenter JF, and Crowe JH. (1987) Modes of interaction of cryoprotectants with membrane phospholipids during freezing. Cryobiology; 24 324-31.

Apperley JF, and Reddy N. (1995) Mechanism and management of treatment-related gonadal failure in recipients of high dose chemoradiotherapy. Blood Rev 9, 93-116.

Arregui L, Rathi R, Megee SO, Honaramooz A, Gomendio M, Roldan ER, and Dobrinski I. (2008) Xenografting of sheep testis tissue and isolated cells as a model for preservation of genetic material from endangered ungulates. Reproduction 136, 85–93.

Avarbock MR, Brinster CJ, and Brinster RL. (1996) Reconstitution of spermatogenesis from frozen spermatogonial stem cells. Nat Med; 2 693-6.

Bagchi A, Woods EJ, and Critser JK. (2008) Cryopreservation and vitrification: recent advances in fertility preservation technologies. Expert Rev Med Devices; 5 359-70.

Benson EE. (2004) Principles of Cryobiology. In Lane, NJ, Benson EE, Fuller BJ (eds). Life in the frozen states, CRC Press, Boca Raton.

Brinster RL and Nagano M. (1998) Spermatogonial stem cell transplantation, cryopreservation and culture. Semin Cell Dev Biol; 9 401-9.

Brinster RL, and Avarbock MR. (1994) Germline transmission of donor haplotype following spermatogonial transplantation. Proc Natl Acad Sci USA 91, 11303-11307.

Brinster RL, and Zimmermann JW. (1994) Spermatogenesis following male germ-cell transplantation. Proc Natl Acad Sci USA 91, 11298-11302.

Brook P, Radford J, Shalet S, Joyce A, and Gosden R. (2001). Isolation of germ cells from human testicular tissue for low temperature storage and autotransplantation. Fertil Steril 75, 269-274.

Courbiere B, Odagescu V, Baudot A, Massardier J, Mazoyer C, Salle B, and Lornage J. (2006) Cryopreservation of the ovary by vitrification as an alternative to slow-cooling protocols. Fertil Steril 41, 1243-1251.

Crowe JH, Carpenter JF, Crowe LM, Anchordoguy TJ (1990) Are freezing and dehydration similar stress vectors? A comparison of modes of interaction of stabilizing solutes with biomolecules. Cryobiology; 27 219-31.

Curaba M, Verleysen M, Amorim AC, Dolmans M-M, Van Langendonckt A, Hovatta O, Wyns C, and Donnez J. (2011) Cryopreservation of prepubertal mouse testicular tissue by vitrification. Fertil Steril 95, 1229-1234.

Deanesly R. (1954) Immature rat ovaries grafted after freezing and thawing. J Endocrinol; 11 197-200.

Fabbri R, Porcu E, Marsella T, Rocchetta G, Venturoli S, and Flamigni C. (2001) Human oocyte cryopreservation: new perspectives regarding oocyte survival. Hum Reprod; 16 411-6.

Fleck RA, Benson EE, Bremner DH, and Day JG. (2000) Studies of free radical-mediated cryoinjury in the unicellular green alga Euglena gracilis using a non-destructive hydroxyl radical assay: a novel approach for developing protistan cryopreservation strategies. Free Radic Res; 32 157-70.

Franca LR, Silva V.A, Jr., Chiarini-Garcia H, Garcia SK, and Debeljuk L. (2000) Cell proliferation and hormonal changes during postnatal development of the testis in the pig. Biology of Reproduction; 63 1629-36.

Frankenhuis MT, Wensing CJG, and Kremer J. (1981) The influence of elevated testicular temperature and scrotal surgery on the number of gonocytes in the newborn pig. International Journal of Andrology; 4 105-10.

Fuller B and Paynter S. (2004) Fundamentals of cryobiology in reproductive medicine. Reprod Biomed Online; 9 680-91.

Fuller BJ. (2004) Cryoprotectants: the essential antifreezes to protect life in the frozen state. Cryo Letters; 25 375-88.

Gañán N, Gomendio M, and Roldan ER (2009). Effect of storage of domestic cat (Felis catus) epididymides at 5 degrees C on sperm quality and cryopreservation. Theriogenology 72, 1268-1277.

Gianaroli L, Selman HA, Magli MC, Colpi G, Fortini D, and Ferraretti AP. (1999) Birth of a healthy infant after conception with round spermatids isolated from cryopreserved testicular tissue. Fertil Steril 72, 539-541.

Goossens E, Frederickx V, Geens M, De Block G, and Tournaye H. (2008) Cryosurvival and spermatogenesis after allografting prepubertal mouse tissue: comparison of two cryopreservation protocols. Fertil Steril; 89 725-7.

Gosden RG, Baird DT, Wade JC, and Webb R. (1994) Restoration of fertility to oophorectomized sheep by ovarian autografts stored at -196 degrees C. Hum Reprod; 9 597-603.

Gouk SS, Loh YFJ, Kumar SD, Watson PF, and Kuleshova LL. (2011) Cryopreservation of mouse testicular tissue: Prospect for harvesting spermatogonial stem cells for fertility preservation. Fertil Steril 95, 2399-2403.

Green SH, Smith AU, and Zuckerman S. (1956) The number of oocytes in ovarian autografts after freezing and thawing. J Endocrinol; 13 330-4.

Honaramooz A, and Yang Y. (2011) Recent advances in application of male germ cell transplantation in farm animals Vet. Med. International. 2011, 657860.

Honaramooz A, Behboodi E, Blash S, Megee SO, and Dobrinski I. (2003a) Germ cell transplantation in goats. Mol. Reprod. Dev. 66, 21-28.

Honaramooz A, Behboodi E, Megee SO, Overton SA, Galantino-Homer H, Echelard Y, and Dobrinski I. (2003b) Fertility and germline transmission of donor haplotype following germ cell transplantation in immunocompetent goats Biol. Reprod. 1260-1264.

Honaramooz A, Cui XS, Kim NH, and Dobrinski I. (2008) Porcine embryos produced after intracytoplasmic sperm injection using xenogeneic pig sperm from neonatal testis tissue grafted in mice. Reprod. Fertil. Dev. 20, 802–807.

Honaramooz A, Li MW, Penedo MC, Meyers S, and Dobrinski I. (2004) Accelerated maturation of primate testis by xenografting into mice. Biol. Reprod. 70, 1500–1503.

Honaramooz A, Megee SO, and Dobrinski I. (2002b) Germ cell transplantation in pigs. Biol. Reprod. 66, 21-28.

Honaramooz A, Snedaker A, Boiani M, Scholer H, Dobrinski I, and Schlatt S. (2002a) Sperm from neonatal mammalian testes grafted in mice. Nature; 418 778-81.

Hughes PE, and Varley MA. (1980) Reproduction in the Pig. United Kingdom: Butterworth and Co.

Hunt CJ, Taylor MJ, and Pegg DE. (1982) Freeze-substitution and isothermal freeze-fixation studies to elucidate the pattern of ice formation in smooth muscle at 252 K (-21 degrees C). J Microsc; 125 177-86.

Jadoul P, Donnez J, Dolmans MM, Squifflet J, Lengele B, and Martinez-Madrid B. (2007). Laparascopic ovariectomy for whole human ovary cryopreservation: technical aspects. Fertil Steril 87, 971-975

Jahnukainen K, Ehmcke J, Hergenrother SD, and Schlatt S. (2007) Effect of cold storage and cryopreservation of immature non-human primate testicular tissue on spermatogonial stem cell potential in xenografts. Hum Reprod; 22 1060-67.

Jezek D, Schulze, W., Kalanj-Bognar, S., Vukelic,Z., Milavec-Puretic, V., Krhen, I. (2001) Effects of various cryopreservation media and freezing-thawing on the morphology of rat testicular biopsies. Andrologia; 33 368-78.

Kasai M and Mukaida T. (2004) Cryopreservation of animal and human embryos by vitrification. Reprod Biomed Online; 9 164-70.

Keros V, Hultenby K, Borgstrom B, Fridstrom M, Jahnukainen K, and Hovatta O. (2007) Methods of cryopreservation of testicular tissue with viable spermatogonia in pre-pubertal boys undergoing gonadotoxic cancer treatment. Hum Reprod; 22 1384-95.

Keros V, Rosenlund B, Hultenby K, Aghajanova L, Levkov L, and Hovatta O. (2005) Optimizing cryopreservation of human testicular tissue: comparison of protocols with glycerol, propanediol and dimethylsulphoxide as cryoprotectants. Hum Reprod; 20 1676-87.

Kishikawa H, Tateno H, and Yanagimachi R. (1999) Fertility of mouse spermatozoa retrieved from cadavers and maintained at 4 degrees C. J Reprod Fertil 16, 217-222.

Leibo SP, Farrant J, Mazur P, Hanna MG, Jr., and Smith LH. (1970) Effects of freezing on marrow stem cell suspensions: interactions of cooling and warming rates in the presence of PVP, sucrose, or glycerol. Cryobiology; 6 315-32.

Lovelock JE and Bishop MW. (1959) Prevention of freezing damage to living cells by dimethyl sulphoxide. Nature; 183 1394-5.

Maksudov GY, Shishova NV, and Katkov II. (2009) In the Cycle of Life: Cryopreservation of Post-Mortem Sperm as a Valuable Source in Restoration of Rare and Endangered Species. In: Endangered Species: New Research. Eds.: A.M. Columbus and L. Kuznetsov. NOVA Science Publishers, Hauppauge NY, pp 189-240.

Martínez AF, Martínez-Pastor F, Alvarez M, Fernandez-Santos MR, Esteso MC, de Paz P, Garde JJ, and Anel L. (2008) Sperm parameters on Iberian red deer: electroejaculation and post-mortem collection. Theriogenology 15, 216-226.

Martinez-Madrid B, Camboni A, Dolmans MM, Nottola SA, Van Langendonckt A, and Donnez J. (2007). Apoptosis and ultrastructural assessment after cryopreservation of whole human ovaries with their vascular pedicle. Fertil Steril 87, 1153-1165.

Mazur P. (1990) Equilibrium, quasi-equilibrium and non-equilibrium freezing of mammalian embryos. Cell Biophys; 17 53-92.

Mazur P. (2004) Principles of Cryobiology. In Fuller, B, Lane N, Benson EE (eds). Life in the frozen states, CRC Press, Boca Raton.

Mc Laren A and Biggers JD. (1958) Successful development and birth of mice cultivated *in vitro* as early as early embryos. Nature; 182 877-8.

Milazzo JP, Vaudreuil L, Cauliez B, Gruel E, Masse L, Mousset-Simeon N, Mace B, and Rives N. (2008) Comparison of conditions for cryopreservation of testicular tissue from immature mice. Hum Reprod; 23 17-28.

Nagano M, Patrizio P, and Brinster RL (2002). Long-term survival of human spermatogonial stem cells in mouse testes. Fertil Steril 78, 1225-1233.

Nakai M, Kaneko H, Somfai T, Maedomari N, Ozawa M, Noguchi J, Ito J, Kashiwazaki N, and Kikuchi K. (2010) Production of viable piglets for the first time using sperm derived from ectopic testicular xenografts. Reproduction 139, 331–335.

Naysmith TE, Blake DA, Harvey VJ, and Johnson NP. (1998) Do men undergoing sterilizing cancer treatments have a fertile future? Hum Reprod 13, 3250-3255.

Oatley JM, de Avila DM, Reeves JJ, and McLean DJ. (2004) Spermatogenesis and germ cell transgene expression in xenografted bovine testicular tissue. Biol. Reprod. 71, 494–501.

Ogawa T, Dobrinski I, Avarbock MR, and Brinster RL. (1999) Xenogeneic spermatogenesis following transplantation of hamster germ cells to mouse testes. Biol Reprod; 60 515-21.

Parkes AS. (1958) Factors affecting the viability of frozen ovarian tissue. J Endocrinol; 17 337-43.

Parrot DMV. (1960) The fertility of mice with orthotophic ovarian grafts derived from frozen tissue. J Reprod Fertil; 1 230-41.

Pegg DE. (2002) The history and principles of cryopreservation. Semin Reprod Med; 20 5-13.

Pegg DE. (2007) Principles of cryopreservation. Methods Mol Biol; 368 39-57.

Pelliniemi LJ. (1975) Ultrastructure of the early ovary and testis in pig embryos. American Journal of Anatomy; 144 89-112.

Polge C, Lovelock, JE. (1952) Preservation of bull semen at -79C. The Veterinary Record; 64 396-97.

Polge C, Smith AU, and Parkes AS. (1949) Revival of spermatozoa after vitrification and dehydration at low temperatures. Nature; 164 666.

Porcu E, Fabbri R, Seracchioli R, Ciotti PM, Magrini O, and Flamigni C. (1997) Birth of a healthy female after intracytoplasmic sperm injection of cryopreserved human oocytes. Fertil Steril; 68 724-6.

Porcu E. (2001) Oocyte freezing. Seminars in Reproductive Medicine; 19 221-30.

Rathi R, Honaramooz A, Zeng W, Schlatt S, and Dobrinski I. (2005) Germ cell fate and seminiferous tubule development in bovine testis xenografts. Reproduction 130, 923-929.

Rathi R, Honaramooz A, Zeng W, Turner R, and Dobrinski I. (2006) Germ cell development in equine testis tissue xenografted into mice. Reproduction 131, 1091-1098.

Rosenlund B, Westlander G, Wood M, Lundin K, Reismer E, and Hillensjö T. (1998) Sperm retrieval and fertilization in repeated percutaneous epididymal sperm aspiration. Hum Reprod 13, 2805-2807.

Rudolph AS and Crowe JH. (1985) Membrane stabilization during freezing: The role of two natural cryoprotectants, trehalose and proline. Cryobiology; 22 367-77.

Ryu BY, Orwig KE, Kubota H, Avarbock MR, and Brinster RL. (2004) Phenotypic and functional characteristics of spermatogonial stem cells in rats. Dev Biol; 274 158-70.

Sato T, Katagiri K, Gohbara A, Inoue K, Ogonuki N, Ogura A, Kubota Y, Ogawa T. (2011) In vitro production of functional sperm in cultured neonatal mouse testes. Nature 471, 504-507.

Schlatt S, Honaramooz A, Boiani M, Scholer HR, and Dobrinski I. (2003) Progeny from sperm obtained after ectopic grafting of neonatal mouse testes Biology of Reproduction, 68: 2331-2335.

Schlatt S, Kim SS, and Gosden R. (2002) Spermatogenesis and steroidogenesis in mouse, hamster and monkey testicular tissue after cryopreservation and heterotopic grafting to castrated hosts. Reproduction; 124 339-46.

Schlatt S, Rosiepen G, Weinbauer GF, Rolf C, Brook PF, and Nieschlag E. (1999) Germ cell transfer into rat, bovine, monkey and human testes Human Reproduction 14, 144-50.

Schoysman R, Vanderzwalmen P, Bertin G, Nijs M, and Van Damme B. (1999) Oocyte insemination with spermatozoa precursors. Curr Opin Urol 9, 541-545.

Schrader M, Müller M, Sofikitis N, Straub B, and Miller K. (2002) Testicular sperm extraction in azoospermic cancer patients prior to treatment--a new guideline? Hum Reprod 17, 1127-1128.

Scott HM, Mason JI, and Sharpe RM. (2009) Steroidogenesis in the fetal testis and its susceptibility to disruption by exogenous compounds. Endocrine reviews; 30 883-925.

Shinohara T, Inoue K, Ogonuki N, Kanatsu-Shinohara M, Miki H, Nakata K, Kurome M, Nagashima H, Toyokuni S, Kogishi K, et al. (2002) Birth of offspring following transplantation of cryopreserved immature testicular pieces and in-vitro microinsemination. Hum Reprod; 17 3039-45.

Smith AU. (1952) Behaviour of fertilized rabbit eggs exposed to glycerol and to low temperatures. Nature; 170 374-5.

Smith AU. (1961) Biological effects of freezing and supercooling. Edward Arnold, London.

Snedaker AK, Honaramooz A, Dobrinski I. (2004) A game of cat and mouse: xenografting of testis tissue from domestic kittens results in complete cat spermatogenesis in a mouse host. J Androl 25, 926-930.

Steven AC and Aebi U. (2003) The next ice age: cryo-electron tomography of intact cells. Trends Cell Biol; 13 107-10.

Stiller CA, Desandes E, Danon SE, Izarzugaza I, Ratiu A, Vassileva-Valerianova Z, and Steliarova-Foucher E. (2006) Cancer incidence and survival in European adolescents (1978-1997). Report from the Automated Childhood Cancer Information System project. Eur J Cancer 42, 2006-2018.

Sutton RL. (1992) Critical cooling rates for aqueous cryoprotectants in the presence of sugars and polysaccharides. Cryobiology; 29 585-98.

Taylor MJ, Pegg DE. (1983) The effect of ice formation on the function of smooth muscle tissue following storage at -21°C and -60°C. Cryobiology; 20 36-40.

Tesarik J, Bah Ceci M, Ozcan C, Greco E, and Mendoza C. (1999) Restoration of fertility by in vitro spermatogenesis. Lancet 353, 555-556.

Tesarik J, Cruz-Navarro N, Moreno E, Cañete MT, and Mendoza C. (2000) Birth of healthy twins after fertilization with in vitro cultured spermatids from a patient with massive in vivo apoptosis of postmeiotic germ cells. Fertil Steril 74, 1044-1046.

Travers A, Milazzo JP, Perdrix A, Metton C, Bironneau A, Mace B, and Rives N. (2011) Assessment of freezing procedures for rat immature testicular tissue. Theriogenology 76, 981-990.

Van Straaten HWM and Wensing CJG. (1977) Histomorphometric aspects of testicular morphogenesis in the pig. Biology of Reproduction; 17 467-72.

Whittingham DG, Leibo SP, and Mazur P. (1972) Survival of mouse embryos frozen to -196 degrees and -269 degrees C. Science; 178 411-4.

Whittingham DG. (1977) Fertilization in vitro and development to term of unfertilized mouse oocytes previously stored at -196 degrees C. J Reprod Fertil; 49 89-94.

Wilhelm D, Palmer S, and Koopman P. (2007) Sex determination and gonadal development in mammals. Physiological reviews; 87 1-28.

Wilmut I. (1972) The low temperature preservation of mammalian embryos. J Reprod Fertil; 31 513-4.

Wyns C, Curaba M, Martinez-Madrid B, Van Langendonckt A, Francois-Xavier W, and Donnez J. (2007) Spermatogonial survival after cryopreservation and short-term orthotopic immature human cryptorchid testicular tissue grafting to immunodeficient mice. Hum Reprod; 22 1603-11.

Wyns C, Van Langendonckt A, Wese F-X, Donnez J, and Curaba M. (2008) Long-term spermatogonial survival in cryopreserved and xenografted immature human testicular tissue. Hum Reprod; 23 2402-14.

Yang Y, Steeg J, and Honaramooz A. (2010). The effects of tissue sample size and media on short-term hypothermic preservation of porcine testis tissue. Cell Tissue Res 340, 397-406.

7

Vitrification of Oocytes and Embryos

Juergen Liebermann

Director of Laboratory, Fertility Centers of Illinois, Chicago, IL
USA

1. Introduction

Currently, controlled ovarian hyperstimulation protocols commonly provide embryos in excess of those needed for fresh transfer. Therefore, techniques have been developed to store these surplus embryos in liquid nitrogen (referred to as cryopreservation) for an indefinite period of time without significant compromise of their quality. Based on data from the Centers for Disease Control and Prevention (CDC) from 2001 to 2004, about 18% of all IVF cycles in the USA used frozen embryos for transfer. In addition, data from the same registry compared live births per transfer using frozen and fresh embryos (25% versus 34% respectively) clearly showing that cryopreservation is an important adjunct to maximize the efficiency of every single patient's oocyte retrieval. The fundamental objectives for successful cryostorage of cells in liquid nitrogen at -196°C can be summarized as follows: **1)** arresting the metabolism reversibly, **2)** maintaining structural and genetic integrity, **3)** achieving acceptable survival rates after thawing, **4)** maintain of developmental competence post thaw and, **5)** the technique has to be reliable and repeatable.

Furthermore, all methods and protocols for cryopreservation should be developed such that ice crystals formation and growth inside the cells or tissues must either be eliminated or massively suppressed. One recent hotly debated topic in the area of reproductive cryobiology is whether slow-cooling or rapid-cooling protocols both satisfy the fundamental cryo-biological principles for reduction of damage by ice crystal formation during cooling and warming, and which approach is better. It is the case nonetheless, that both methods of cryopreservation of biological material include six principal steps: **1)** initial exposure to the cryoprotectant (intracellular water has to be removed by gradual dehydration, **2)** cooling (slow/rapid) to subzero temperatures (-196°C), **3)** storage at low temperature, **4)** thawing/warming by gradual rehydration, **5)** dilution and removal of the cryoprotectant agents and replacement of the cellular and intracellular fluid at precise rate and, **6)** recovery and return to a physiological environment.

Although initially reported in 1985 as a successful cryopreservation approach for mouse embryos, vitrification has taken a backseat in human assisted reproduction. However, the practical advantages of this cryopreservation method have more recently caught the attention of many ART laboratories as a feasible alternative to traditional slow freezing methods. Since 1985 more than 2,100 publications can be found referring to the topic of "vitrification", which is further evidence of the burgeoning growth of interest in this cryopreservation technology. One "drawback" considered by embryologists who are not familiar with the vitrification technique, is the use of higher concentration of cryoprotectants, which does potentially mean that the vitrification solutions are more toxic than their counterpart solutions used for conventional slow freezing. However, with better understanding of the physical and biological principles of vitrification this has lead to numerous successful clinical applications of this technique within the field of assisted reproduction. As of today, all developmental stages of human embryos cultured in vitro have been successfully vitrified and warmed, with resulting offspring. Today, slow freezing technology still has the longest clinical track record, and greater 'comfort level' amongst embryologists. Nevertheless, vitrification with its increasing clinical application is showing a trend of greater consistency and better outcomes when compared to slow freezing technology. Therefore, when (not if) IVF programs overcome the fear of the 'unknown', and take on the challenge of the short learning curve with vitrification, then at that point vitrification will become the clinical standard for human embryo cryopreservation.

Cryopreservation at low temperature slows or totally prevents unwanted physical and chemical change. The major disadvantage to using low temperature cryostorage is that it can lead to the crystallization of water, and thereby this approach can create new and unwanted physical and chemical events that may injure the cells that are being preserved. Although the results achieved by slow freezing in many cases seem quite successful (Gardner *et al.*, 2003; Van den Abbeel *et al.*, 2005), ice crystal formation still renders traditional slow-freezing programs generally less consistent in their clinical outcomes. Another downside to the slow freezing approach is the time to complete such freezing procedures for human embryos, which can range from 1.5 to 5hrs. This is due to the fact that the slow rate of cooling attempts to maintain a very delicate balance between multiple factors that may result in cellular damage by ice crystallization and osmotic toxicity. Traditionally slow-freeze embryo cryopreservation has been a positive contributor to cumulative patient pregnancy rates, but ultimately the limitations of current slow-rate freezing methods in ART have become more evident in the shootout with vitrification-based cryostorage.

Vitrification is one of the more exciting developments in ART in recent years that attempts to avoid ice formation altogether during the cooling process by establishing a glassy or vitreous state rather than an ice crystalline state, wherein molecular translational motions are arrested without structural reorganisation of the liquid in which the reproductive cells are suspended. To achieve this glass-like solidification of living cells for cryostorage, high cooling rates in combination with high concentrations of cryoprotectants are used. A primary strategy for vitrifying cells and tissue is to increase the speed of thermal conductivity, while decreasing the concentration of the vitrificants to reduce their potential toxicity. There are two main ways to achieve the vitrification of water inside cells efficiently: **a)** to increase the cooling rate by using special carriers that allow very small volume sizes containing the cells to be very rapidly cooled; and **b)** to find materials with rapid heat transfer. However, one has to take into account that every cell seems to require its own optimal cooling rate, e.g., mature unfertilized oocytes are much more sensitive to chilling injury than any of the cell stages of the pre-implantation embryo.The earliest attempts using vitrification as an ice-free cryopreservation method for embryos were first reported in 1985 (Rall & Fahy, 1985). In 1993 successful vitrification of mouse embryos was demonstrated (Ali & Shelton, 1993). Furthermore, bovine oocytes and cleavage-stages were vitrified and warmed successfully a few years later (Vajta *et al.* 1998). In 1999 and 2000 successful pregnancies and deliveries after vitrification and warming of human oocytes were reported (Kuleshova *et al.*, 1999; Yoon *et al.*, 2000). Since that time, and because it seems to be that both entities appear to be especially chill-sensitive cells in ART, oocytes and blastocysts seem to receive a potentially significant boost in survival rates by avoiding ice-crystallization using vitrification (Walker *et al.*, 2004). In general, vitrification solutions are aqueous cryoprotectant solutions that do not freeze when cooled at high cooling rates to very low temperature. Interest in vitrification has clearly risen as evinced by the almost exponential growth of scientific publications about vitrification. Vitrification is very simple, requires no expensive programmable freezing equipment, and relies especially on the placement of the embryo in a very small volume of vitrification medium (refered also as "minimal volume approach") that must be cooled at extreme rates not obtainable in traditional enclosed cryo-storage devices such as straws and vials. The importance of the use of a small volume, also referred to „minimal volume approach" was described and published in 2005 (Kuwayama *et al.*, 2005; Kuwayama, 2007). In general, the rate of cooling/warming and the concentration of the cryoprotectant required to achieve vitrification are in inversely related. In addition, recent publications have shown the dominance of warming rate over cooling rates in the survival of oocytes subjected to a vitrification procedure (Serki & Mazur, 2009; Mazur & Seki, 2011).

During vitrification, by using a cooling rate in the range of 2,500 to 30,000°C/min or greater, water is transformed directly from the liquid phase to a glassy vitrified state. The physical definition of vitrification is the solidification of a solution at low temperature, not by ice-crystallization but by extreme elevation in viscosity during cooling (Fahy *et al.*, 1984; Fahy 1986). Vitrification of the aqueous solution inside cells can be achieved by increasing the speed of temperature change, and by increasing the concentration of the cryoprotectant used. However, a major potential drawback of vitrification is the use of high concentration of cryoprotectant, and an unintentional negative impact of these cryoprotectants in turn can be their toxicity, which may affect the embryo and subsequent development in utero. It is therefore essential to achieve a fine balance between the speed of cooling and the concentration of the vitrifying cryoprotectants. This is necessitated by the practical limit for the rate of cooling, and the biological limit of tolerance of the cells for the concentration of toxic cryoprotectants being used to achieve the cryopreserved state. It is important to note that recently published papers (Takahashi *et al.*, 2005; Liebermann & Tucker, 2006; Liebermann, 2009, 2011) have shown that the use of relatively high concentration of cryoprotectants such as 15% (vol/vol) ethylene glycol (EG) used in an equimolar mixture with dimethyl sulphoxide (DMSO) had no negative effect on the perinatal outcomes from blastocyst transfers following vitrification when compared with those from fresh blastocyst transfers.

Vitrification in principle is a simple technology, that is potentially faster to apply, and relatively inexpensive; furthermore, it is becoming clinically established, and is seemingly more reliable and consistent than conventional cryopreservation when carried out appropriately (Tucker *et al.*, 2003; Liebermann & Tucker, 2004).

Cryoprotectant agents are essential for the cryopreservation of cells. Basically two groups of cryoprotectants exist: **1)** permeating (*glycerol, ethylene glycol, dimethyl sulphoxide*); and **2)** non-permeating (*saccharides, protein, polymers*) agents. The essential component of a vitrification solution is the permeating agent. These compounds are hydrophilic non-electrolytes with a strong dehydrating effect. Furthermore, these CPAs are able to depress the "freezing point" of the solution. Regarding the high concentration of cryoprotectant used for vitrification, and in view of the known biological and physiochemical effects of cryoprotectants, it is suggested that the toxicity of these agents is a key limiting factor in cryobiology. Not only does this toxicity prevent the use of fully protective levels of these additives, but it may also be manifested in the form of cryo-injury above and beyond that seen occurring due to classical causes of cell damage (osmotic toxicity and ice formation) during cryopreservation. In spite of this, the permeating CPA should be chosen firstly by their permeating property, and secondly on the basis of their potential toxicity. Because the permeating CPA is responsible for the toxicity (*the key limiting factor in cryobiology*), different cryoprotectants have been tested for their relative toxicity, and the results indicate that ethylene glycol (EG; MW 62.02) is the least toxic followed by glycerol. Additionally, these highly permeating cryoprotectants are also more likely to diffuse out of the cells rapidly and the cells regained their original volume more quickly upon warming, thus preventing osmotic injury. Therefore, the most common and accepted cryoprotectant for vitrification procedures is ethylene glycol (EG). Today EG is more commonly used in an equimolar mixture with DMSO. Often additives are added to the vitrification solution such as disaccharides. Disaccharides, for example sucrose, do not penetrate the cell membrane, but they help to draw out more water from cells by osmosis, and therefore lessen the exposure time of the cells to the toxic effects of the cryoprotectants. The non-permeating sucrose also acts as an osmotic buffer to reduce the osmotic shock that might otherwise result from the dilution of the cryoprotectant after cryostorage. In addition, permeating agents are able to compound with intracellular water and therefore water is very slowly removed from the cell. Hence the critical intracellular salt concentration is reached at a lower temperature. Removal of the cryoprotectant agent during warming can present a very real problem in terms of trying to reduce toxicity to the cells. Firstly, because of the toxicity of the vitrification solutions, quick dilution of them after warming is necessary; and secondly, during dilution water permeates more rapidly in to the cell than the cryoprotective additive diffuses out. As a consequence of the excess water inflow the cells are threatened by injury from osmotic swelling. In this situation the non-permeating sucrose acts as an osmotic buffer to reduce the osmotic shock. During warming using a high extracellular concentration of sucrose (e.g., 1.0M) counterbalances the high concentration of the cryoprotectant agents in the cell, as it reduces the difference in osmolarity between the intra- and extracellular compartments. The high sucrose concentration cannot totally prevent the cell from swelling, but it can reduce the speed and magnitude of swelling (Liebermann and Tucker, 2002; Liebermann *et al.*, 2002a; 2003).

2. Oocytes

The cryopreservation of human oocytes constitutes a important step forward in Assisted Reproductive Technology (ART) despite the fact that for more than 2 decades oocyte cryopreservation has long been the focus of unsuccessful efforts to perfect its clinical application. More recently, vitrification as an alternative to traditional slow freezing prootcols has been shown to provide high degrees of success in vitrified metaphase-II human oocytes. Although oocyte cryopreservation historically has low efficiency mainly because of low rates of survival, fertilization, and cleavage, data on ~2000 "frozen oocyte" babies born worldwide since 1986 exists. The question arises as to what makes oocytes so unique compared to embryos, besides differences in cell size and membrane permeability? Oocytes have a low volume-to-surface ratio; hence they are less efficient at taking up cryoprotectant and at loosing water. Other differences to be considered are **a)** that the maternal DNA is held suspended in the cytoplasm on the meiotic spindle & not within the protective confines of the nuclear membrane, therefore damage in the DNA and microtubules could explain the limited success of oocytes, **b)** the oocyte is arrested in a state primed for activation, and **c)** the changes in its environment can cause parthenogenetic activation. What are the applications then for oocyte cryopreservation in the US? One application would be to preserve fertility in women with malignant/premalignant conditions who would have to undergo treatment that might negatively impact their future ability to have children (50,000 per year <40 yr old), also in women who may want to delay childbearing ('clock-tickers') because of their careers, partnership status or psychological/ emotional reasons. A very interesting approach is donor oocyte banking, which makes the donor-recipient cycle more convenient by facilitating the "egg donation" and allows quarantining of the oocytes, which provides a unique advantage in economy as well as feasibility. Other applications are if a male is unable to produce a semen sample on the day of egg retrieval and or it could also eliminate ethical/moral questions of producing extra embryos. Overall, oocyte cryostorage offers an opportunity to reduce number of embryos generated per IVF cycle, and therefore lessening the pressure on the patient to increase the number of fresh embryos transferred. In addition, while also reducing embryo cryostorage it has the benefit of helping women "retain ownership" of their ability to be genetic parents at a time of their choosing, a time of greater convenience & health. The live born babies from cryopreserved oocytes have shown no apparent increase in congenital anomalies. Although 13 years later after the first slow-freeze birth, the number of reported babies born as a result of vitrified oocytes is now approaching that of slow-frozen oocytes without any increasing risk in congenital abnormalities (Noyes *et al.*, 2009). Vitrification of oocytes does not appear to increase risks of abnormal imprinting or disturbances in spindle formation or chromosome segregation (Trapphoff *et al.*, 2010). It has the greatest potential for successful oocyte cryopreservation and with its increased clinical application is showing a trend to greater consistency and better outcomes (similar to outcomes between fresh or warmed oocytes). Vitrification of oocytes, when applied to properly screened patients, will be a useful technology in reproductive medicine practice and will constitute a major step forward in ART.

Fortunately to date, no significant increase in abnormalities has been reported from these cryostored oocyte pregnancies (Chian *et al.*, 2009), regardless of the historical concerns that cryopreservation of mature oocytes might disrupt the meiotic spindle and thus increase the potential for aneuploidy in the embryos arising from such eggs. These concerns have mostly been allayed by publications that show no abnormal or stray chromosomes from previously frozen oocytes (Gook & Edgar, 1999), and FISH comparison of embryos from fresh and thawed oocytes show no increase in anomalies (Cobo *et al.*, 2001). There also appears to be adequate recovery of the meiotic spindle post-cryopreservation whether using conventional or vitrification technology (Chen *et al.*, 2004; Bianchi *et al.*, 2005; Larman *et al.*, 2007). The scientific literature on oocyte cryopreservation grows daily it seems. Most reports focus on clinical pregnancy rates (Boldt *et al.*, 2003; Boldt *et al.*, 2006), and as such while this data is helpful to increase our confidence in the technology, it does little to research new directions for oocyte cryopreservation.

3. Zygotes

Conventional cryopreservation of pronuclear zygotes (2PN) is well established in countries such as Germany where freezing of later stage human embryos is by law or by ethical reasons not allowed. The time to complete the conventional protocol to cryopreserved zygotes is 98min. In Germany the clinical pregnancy outcomes arising from the frozen/thawed 2PN cycles is about 18%, with an implantation of around 10% per embryo transferred. The time to complete vitrification of zygotes requires approximately 12min. Recently successful vitrification of 2PN with high survival (~ 90%), cleavage rates on day-2 (>80%), and blastocyst formation of 31% and pregnancies were reported (Park *et al.*, 2000; Jelinkova *et al.*, 2002; Liebermann *et al.*, 2002b; Al-Hasani *et. al.*, 2007). Zygote vitrification implemented as a clinical setting can provide a clinical pregnancy rate of close to 30%, with an implantation rate of 17% (Al-Hasani *et al.*, 2007). The pronuclear stage appears well-able to withstand the vitrification and warming conditions, which is probably due to the significant membrane permeability changes that occur post-fertilization; such changes to the oolemma may also make it more stable and able to cope with the vagaries of the cold-shock and striking osmotic fluctuations that occur during the vitrification process.

4. Cleavage stage embryo

Reports of human embryo vitrification have been more frequent. Liebermann and Tucker (2002) using either the cryoloop or the hemi-straw system (HSS) showed post-warming survival rates (after 2 hours of culture of day-3 embryos where more than half of their blastomeres were intact) from 84 to 90% which was dependent on the carrier system used. There was a reasonable further cleavage and compaction rate of 34%. This finding supports previous reports in which high survival rates of eight-cell human embryos using 40% EG were documented (Mukaida *et al.*, 1998). In comparison to traditional slow-rate cryopreservation, a survival rate of cleavage stage embryos of 76% was reported with vitrification (Jericho *et al.*, 2003). Recently reported successful pregnancies and deliveries after vitrification of day-3 human embryos using the OPS have been reported (El-Danasouri and Selman, 2001; Selman and El-Danasouri, 2002). Their results showed a negative correlation between stage of development and survival, eight-cell embryos showed a higher survival rate (79.2%; 62/78) than did embryos with fewer than six cells (21.1%; 11/53) after vitrification (El-Danasouri and Selman, 2001). Despite the fact, that Liebermann and Tucker (2002) achieved a promising post-warming survival rate, overall only about 34% of the surviving embryos had the developmental potential to reach the compaction stage. Recently publications on cleavage stage vitrification provided good outcome data. Loutradi *et al.* (2008) were performing a meta-analysis and systematic review by comparing traditional and vitrification protocols for cleavage stage embryos, and found a survival rate of 84.0% versus 97.0%. In addition, clinical pregnancy rates between 35 and 48%, with implantation rates between 15 to 39% have been reported (Rama Raju *et al.*, 2005; Desai *et al.*, 2007; Li *et al.*,

2007; Balaban *et al.*, 2008). So clearly vitrification appears to have a positive impact on overall embryo utilization. A study on the neonatal outcome of 907 vitrified/warmed cleavage stage embryos found no significant increase in the congenital birth defect rate when compared with pregnancies using fresh cleavage stage embryos (Rama Raju *et al.*, 2009).

5. Blastocyst stage

Vitrification of human blastocysts using different carriers shows survival rates of 70% to 90%, with clinical pregnancy rates of 37% to 53% and implantation rates of 20% to 30% ((Yokota *et al.*, 2000, 2001; Reed *et al.*, 2002; Mukaida *et al.*, 2001; 2003; Hiraoka *et al.*, 2004; Vanderzwalmen *et al.*, 2002; 2003; Huang *et al.*, 2005; Liebermann & Tucker, 2006; Liebermann, 2009, 2011).

6. The advantage of blastocyst cryopreservation

Activation of the embryonic genome occurs after the 8-cell stage (3 days postoocyte retrieval) is reached (Braude *et al.*, 1988). If the activation does not occur, the embryo will not survive further. Therefore, the improvement of human IVF outcomes requires identification of embryos that will progress beyond the 8-cell stage. Blastocyst culture (5 days postoocyte retrieval) allows for the transfer of embryos that clearly have an activated embryonic genome. This requires that the elimination of embryos in extended culture from day 3 to day 5 should depend solely on their inherited survival potential and not be a consequence of an adverse effect exerted by the sequential media used for culture beyond day 3. Additional advantages in cryopreserving at the blastocyst stage are: 1) At this stage a lower numbers of embryos can be transferred in fresh cycles, resulting in less high order multiple pregnancies, 2) The same is true for cryopreserved blastocysts showing higher pregnancy rates and implantation per thawed embryo transferred, 3) Approximately 120 hours (day five) into development the healthy human embryo should be at the blastocyst stage comprised of some 50 to 150 cells, of which about 20 to 30% make up the inner cell mass (ICM), the remainder making up the trophectoderm (TE), 4) the higher cell number allows better compensation for cryo-injuries, which results in greater viability and faster recovery, 5) the cytoplasmatic volume of the cells is lower, thus the surface-volume ratio is higher, and that in turn makes the penetration of the cryoprotectant faster, and 6) on average fewer embryos per patient were frozen-stored, but each one when thawed has a greater potential for implantation.

Both natural and hormone replacement cycles seem to provide comparable levels of receptivity in naturally cycling women, though they differ in level of convenience. Regardless of the day of cryopreservation of the embryo (whether day 5, 6 or 7), at thawing/warming blastocysts should be treated as if they had been frozen on the fifth day of development. Vitrification of blastocysts has been undertaken utilizing an "open system" (Cryotop; Kitazato Bio Pharma Co. Ltd., Fuji-shi, Japan), and since 2007 on a "closed system" (HSV [High Security Vitrification Kit]; CryoBio System, L'Aigle, France) after a two-step loading with cryoprotectant agents at 24°C. Briefly, blastocysts were placed in equilibration solution, which is the base medium (Hepes-buffered HTF with 20% Serum Supplement Substition (SSS) containing 7.5% (v/v) ethylene glycol (EG) and 7.5% (v/v) dimethyl sulfoxide (DMSO). After 5-7 min, the blastocysts were washed quickly in vitrification solution, which is the base medium containing 15% (v/v) DMSO, 15% (v/v) EG, and 0.5M sucrose, for 45-60sec and transferred onto the Cryotop or HSV using a micropipette. Immediately after the loading of not more than two blastocysts in a 1μl drop on the Cryotop, the carrier was plunged into fresh clean liquid nitrogen (LN2). After loading the embryos, the Cryotop was capped under the LN2 to seal and protect the vitrified material prior to cryo storage. In contrast, after loading the HSV, the straw was heat sealed and then plunged in LN2, and stored the same way as the cryotop (Liebermann & Tucker, 2006; Liebermann, 2009, 2011).

To remove the cryoprotectants, blastocysts were warmed and diluted in a two step process. With the Cryotop or HSV submerged in LN2, the protective cap (Cryotop) or inner straw (HSV) were removed, and then both carriers with the blastocysts were removed from the LN2 and placed directly into a pre-warmed (~35-37°C) organ culture dish containing 1ml of 1.0M sucrose. Blastocysts were picked up directly from the Cryotop and placed in a fresh drop of 1.0M sucrose at 24°C. After 5min blastocysts were transferred to 0.5M sucrose solution. After an additional 5min, blastocysts were washed in the base medium and returned to the culture medium (SAGE Blastocyst Medium, Trumbull, CT, USA) until transfer.

Between January 2004 and July 2011 the *Fertility Centers of Illinois* "IVF Laboratory River North"(Chicago) has vitrified 13,568 blastocysts *without artificial shrinkage* before the cryopreservation procedure (Table 1). After 2562 frozen embryo transfers (FET) including day 5 and day 6 blastocysts with a mean age of the patients of 34.9 ± 5.1 years, to date we have seen a survival rate, implantation, and clinical pregnancy rate per transfer (cPR) of 97.3%, 30.2%, and 42.5%, respectively (Table 2). After 7 1/2years of vitrifying blastocysts the perinatal outcome is as follow: from 687 deliveries with vitrified blastocysts, 852 babies (422 boys and 430 girls) were born (Table 2). No abnormalities were recorded. The singleton, twin and triplet pregnancy rates were 71%, 27%, and 2%, respectively.

Day of Development	Day 5	Day 6	Day 7	Total
Number of Blastocysts vitrified (%)	6220 (46%)	6988 (51%)	360 (3%)	13568

Table 1. Retrospective data from 3,712 patients (average age 33.8±4.9) with blastocyst cryopreservation by vitrification from January 2004 till July 2011.

Technique	VIT
Patient's age (y)	34.9 ± 5.1
No. of warmed cycles	2580
No. of transfers	2562
No. of blastocysts warmed	4965
No. of blastocysts survived (%)	4829 (97.3)
No. of blastocysts transferred	4752
Mean no. of blastocysts transferred	1.8
No. of implantations (%)	1432 (30.2)
No. of positive pregnancy/warm (%)	1255 (48.6)
No. of positive pregnancy/VET (%)	1255 (49.0)
No. of clinical pregnancy/warm (%)	1089 (42.2)
No. of clinical pregnancy/VET (%)	1089 (42.5)
Ongoing pregnancies/VET (%)	875 (34.2)
No. of livebirths	852 (422 boys & 430 girls)

Table 2. Retrospective data from the blastocyst cryopreservation program (Fertility Centers of Illinois, Chicago) where vitrification (VIT) technology was applied from January 2004 till July 2011.

When the vitrified-warmed blastocysts were divided into day 5 and day 6 groups, the following data was gather (Table 3 & 4). In 1265 FETs transferring day 5 blastocysts, the survival, implantation, and cPR were 97.6%, 34.8%, and 48.3% compared to 97.2%, 25.3%, and 36.5% of in 1204 day 6 FETs.

Patient's Age	< 35	35-37	38-40	> 40	Donor	Total
Ø Age	30.8±2.6	35.8±0.8	38.8±0.8	42.8±2.0	43.5±4.7	34.6±5.2
Day 5 Cycles	678	248	157	74	112	1269
Day 5 Transfers	677	247	155	74	112	1265
Embryos survived (%)	97.5%	96.9%	98.7%	96.0%	97.7%	97.6
Embryos transferred (MEAN)	1.9	1.8	1.9	1.9	1.8	1.9
Positive Pregnancies/Transfer	56.1%	56.7%	56.1%	51.4%	54.5%	55.8%
Clinical Pregnancies/Transfer	50%	49%	43%	43%	50%	48.3%
Ongoing Pregnancies/Transfer	43%	37%	30%	30%	39%	38.8%
# Sacs	478	152	87	42	69	828
Implantation Rate	37.3%	33.3%	29.7%	29.2%	33.7%	34.8%

Table 3. Retrospective outcome data at FCI from vitrified day 5 blastocysts in regards to the patients age between June 2007 till July 2011.

Patient's Age	< 35	35-37	38-40	> 40	Donor	Total
Ø Age	31.0±2.4	36.0±0.8	38.8±0.8	42.5±1.8	43.9±4.9	35.1±5.0
Day 5 Cycles	586	271	177	103	83	1220
Day 5 Transfers	579	266	176	101	82	1204
Embryos survived (%)	96.7%	97.8%	97.9%	95.2%	99.4%	97.2%
Embryos transferred (MEAN)	1.8	1.8	1.8	1.7	1.8	1.8
Positive Pregnancies/Transfer	43.0%	43.6%	39.2%	31.1%	50.0%	42.1%
Clinical Pregnancies/Transfer	37%	38%	35%	29%	42%	36.5%
Ongoing Pregnancies/Transfer	31%	30%	27%	20%	32%	29.2%
# Sacs	276	128	80	34	41	559
Implantation Rate	25.5%	26.5%	24.8%	19.0%	28.1%	25.3%

Table 4. Retrospective outcome data at FCI from vitrified day 6 blastocysts blastocyst in regards to the patients age between June 2007 till July 2011.

In addition, in 1128 FET using aseptic vitrification, 2041 blastocysts were transferred with a survival, implantation, and cPR of 98.4%, 31.8%, and 44.2%, respectively (Table 5). After 4 1/2 years of vitrifying blastocysts using a closed system the perinatal outcome is as follow: 313 babies (165 boys and 148 girls) were born (Table 5). No abnormalities were recorded.

Technique	aVIT
Patient's age (y)	34.5 ± 5.0
No. of warmed cycles	1132
No. of transfers	1128
No. of blastocysts warmed	2102
No. of blastocysts survived (%)	2069 (98.4)
No. of blastocysts transferred	2041
Mean no. of blastocysts transferred	1.8
No. of implantations (%)	650 (31.8)
No. of positive pregnancy/warm (%)	572 (50.7)
No. of positive pregnancy/VET (%)	572 (53.9)
No. of clinical pregnancy/warm (%)	499 (44.1)
No. of clinical pregnancy/VET (%)	499 (44.2)
Ongoing pregnancies/VET (%)	423 (37.5)
No. of livebirths	313 (165 boys & 148 girls)

Table 5. Retrospective data from the blastocyst cryopreservation program (*Fertility Centers of Illinois, Chicago*) where aseptic vitrification (aVIT) technology was applied from June 2007 till July 2011.

Our data has shown that freezing at the blastocyst stage provides excellent survival, implantation and clinical pregnancy (Liebermann & Tucker, 2006; Liebermann, 2009, 2011). To achieve this data the following points should be considered: a) without a successful blastocyst vitrification storage program, extended culture should never be attempted, b) the blastocyst is composed of more cells and therefore better able to compensate for cryo-injury, c) the cells are smaller thus making cryoprotectant penetration faster, and d) on average fewer embryos per patient are cryo-stored, but each one when thawed, has a greater potential for implantation, often with an opportunity for an ET with a single blastocyst.

Furthermore, a vitrification solution with a mixture of 7.5% EG/DMSO, followed by a 15% EG/DMSO with 0.5M sucrose step is safe for clinical use, giving rise to healthy babies without abnormalities. Vitrification of blastocysts using and open or closed system (Cryotop or HSV) is effective for achieving high implantation and pregnancy rates as seen in fresh embryo transfers. Although the outcome in terms of implantation and clinical pregnancy is significantly different when comparing day 5 blastocyst to day 6 blastocysts, our data should encourage cryopreservation of day 6 blastocysts as well. Based on the data presented, it is clear that the vitrification of Day 6 blastocysts is of clinical value since it can result in live births. This observation is confirmed by Saphiro et al. (2001) and Levens et al. (2008); they found that blastocyst development rate impacts outcome in slow cryopreserved blastocyst transfer cycles.

In conclusion, vitrification of human blastocysts is a viable and feasible alternative to traditional slow freezing methods. The key to this success lies in the more optimal timing of embryo cryopreservation, e.g. individual blastocysts may be cryopreserved at their optimal stage of development and expansion. In addition, the repeatedly discussed topic of using open systems (direct contact between cells and LN2) and the possible danger of contamination by bacteria, fungus or different strains of virus from LN2, can be avoided by moving forward to a closed system providing lower cooling rates, but without a negative impact on the outcome.

7. Contamination of LN2: Open versus closed systems

There are many potential advantages of vitrification in that it is an easy, cheap, fast and an apparently successful cryopreservation method; however, there is one issue that is still up for debate. It has been shown that fungi, bacteria and viruses are able to survive in liquid nitrogen (LN2) (Tedder et al, 1995; Fountain et al, 1997; Bielanski et al, 2000; 2003; Kyuwa et al, 2003; Letur-Konirsch et al, 2003). Given the direct exposure of the human cells as they are directly plunged into LN2 during the vitrification process, this therefore raises the question as to whether the LN2 has to be sterilized, as it may be a possible source of contamination for those cells. To this point there has been no fungal, viral or bacterial contamination that has been described from about 400 publications related to vitrification since the first report in 1985. Bielanski and colleagues (2000) demonstrated a viral transmission rate of 21 % to human embryos stored in open freezing containers under experimental conditions of extremely elevated viral presence; while in contrast all embryos stored in sealed freezing containers were free from contamination. Based on this observation they proposed that the sealing of freezing containers appears to prevent exposure to potential contaminants. Commercial systems to purify LN2 by filtration have been developed, however this technology to date has received little practical application in IVF laboratories that have active cryopreservation programs. While it is not totally clear that contamination is a real risk in everyday use of LN2, nevertheless it may be prudent to consider routine sterilization of LN2 when open carrier systems are used for vitrification, followed by a sealing of that system for cryo-storage. Further there are currently at least three 'closed' sealed vitrification systems that are commercially available, with FDA clearance, that represent successful alternatives to open systems for embryo vitrification (Liebermann, 2009, 2011)

8. Conclusions and future directions

Vitrification is a very promising cryopreservation method with many advantages, and an ever increasing clinical track record. A standardized vitrification protocol applicable to all stages of the pre-implantation embryo may not be realistic because of: **a)** different surface-to-volume ratios; **b)** differing cooling rate requirements between oocytes, zygotes, cleavage stage embryos and blastocysts; and **c)** variable chill-sensitivity between these different developmental stages. Currently however, the most widely used protocol applied to any embryo stage is the two-step equilibration in an equi-molar combination of the cryoprotectants ethylene glycol and DMSO, at a concentration of 15% each (v/v) supplemented with 0.5 mol/l sucrose.

For the adoption of vitrification in ART, as with all new technologies, there has been initial resistance; but as clinical data has been accrued, this technology is becoming more commonly adopted as standard procedure in many IVF programs worldwide. With this increased use in human assisted reproduction will come evolution of the vitrification process as it is fine tuned to clinical needs, so pushing forward its development to higher levels of clinical efficiency, utilization and universal acceptance.

9. Acknowledgement

The author thank the *Fertility Centers of Illinois* (FCI) and the embryologists at the FCI IVF Laboratory River North for their invaluable contributions and support in pushing vitrification to become our standard protocol for cryopreservation of human oocytes and blastocysts within our program.

10. References

[1] Ali and Shelton (1993). Vitrification of preimplantation stages of mouse embryos. J Reprod Fertil 98:459–465.

[2] Al-Hasani S, Ozmen B, Koutlaki N, Schoepper B, Diedrich K, Schultze-Mosgau A. (2007) Three years of routine vitrification of human zygotes: is it still fair to advocate slow-rate freezing? Reprod Biomed Online 14:288-93.

[3] Balaban B, Urman B, Ata B, Isiklar A, Larman MG, Hamilton R, Gardner DK (2008) A randomized controlled study of human Day 3 embryo cryopreservation by slow freezing or vitrification: vitrification is associated with higher survival, metabolism and blastocyst formation. Hum Reprod.23:1976-82.

[4] Bianchi V, Coticchio G, Fava L, Flamigni C, Borini A (2005) Meiotic spindle imaging in human oocytes frozen with a slow freezing procedure involving high sucrose concentration. Hum Reprod. 20:1078-83.

[5] Bielanski et al. (2000). Viral contamination of embryos cryopreserved in liquid nitrogen. Cryobiology 40:110-116.

[6] Bielanski et al. (2003). Microbial contamination of embryos and semen during long term banking in liquid nitrogen. Cryobiology 46:146-152.

[7] Boldt J, Cline D, McLaughlin D (2003) Human oocyte cryopreservation as an adjunct to IVF - embryo transfer cycles. Hum. Reprod. 18:1250-5

[8] Boldt J, Tidswell N, Sayers A, Kilani R, Cline D (2006) Human oocyte cryopreservation: 5-year experience with a sodium-depleted slow freezing method. Reprod Biomed Online 13:96-100.

[9] Braude P, Bolton V, Moore S (1988) Human gene expression first occurs between the four- and eight-cell stages of preimplantation development. Nature 332:459-461.

[10] Chen CK, Wang CW, Tsai WJ, Hsieh LL, Wang HS, Soong YK (2004) Evaluation of meiotic spindles in thawed oocytes after vitrification using polarized light microscopy. Fertil Steril 82:666-72.

[11] Chian RC, Gilbert L, Huang JY et al. 2009 Live birth after vitrification of in-vitro matured human oocytes. Fertil Steril 91, 372–376.

[12] Cobo A, Rubio C, Gerli S, Ruiz A, Pellicer A, Remohi J (2001) Use of fluorescence in situ hybridisation to assess the chromosomal status of embryos obtained from cryopreserved oocytes. Fertil Steril 75:354-60.

[13] Desai N, Blackmon H, Szeptycki J, Goldfarb J. (2007) Cryoloop vitrification of human day 3 cleavage-stage embryos: post-vitrification development, pregnancy outcomes and live births. Reprod Biomed Online, 14:208-13.

[14] El-Danasouri and Selman (2001). Successful pregnancies and deliveries after a simple vitrification protocol for day 3 human embryos. Fertil Steril 76:400-402.

[15] Fahy et al. (1984). Vitrification as an approach to cryopreservation. Cryobiology 21:407-426.

[16] Fahy (1986). Vitrification: a new approach to organ cryopreservation. In: Merryman HT (ed.), Transplantation: approaches to graft rejection, New York: Alan R Liss, 305-335.

[17] Fountain et al. (1997). Liquid nitrogen freezer: a potential source of microbial contamination of hematopoietic stem cell components. Transfusion 37:585-591.

[18] Gardner et al. (2003). Changing the start temperature and cooling rate in a slow-freezing protocol increases human blastocyst viability. Fertil Steril 79:407-410.

[19] Gook DA, Edgar DH (199) Cryopreservation of the human female gamete: current and future issues. Hum Reprod 14:2938-40.

[20] Huang CC, Lee TH, Chen SU, Chen HH, Cheng TC, Liu CH, et al., (2005) Successful pregnancy following blastocyst cryopreservation using super- cooling ultra-rapid vitrification. Hum Reprod 20:122–128.

[21] Jelinkova et al. (2002). Twin pregnancy after vitrification of 2-pronuclei human embryos. Fertil Steril 77:412-414.

[22] Kuleshova et al. (1999). Birth following vitrification of a small number of human oocytes: case report. Hum Reprod 14:3077-3079.

[23] Kuwayama M, Vajta G, Ieda S, Kato O (2005) Comparison of open and closed methods for vitrification of human embryos and the elimination of potential contamination. Reprod Biomed Online; 11: 608-14.

[24] Kuwayama M (2007) Highly efficient vitrification for cryopreservation of human oocytes and embryos: the Cryotop method. TheriogenologY 67:73–80.

[25] Kyuwa et al. (2003). Experimental evaluation of cross-contamination between cryotubes containing mouse 2-cell embryos and murine pathogens in liquid nitrogen tanks. Exp Anim 52:67-70.

[26] Larman MG, Minasi MG, Rienzi L, Gardner DK (2007)   Maintenance of the meiotic spindle during vitrification in human and mouse oocytes. Reprod Biomed Online 15:692-700.

[27] Levens ED, Whitcomb BW, Henessy S, James AN, Yauger BJ, Larsen FW 2008 Blastocyst development rate impacts outcome in cryopreserved \blastocyst transfer cycles. Fertility and Sterilit, 90, 2138-2143.

[28] Letur-Kornisch et al. (2003). Safety of cryopreservation straws for human gametes or embryos: a study with human immunodeficiency virus-1 under cryopreservation conditions. Hum Reprod 18:140-144.

[29] Li Y, Chen ZJ, Yang HJ, Zhong WX, Ma SY, Li M. 2007 Comparison of vitrification and slow-freezing of human day 3 cleavage stage embryos: post-vitrification development and pregnancy outcomes. Zhonghua Fu Chan Ke Za Zhi. 42:753-5.

[30] Liebermann et al. (2002a). Potential importance of vitrification in reproductive medicine. Biol Reprod 67:1671-1680.

[31] Liebermann et al. (2002b). Blastocyst development after vitrification of multipronucleate zygotes using the flexipet denuding pipette (FDP). RBMOnline 4:146-150.

[32] Liebermann and Tucker (2002). Effect of carrier system on the yield of human oocytes and embryos as assessed by survival and developmental potential after vitrification. Reproduction 124:483-489.

[33] Liebermann et al. (2003). Recent developments in human oocyte, embryo and blastocyst vitrification: where are we now? RBMOnline 7:623-633.

[34] Liebermann and Tucker (2004). Vitrifying and warming of human oocytes, embryos, and blastocysts: vitrification procedures as an alternative to conventional cryopreservation methods Mol Biol 254:345-364.

[35] Liebermann J, Tucker MJ 2006 Comparison of vitrification versus conventional cryopreservation of day 5 and day 6 blastocysts during clinical application. Fertility and Sterility 86, 20-26.

[36] Liebermann J 2009 Vitrification of human blastocysts: an update. Reproductive BioMedicine Online 19:Suppl 4, 105-114.

[37] Liebermann J (2011) More than six years of Blastocyst Vitrification – What is the verdict? US Obstetrics and Gynecology, 5: 14-17.

[38] Loutradi KE, Kolibianakis EM, Venetis CA, Papanikolaou EG, Pados G, Bontis I, Tarlatzis BC 2008 Cryopreservation of human embryos by vitrification or slow freezing: a systematic review and meta-analysis. Fertil Steril. 90:186-93.

[39] Mazur P, Seki S (2011) Survival of mouse oocytes after being cooled in a vitrification solution to -196°C at 95° to 70,000 °C/min and warmed at 610° to 118,000 °C/min: A new paradigm for cryopreservation by vitrification. Cryobiology 62: 1-7.

[40] Mukaida et al. (1998). Vitrification of human embryos based on the assessment of suitable conditions for 8-cell mouse embryos. Hum Reprod 13:2874-2879.

[41] Mukaida T, Nakamura S, Tomiyama T, Wada S, Kasai M (2001) Takahashi K Successful birth after transfer of vitrified human blastocysts with use of a Cryoloop containerless technique. Fertil Steril 2001;76:618 –623

[42] Mukaida T, Takahashi K, Kasai M (2003) Blastocyst cryopreservation: ultrarapid vitrification using Cryoloop technique. Reprod Biomed Online 6:221–225

[43] Noyes N, Porcu E, Borini A. (2009) Over 900 oocyte cryopreservation babies born with no apparent increase in congenital anomalies. Reprod Biomed Online 18:769-76.

[44] Park et al. (2000). Ultra-rapid freezing of human multipronuclear zygotes using electron microscope grids. Hum Reprod 15:1787-1790.

[45] Rall and Fahy (1985). Ice-free cryopreservation of mouse embryos at –196 °C by vitrification. Nature 313:573–575.

[46] Rama Raju GA, Haranath GB, Krishna KM, Prakash GJ, Madan K. (2005) Vitrification of human 8-cell embryos, a modified protocol for better pregnancy rates. Reprod Biomed Online. 2005:11:434-7.

[47] Rama Raju GA, Jaya Prakash G, Murali Krishna K, Madan K. (2009) Neonatal outcome after vitrified day 3 embryo transfers: a preliminary study. Fertil Steril. 92:143-8

[48] Reed ML, Lane M, Gardner DK, Jensen NL, Thompson J (2002) Vitrification of human blastocysts using the Cryoloop method: successful clinical application and birth of offspring. J Assist Reprod Genet 2002;19:304–306.

[49] Selman and El-Danasouri (2002). Pregnancies derived from vitrified human zygotes. Fertil Steril 77:422-423.

[50] Seki S , Mazur P (2009) The dominance of warming rate over cooling rate in the survival of mouse oocytes subjected to a vitrification procedure. Cryobiology 59: 75-82

[51] Shapiro B, Richter K, Harris D, Daneshmand S 2001 A comparison of day 5 and 6 blastocysts transfers. Fertility and Sterility 75, 1126 –1130.

[52] Takahashi et al. (2005) Perinatal outcome of blastocyst transfer with vitrification using cryoloop: a 4-year follow-up study. Fertil Steril 84:88-92.

[53] Tedder, R.S., Zuckerman, M.A., Goldstone, A.H., Hawkins AE, Fielding A, Briggs EM. et al. (1995) Hepatitis-B transmission from contaminated cryopreservation tank. *Lancet* 346, 137-140.

[54] Trapphoff et al. (2010) DNA integrity, growth pattern, spindle formation, chromosomal constitution and imprinting patterns of mouse oocytes from vitrified pre-antral follicles. Human Reproduction 25, 3025-3042.

[55] Tucker et al. (2003). Cryopreservation protocols. In Color Atlas of Human Assisted Reproduction: Laboratory and Clinical Insights, eds. P Patrizio, V Guelman and MJ Tucker, pp 257-276, Lippincott, Williams and Wilkins, Philadelphia.

[56] Vajta et al. (1998). Open pulled straws (OPS) vitrification: a new way to reduce cryoinjuries of bovine ova and embryos. Mol Reprod Dev 51: 53-58.

[57] Van den Abbeel et al. (2005). Slow controlled-rate freezing of sequentially cultured human blastocysts: an evaluation of two freezing strategies. Hum Reprod 20: 2939-45.

[58] Vanderzwalmen P, Bertin G, Debauche Ch, Standaert V, van Roosendaal E, Vandervorst M, et al., (2002) Births after vitrification at morula and blastocyst stages: effect of artificial reduction of the blastocoelic cavity before vitrification. Hum Reprod 17:744 –751.

[59] Vanderzwalmen P, Bertin G, Debauche Ch, Standaert V, Bollen N, van Roosendaal E, et al., (2003) Vitrification of human blastocysts with the hemistraw carrier: application of assisted hatching after thawing. Hum Reprod 18:1501–1511.

[60] Walker et al. (2004). Vitrification versus programmable rate freezing of late stage murine embryos: a randomized comparison prior to application in clinical IVF. RBMOnline 8:558-568.

[61] Yokota Y, Sato S, Yokota M, Ishikawa Y, Makita M, Asada T, et al. (2000) Successful pregnancy following blastocyst vitrification. Hum Reprod 15:1802–1803.

[62] Yokota Y, Sato S, Yokota M, Yokota H, Araki Y (2001) Birth of a healthy baby following vitrification of human blastocysts. Fertil Steril 75:1027–1029.

[63] Yoon et al. (2000). Pregnancy and delivery of healthy infants developed from vitrified oocytes in a stimulated in vitro fertilization-embryo transfer program [letter]. Fertil Steril 74:180-181.

8

Oocyte Cryopreservation for the Elective Preservation of Reproductive Potential

Catherine Bigelow and Alan B. Copperman
Mount Sinai School of Medicine, New York, NY
USA

1. Introduction

Cryopreservation has been a technique used in reproductive endocrinology and infertility medicine since the early 1980s. Embryo cryopreservation, specifically, has been widely used in many in vitro fertilization (IVF) programs worldwide. This method has been well studied and is a common strategy employed for storing supernumerary embryos after IVF cycles, among other applications. Oocyte cryopreservation, which involves cryopreservation of unfertilized human ova, is a newer procedure that is gaining popularity due to its many benefits, including delay of childbearing, fertility preservation for cancer patients, and avoidance of ethical, religious or legal dilemmas surrounding embryo cryopreservation. While this technique is still considered "experimental," oocyte cryopreservation is rapidly gaining acceptance in the field of fertility preservation. The current chapter discusses the multifaceted reasons for delayed childbearing, the applications of oocyte cryopreservation, and technical aspects of this procedure. Additionally, arguments are presented to counter the "experimental" label of oocyte cryopreservation and obstetric and perinatal outcome data are analyzed.

2. The history of human oocyte cryopreservation

The first human pregnancy after cryopreservation and thaw of an 8-cell embryo occurred in 1983 (Trounson & Mohr, 1983). The first reports of mature human oocyte cryopreservation also occurred in the 1980s, with the first live birth after oocyte cryopreservation using a slow-freeze method being reported in 1986 (Chen, 1986). In this early case report, mature oocytes were cryopreserved using a slow-freeze, rapid-thaw method and DMSO was used as a cryopreservant (Figure 1). Chen achieved an egg survival rate of 80%, with an 83% fertilization rate in the thawed surviving oocyte population. This cohort of 40 oocytes ultimately resulted in one viable twin gestation. Despite this promising early work, significant advances in the field of oocyte cryopreservation did not occur until decades later.

Since the first reports of successful egg freezing, there have been many changes and advances in the protocols and techniques utilized to maximize post-thaw success rates. Alterations in cryopreservants and media have been tested and improved in the past three decades. Replacement of sodium with choline in the cryopreservation media has been shown to improve cryopreservation outcomes (Quintans et al., 2002; Stachecki et al., 1998). Alternative strategies, including trehalose injection have also been introduced in attempts to improve survival of cryopreserved oocytes (Eroglu et al., 2000; Jain & Paulson, 2006).

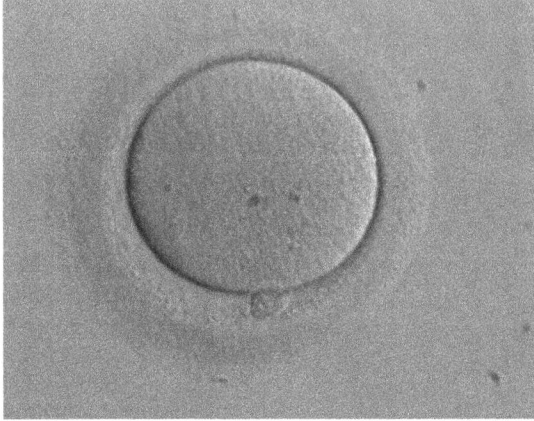

Courtesy of Reproductive Medicine Associates of New York, LLP.
Fig. 1. Mature human oocyte.

The introduction of vitrification led to the first live birth after this technique in 1999 (Kuleshova et al., 1999). Vitrification of embryos has been shown to be reliable (Kolibianakis et al., 2009) and vitrification is now routinely applied to oocytes. The efficiency of both oocyte cryopreservation methods has been studied and shows similar trends in improvement. While more data exists for slow-freeze cryopreservation, due solely to the number of years that this method has been available, vitrification data is equally promising. In a 2006 meta-analysis, Oktay et al. demonstrated live birth rates increasing after slow-freezing from 21.6% per transfer (from 1996 to 2004) to 32.4% (from 2002 to 2004). The vitrification data showed a similar trend, with a live birth rate of 29.4% before 2005 and 39% after 2005 (Oktay et al., 2006). Through the use of intracytoplasmic sperm injection (ICSI), many of the concerns about zona pellucida hardening from cryopreservation were bypassed. This ability to augment fertilization of thawed oocytes altered the outlook on oocyte cryopreservation and made this a more viable option for fertility preservation.

More recently, experimentation with cryopreservation of ovarian tissue through orthotopic or heterotopic transplantation has been attempted. The first child born after ovarian tissue cryopreservation was documented in 2004, to a woman who had a history of chemotherapy and radiation treatment for lymphoma (Donnez et al., 2004). Due to fewer studies on optimal cryopreservation protocols and methods of tissue selection, this technique has been considered more experimental than oocyte cryopreservation. In addition, optimal strategies for enhancing graft revascularization are limited in the literature and ideal tissue size has not been established. In a case series of 13 patients who underwent ovarian tissue cryopreservation due to various diseases requiring chemotherapy, large strips (8-10mm x 5mm) and small cubes (2mm x 2mm) of ovarian tissue were both effective in restoring ovarian function (Donnez et al., 2011). However, due to small numbers of human patients having undergone this procedure and a lack of standardized protocols, these outcomes are difficult to interpret. Furthermore, appropriate candidate selection for ovarian tissue cryopreservation has note been defined. Due to waning primordial follicle counts as women age, it has been suggested that ovarian tissue cryopreservation should be limited, at the very least, to women <40 years of age (Oktay, 2002). Some concern also exists about the risk of ovarian metastasis and the reintroduction of malignant cells upon transplantation of thawed ovarian tissue. Minimal residual disease (MRD) in cryopreserved ovarian tissue of patients with leukemia has been demonstrated in humans, with the prevalence of MRD in chronic myeloid leukemia and acute lymphoblastic leukemia as high as 33% and 70%, respectively (Dolmans et al., 2010). Given these risks, strategies to effectively test cryopreserved ovarian tissue for evidence of MRD are required before this technique can be widely utilized in clinical practice. In light of these uncertainties, ovarian tissue cryopreservation is still in its infancy with regard to fertility preservation. Additionally, immature oocyte cryopreservation is being studied but is also in early experimental stages, according to the American Society for Reproductive Medicine (ASRM) (ASRM Practice Committee, 2008). This potentially new frontier is still being studied in primate models and preliminary human studies are ongoing.

These changes in cryopreservants, rates of freezing, fertilization, and protocols for cryopreservation have improved outcomes. As research funding, referring provider knowledge, and patient interest in oocyte cryopreservation increase, we can anticipate continued advancements in the field of fertility preservation.

3. The significance of human aging

Human aging has been well-studied and is a known contributor to the decline in fertility experienced by women. Female fecundity, or the ability to produce offspring, declines with advancing age. This is partially due to decreased numbers of oogonia, which have a steady rate of atresia from birth, with a more rapid decline around the age of 37.5 years. Numbers of oogonia, or primordial fetal oocytes, are maximal at 20 weeks' gestation, totaling between six and seven million. At birth, this number has already declined to one to two million; a mere 400,000 oocytes remain at the beginning of puberty. While this number still seems rather high, only around 500 of these oocytes are destined for maturation and ovulation. The remainder will be lost through a highly controlled system of follicular atresia and apoptosis (Williams Gynecology, 2008 Ed.), until around 1000 oocytes remain at the time of menopause (Figure 2). Since women now live longer, a larger portion of their lives are spent in reproductive senescence, and the need for reproductive assistance due to challenges associated with diminished ovarian reserve has increased (Faddy et al., 1992).

Recent research has suggested that there may be a population of oogonial stem cells, similar to that seen in males for lifelong spermatocyte production. Several studies have pointed toward the presence of mitotically-active germline stem cells in the mammalian ovary (Johnson et al., 2004; Pacchiarotti et al., 2010; Parte et al., 2011; Zou et al., 2009). Many groups have conducted experiments which have isolated stem cells capable of sustaining oocyte and follicle production *in vitro*. While these results are controversial and disputed by some (Byskov et al., 2005), the potential for regeneration of oocytes and follicular development throughout the female life span is an exciting and promising future area in assisted reproduction.

Mathematical models have been developed in order to generate prediction rules for numbers of remaining oocytes and reproductive capacity. Oocyte atresia appears to follow a bi-exponential pattern, with a more rapid decline in oocyte number occurring after a critical number of 25,000 follicles remain around the age of 37.5 years (Faddy et al., 1992). According to this model, around 1000 follicles remain at the age of 51, which corresponds to the median age of menopause in the general population. Other authors have studied histological samples to identify the rate of recruitment of non-growing follicles (NGF) in human ovaries from prenatal samples through menopause (Wallace & Kelsey, 2010). This model suggests that up to 81% of the variance in non-growing follicles is due to age alone. Interestingly, the authors' mathematical model demonstrates an increased rate of non-growing follicle recruitment until the age of 14 years old, after which NGF recruitment decreases until the menopause. Using this best-fitting asymmetric peak mathematical model, it may be possible to predict ovarian reserve in women based on age and guide discussions of fertility preservation in women seeking information about oocyte cryopreservation.

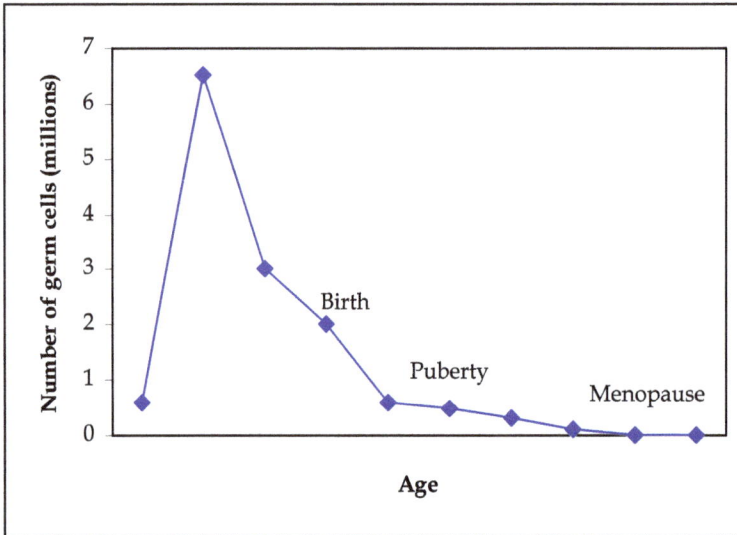

Fig. 2. Number of germ cells across the human female lifespan. Germ cells peak around 6 months post-conceptional age at a level of 6-7 million. At birth, this number has declined to around 2 million germ cells remaining in the infant ovary. Further decline occurs during the rest of the lifespan, with ~500,000 remaining at puberty and only ~1000 oocytes left at menopause.

Though we have some clinical tools to help predict a woman's reproductive capacity, including hormonal tests and the basal antral follicle count, the ramifications of human aging on reproduction are still variable and difficult to predict. Traditionally, elevated levels of basal follicle stimulating hormone (FSH) and abnormal estradiol (E2) levels have been used to guide physicians who are assessing ovarian reserve. FSH is measured in the early follicular phase of the menstrual cycle, when luteal inhibin levels decrease. Classically, it is measured on day 3 after the onset of menses. Studies have shown that a day 3 FSH level above 15 mIU/mL predicts significantly lower rates of pregnancy (Scott, 1995). Concomitant measure of E2 levels may decrease the rate of false negatives when FSH values are used alone. Estradiol should be thus be measured concurrently with day 3 FSH testing. The basal antral follicle count (BAFC) has also been used widely in the field of reproductive endocrinology and infertility to help predict ovarian reserve. BAFC <4 has a specificity of 98.7% when predicting non-pregnancy following IVF (Gibreel et al., 2009). BAFC may therefore be an appropriate measure of ovarian reserve in women undergoing infertility evaluation. Meta-analysis has also shown that BAFC of less than 4 has a sensitivity and specificity to predict cycle cancellation of 66.7% and 94.7%, respectively. Additionally, women with a BAFC of less than 4 are 37 times more likely to have their cycle cancelled (Gibreel et al., 2009).

A newer marker for predicting ovarian reserve is anti-Müllerian hormone, or AMH, which has been in the literature since the early 2000s (Gruijters et al., 2003). Serum AMH levels are constant throughout the menstrual cycle, unlike FSH or E2, and are not affected by other hormone levels. Because of these relatively constant levels, AMH may be useful for predicting ovarian response to stimulation cycles for IVF; its predictive power seems to be similar to that of the BAFC (La Marca et al., 2009). Additionally, AMH is secreted in primary, preantral, and small antral follicles, which are thought to comprise the pool of ovarian reserve (Figure 3). This endocrine marker is secreted by granulosa cells and reflects the transition of resting primordial follicles to growing follicles (Sowers et al., 2008). Additionally, AMH levels diminish as an FSH-dependent dominant follicle begins to develop (Broekmans et al., 2008), reinforcing its role as a marker of preantral and small antral follicles in the pool of ovarian reserve. AMH is not, on the other hand, expressed in atretic follicles. Therefore, its levels are directly correlated to the number of viable, growing follicles that remain in the ovary. Levels of AMH decline in a predictable fashion as women near the menopausal transition, which has been studied in concordance with declining levels of inhibin-B and increasing levels of FSH (thus reinforcing the soundness of this marker as a predictor of declining ovarian reserve) (Sowers et al., 2008). There is a statistical association between AMH and FSH levels in assessing ovarian reserve. Singer et al. compared the correlation between these two hormones and found that serum AMH level is highly predictive of baseline FSH level. Using these two serum marker levels in combination may prove to be a useful predictor of ovarian reserve (Singer et al., 2009).

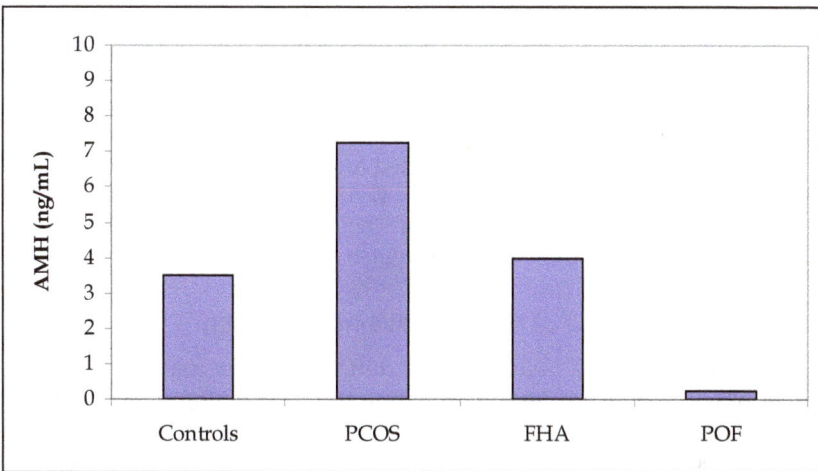

Fig. 3. Mean AMH plasma levels in patients and controls. Women with PCOS have significantly higher levels of plasma AMH and women with POF have significantly lower levels, when compared to controls and women with FHA ($p<0.05$). PCOS: Polycystic Ovarian Syndrome; FHA: Familial Hypothalamic Amenorrhea; POF: Premature Ovarian Failure. *Adapted from Broekmans et al., 2008.*

In our NYC-based infertility clinic, women presenting for new oocyte cryopreservation consultations were retrospectively evaluated. Of the 519 women presenting for new patient consultation, approximately 1/3 initiated oocyte cryopreservation cycles. The best predictors of successful oocyte cryopreservation cycles were (in order) BAFC, day 3 FSH, and age (all p<0.05) (Barritt et al., 2010). Importantly, providers must remember that all of these tests and models attempt to predict the *quantity* of oocytes available for future reproduction. Unfortunately, tests to predict oocyte *quality* are still lacking. Models incorporating multiple variables may end up being the best predictor of ovarian reserve and ART cycle success, though many still consider age the best predictor of ovarian reserve and reproductive potential.

The risk of aneuploidy is increased in older oocytes, which leads to higher rates of chromosomally abnormal fetuses and spontaneous abortion. Approximately 15-20% of pregnancies end in spontaneous abortion, or miscarriage (Barron, 1968). Maternal age has long been recognized as a risk factor for pregnancy loss. Risk of chromosomal abnormalities, decreased fecundity, and prevalence of comorbid medical illnesses rise with increasing age – all of which may lead to spontaneous abortion (Barron, 1968). Aneuploidy is thought to affect around 20% of human oocytes (Jones, 2008). Some hypothesize that rates of aneuploidy increase with age through a "two-hit" pathway: nondisjunction followed by an inability of the oocyte to detect the chromosomal abnormality. Nondisjunction, or inappropriate chromosomal separation during meiosis I, is a leading cause of aneuploidy and increases with maternal age. Oocytes from older women may have decreased cohesive bonds between chromosomes, further predisposing them to meiotic errors (Jones, 2008). Additionally, as oocytes age, they may be unable to detect errors in recombination and sister chromatid separation.

It has been well-documented that infertility rates increase with age and that reproductive aging is primarily related to oocyte age. One prospective study demonstrated infertility rates increasing from 8% in women aged 19-26 years to 13-14% in women aged 27-34 years, and ultimately to 18% for women aged 35-39 years (Dunson, 2004). Similarly, there is a decline in success rates of fresh-cycle, non-donor oocyte IVF as a woman ages. Live birth rates per embryo transfer have been documented around 47.5% for women <35 years old, with a progressive decline to 17.0% in women 41-42 years of age, according to 2009 data from the Society of Assisted Reproductive Technologies (SART) (SART, 2009). In light of this data, strategies to preserve fertility for young women are paramount.

4. Changing demographics of reproductive-aged women

In the United States, there has been a notable shift in the demographics of reproductive-aged women. Many women are delaying childbearing in the setting of career pursuits and shifting societal expectations of gender roles. An analysis of Danish fertility rates from 1980-2001 showed an increase in the mean age of childbearing of 3 years over the 21 year period (Hvidtfeldt et al., 2010). There appears to be a global shift in reproduction leading to delayed childbearing and increased maternal age.

Reasons for delaying childbearing are multifaceted and complex. Perceived career threats are a very real and prevalent issue in young women of reproductive age (Willett et al., 2010). For women in professional programs, fear about extension of training, loss of future career opportunities and concern for pregnancy complications are all significantly higher than in men. These concerns lead to choosing between career training and childbearing, thus risking subfertility by delaying reproduction for the sake of a woman's profession. Studies at our center have evaluated motivations for and trends in elective preservation of fertility in women seeking care at a New York City infertility clinic. Women seeking elective egg freezing were likely to have a high level of education, with all women having at least a bachelor's degree and 75% holding a master's or professional degree. These women were all single, nulliparous, and the majority expressed a desire to be sure they had taken advantage of all reproductive opportunities (Gold et al., 2006). Half of women interviewed described being pressured by their "biological clock" and many wanted to freeze eggs as an "insurance policy," though did not anticipate needing to use them. Interestingly, the mean patient age was 39 years old and 65% of women had reported only recently learning about egg freezing technology. In a multicenter analysis, more than 3000 women called to inquire about fertility preservation. Of these women, those who actually completed a cycle had a significantly higher average age of 37.1 years; patients who were older than 35 had fewer cycles that resulted in the recommended number of metaphase II oocytes for cryopreservation (Frank Sage et al., 2008). This may suggest an inadequate awareness of the age-related decline in fertility that occurs as part of normal human aging. Most studies on reproductive outcomes after oocyte cryopreservation (including oocyte survival rates, fertilization rates, and number of pregnancies) have analyzed women under the age of 35 (Jain & Paulson, 2006). Because of this limitation in the body of literature on oocyte cryopreservation, providers should ideally cryopreserve oocytes in women <35 years of age. As oocyte cryopreservation becomes more publicized and accurate information about declining female fertility is disseminated, the mean age of cryopreservation may decrease.

Trends in the local and national economy have been studied in relation to elective medical procedures, including oocyte cryopreservation. Costs of oocyte and embryo cryopreservation have been evaluated through the LIVESTRONG database of 154 participating reproductive centers. For the average patient, the cost of oocyte cryopreservation is around $7,800, compared to an average of $9,300 for embryo cryopreservation (Beck et al., 2010). The costs of fertility preservation are variable based on geography and center. In a New York City private IVF program, annual per capita income showed significant positive correlation with new consults for oocyte cryopreservation. Additionally, as annual unemployment rates increased, the number of new consults significantly decreased (Flisser et al., 2009).

Oocyte cryopreservation has many social and ethical advantages over embryo cryopreservation. Embryo cryopreservation remains the standard recommendation for fertility preservation according to ASRM guidelines, mainly due to the amount of literature studying this technique. Single women, however, may encounter social issues with freezing embryos. The option to extend fertility without the need for a male partner or sperm donor is frequently appealing to women who are not in a long-term relationship. The discomfort of anonymity associated with sperm donors is eliminated with egg freezing. Other potential issues include decisions regarding paternity and legal obligations for patients who undergo directed sperm donation, strategies for disposing of embryos if a woman gets married later in life, and how to handle the disposition of embryos if the egg donor dies and does not have explicit advanced directives in place (Jain & Paulson, 2006). These dilemmas are all circumvented with oocyte cryopreservation. Additionally, infertility centers avoid the often difficult task of synchronizing cycles between oocyte donors and recipients, in the case of third party reproduction (Oktay et al., 2010). Improvement in coordination of care, costs, and the ability to quarantine oocytes for infectious disease testing are benefits of oocyte cryopreservation for egg donors.

Fertility preservation for cancer patients undergoing potentially sterilizing chemotherapy and radiation has been a widely accepted application of oocyte cryopreservation. Management of all of the gynecologic cancers has the potential to affect ovarian reserve. Cervical cancer often requires pelvic radiation and endometrial cancer is frequently treated with hysterectomy and bilateral salpingoophorectomy. Therapy for breast cancer, the most common cancer in women in the United States, commonly utilizes cyclophosphamide, which has well-known ovary-toxic effects and leads to premature ovarian failure (Oktay & Sönmezer, 2007). Ovarian stimulation is necessary for both oocyte and embryo cryopreservation for these patients; stimulation protocols have been developed to avoid excessive estrogen exposure in women with estrogen-responsive cancers. For patients who do not need to immediately initiate chemotherapy (or other therapies that may affect the ovary), cryopreservation is a viable option for fertility preservation. In a retrospective data analysis of a NYC infertility clinic from 2005-2007, women presenting for pre-cancer treatment oocyte cryopreservation cycles were evaluated. The average time between initial consultation and completion of the cryopreservation cycle was 37.2 ± 22.5 days, and a mean number of 17.8 oocytes were retrieved across the 4 patients studied (Barritt et al., 2008). Early referral to a fertility center is vital, as patients will require 2 weeks of stimulation after menses in order to retrieve oocytes for cryopreservation. Many oncologists are supportive of their patients' desire to preserve fertility, even in light of the potential delay of chemotherapy and need for gonadotropin stimulation. Women who require immediate initiation of chemotherapy or pediatric cancer patients may benefit from ovarian tissue cryopreservation, though studies of this technique are still quite small and this strategy has not yet been widely used (Oktay & Sönmezer, 2007).

Reproductive endocrinologists approaching the patient interested in elective fertility preservation need to recognize the demographic shifts and societal attitudes toward oocyte cryopreservation. The wide variety of applications of oocyte cryopreservation, including delayed childbearing, ethical opposition to embryo cryopreservation, improvement in third party oocyte donation and fertility preservation for cancer patients, all highlight the advantages of this emerging reproductive technology.

5. Technical aspects of oocyte cryopreservation

Oocyte cryopreservation is a delicate and complex process. Mammalian cells are generally stored at a temperature of -196°C, at which no biological activity takes place. Cryopreservation must transform human oocytes from a biologically active system at 37°C to an inert structure at -196°C; oocytes are most vulnerable during this temperature transition. Membrane permeability and kinetics vary throughout the developmental cycle of the oocyte; metaphase II oocytes have demonstrated higher post-cryopreservation survival in a mouse model (Gook & Edgar, 2007). There are three main goals of oocyte cryopreservation: avoidance of ice crystal formation, avoidance of solution effect, and avoidance of osmotic shock (Jain & Paulson, 2006). As water freezes and expands to form ice, *crystal formation* causes shearing forces on organelles and increases intracellular pressure. Additionally, as water transitions from its liquid to solid form, any solutes dissolved in liquid water are excluded from the ice. This can lead to very high, if not toxic, levels of non-liquid solutes and electrolytes, known as *solution effect*. Further damage to intracellular proteins can occur in the presence of these toxic levels of intracellular substances during cryopreservation. Finally, *osmotic shock* can occur in the setting of rapid rewarming, during which rapid free water shifts lead to cell shrinking and swelling to accommodate alterations in extracellular osmotic pressure. These three goals are achieved through the use of different cryoprotectant chemicals. Cryoprotectants facilitate oocyte cryopreservation by generating an osmotic gradient by which water can exit the oocyte. Permeating cryoprotectants are able to enter the oocyte, thereby preventing cell shrinkage during osmosis of water to the extracellular space.

Two protocols for oocyte cryopreservation exist, slow-freeze methods and vitrification. While slow-freezing is the most widely used and has been studied more in the literature, recent studies in embryos suggest that vitrification may have improved post-thaw survival rates, though it is still not clear whether there are significant differences in clinical pregnancy rates. These methods are discussed here and are analyzed in light of recent evidence of comparative efficacy. In addition, methods for ovarian tissue cryopreservation are briefly discussed.

5.1 Slow freeze

Slow freezing has traditionally been the more widely-used technique for mature oocyte cryopreservation. This technique was first described in 1972 by Whittingham et al. after successful slow freeze and post-thaw survival of mouse embryos (Whittingham et al., 1972). The technology was first applied to human embryos in 1983, and resulted in successful post-thaw survival and pregnancy after cryopreservation (Trounson & Mohr, 1983), followed by live birth after mature oocyte cryopreservation in 1986 (Chen, 1986).

Slow freeze cryopreservation is achieved using initial low cryoprotectant concentrations to reduce toxicity while the oocyte is still metabolically active (Jain & Paulson, 2006). The temperature is lowered gradually, at rates between 0.3-2°C/minute. This slow rate of cooling allows retardation of the metabolic rate in the oocyte without accumulating toxic levels of cryoprotectant. Propanediol (PROH) and dimethylsulfoxide (DMSO) are permeating cryoprotectants which form hydrogen bonds with intracellular water molecules and prevent ice crystal formation, thus achieving the first goal of successful cryopreservation. Additionally, the presence of PROH dilutes electrolyte concentrations by remaining in solution (due to its low freezing point); this prevents solution effect, which is the second goal of cryopreservation. PROH is preferred to DMSO as a cryoprotectant, as it is thought to be less toxic to the oocyte (Renard & Babinet, 1984). Additionally, using 0.2-0.3M sucrose as a nonpermeating cryoprotectant during oocyte dehydration seems to improve post-thaw survival (Fabbri et al., 2001). "Seeding" the extracellular solution with an ice crystal occurs around -6°C, during which an ice front grows and excludes solutes, thereby increasing their concentration around the oocyte. This ice front can potentially cause intracellular damage to the oocyte if it comes in contact with the cell or can lead to gas bubble formation (Ashwood et al., 1988). The oocyte is maintained at -6°C for 10 to 30 minutes before being further cooled to -32°C. At this point, metabolic activity in the oocyte is extremely low and the cell is plunged into a Dewar vessel of liquid nitrogen to vitrify any remaining cryoprotectant solution (Figure 4). The Dewar vessel is capable of maintaining a near constant temperature for the frozen oocytes during storage.

Fig. 4. Liquid nitrogen Dewar vessel, used for freezing and storing cryopreserved oocytes. *Courtesy of Reproductive Medicine Associates of New York, LLP.*

Thawing of embryos occurs at a rate of 4-25°C/minute. A relatively rapid temperature transition is needed to prevent recrystallization of water in the cell. Nonpermeating cryoprotectants, such as sucrose or other disaccharides, are utilized to help prevent osmotic shock during thawing, as high levels of permeating cryoprotectants are present intracellularly (Jain & Paulson, 2006). This helps achieve the third goal of cryopreservation (Figure 5).

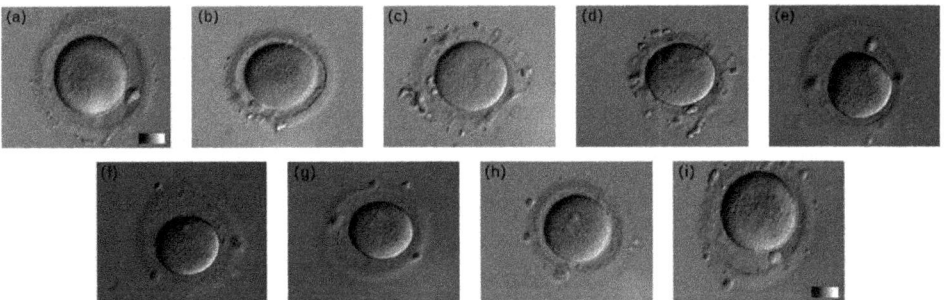

Fig. 5. Oocyte volume changes during freezing and thawing with slow-freeze protocol. (a) De-cumulated oocyte. (b)-(d) Oocyte undergoing freezing stages. (e)-(h) Oocyte undergoing thawing stages of slow-freeze protocol. (i) Oocyte after cryopreservation. *Images courtesy of Herrero et al., 2011.*

Slow freezing has limitations. First, this method is expensive and requires programmable freezing equipment that must be purchased by the IVF laboratory. This poses a substantial cost to many centers. Additionally, this method is extremely time-consuming, taking embryologists at least 90 minutes to successfully cryopreserve oocytes. Despite these drawbacks, this technique is still the most widely used and has the most literature available about tested protocols and outcomes.

5.2 Vitrification

Vitrification, which literally means "the act or process of converting into glass," is an alternative method to slow freeze cryopreservation. This technique uses high concentrations of cryoprotectants and a rapid cooling rate to convert liquid intracellular water directly into a glassy, vitrified state.

This method of oocyte cryopreservation was first described in humans in 1986 (Fahy et al., 1986), with the first live birth after vitrification occurring in 1999 (Kuleshova et al., 1999). Oocytes are directly exposed to liquid nitrogen which practically eliminates ice crystal formation, due to the rapid cooling rate, around 20,000°C/minute. This minimizes the risk of physical damage to the oocyte from shearing of organelles or increased intracellular pressure. The oocyte is converted rapidly into an amorphous state (Figure 6). A plastic straw containing cryoprotectants and the oocyte is directly plunged into the liquid nitrogen. While initial studies of these "cryoprotectant cocktails" found them to be incredibly toxic, extensive evaluation has indicated that the combination with minimal toxicity is a combination of a high concentration of ethylene glycol (5.5M) and sucrose (1.0M) (Ali & Shelton, 1993). Further modification of the cryoprotectant protocols has decreased the concentration of ethylene glycol to 5.0M (Kuwayama et al., 2005a). Other groups have had high success with vitrification using 2.5M ethylene glycol, 0.5M sucrose and 2.1M DMSO (Gook & Edgar, 2007). These changes in methodology have led to continued improvement in vitrification outcomes, including improved oocyte post-thaw survival, fertilization rates, and pregnancy outcomes.

Some studies have reported the potential for disease transmission, especially viral illnesses, through direct contact with contaminated liquid nitrogen using open-carrier systems for vitrification (Bielanski et al., 2000, 2003), in which there is direct contact between the cryoprotectant media and liquid nitrogen. Closed-carrier system vitrification, in which oocytes are not in direct contact with liquid nitrogen, have been shown to have similar blastocyst survival, pregnancy rates, and live birth rates as open-carrier systems (Kuwayama et al., 2005b), without the theoretical risk of horizontal viral transmission (Bielanski et al., 2000). Closed-carrier systems cool at a slower rate (around 200°C/minute) but have similar rates of post-thaw embryo development, and may demonstrate similar efficacy (Jain & Paulson, 2006).

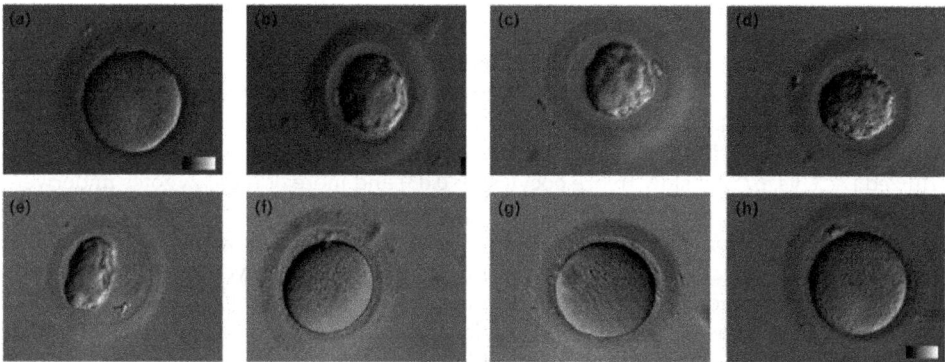

Fig. 6. Oocyte volume changes during vitrification and thaw. (a) De-cumulated oocyte before cryopreservation. (b)-(e) Oocyte undergoing vitrification. (f)-(g) Oocyte during warming phase of vitrification protocol. (h) Oocyte after cryopreservation. *Images courtesy of Herrero et al., 2011.*

Vitrification has its own drawbacks. This method, too, is very expensive for IVF centers to implement in terms of costs of freezing and thawing media. Additionally, this technique has a high learning curve, which must be considered. On the other hand, vitrification does not take as much time as slow freezing due to the rapid cooling procedure and does not require expensive embryology lab equipment. As vitrification continues to be used and data accrued about the success of this method, it is likely to alter the choice of cryopreservation protocols worldwide.

5.3 Slow freeze versus vitrification for oocyte cryopreservation

Slow freezing reports first began 13 years before literature on vitrification emerged. Both of these methods have demonstrated increasing efficiency over time, with continually improving live birth and ongoing pregnancy rates per transfer (Oktay et al., 2006). Although there is a lag in the data for vitrification outcomes, the number of babies born after vitrification is approaching that of slow freeze methods for oocyte cryopreservation (Noyes et al., 2009).

In a recent meta-analysis of randomized controlled trials (RCTs) comparing these two methods, vitrification was found to have better post-thawing survival rates for cleavage stage embryos (odds ratio [OR] 6.35, 95% confidence interval [CI] 1.14, 35.26) and for blastocysts (OR 4.09, 95% CI 2.45, 6.84) (Kolibianakis et al., 2009). A significantly higher number of embryos cryopreserved in the cleavage stage developed into blastocysts following vitrification. Clinical pregnancy rates, however, demonstrated no significant difference between slow freeze and vitrification protocols. This meta-analysis was undertaken to evaluate and summarize the available evidence for cryopreservation of human embryos, not oocytes. Additionally, the data amassed for this meta-analysis came from only 6 RCTs, only one of which commented on live birth rates. The authors, in light of this limited data, call for well-designed randomized controlled trials to further study differences between and advantages or disadvantages of these cryopreservation techniques.

Recently, a prospective randomized comparison of slow freeze versus vitrification for mature human oocyte cryopreservation was performed in Brazil (Smith et al., 2010). In this study, women with supernumerary oocytes retrieved (more than nine) were consented and randomized to either slow freeze or vitrification of these supernumerary oocytes. Demographic characteristics between the two groups of women were similar, including patient age, baseline laboratory values, and number of oocytes collected. Semen parameters were also similar between the groups and all oocytes were inseminated by intracytoplasmic sperm injection (ICSI). Oocyte survival after thawing was significantly higher in those having undergone vitrification. Additionally, a higher percentage of vitrified oocytes were fertilized (77% vs. 67% of slow freeze oocytes; $p<0.03$) and more of these zygotes underwent cleavage from day 1 to day 2 (84% vs. 71%, respectively; $p<0.01$). Perhaps the most important outcome for any assisted reproductive technology, however, is the rate of pregnancy. Biochemical and clinical pregnancy rates per thaw cycle were significantly higher in the vitrification group compared to the slow freeze group (46% vs. 17% and 38% vs. 13%, respectively; $p<0.01$ and $p<0.02$) (Table 1). Additionally, the two groups had similar rates of spontaneous abortion following embryo transfer. Perinatal outcomes were not evaluated by these authors. From case reports evaluating live births following oocyte cryopreservation, the average gestational age at delivery for slow freeze was 36.9 weeks, compared to 39 weeks' gestational age at delivery after vitrification (Noyes et al., 2009). This data suggests improved efficiency and a clinical advantage of oocyte vitrification for elective fertility preservation. Reproductive endocrinologists should be aware of this recent data when considering the implementation of oocyte cryopreservation into their clinical practice and when counseling patients seeking fertility preservation.

	Slow-Freeze	Vitrification	p-value
Number of cycles	30	48	NA
Oocytes thawed	238	349	NA
Oocytes thawed per treatment (mean ± SE)	7.9 ± 0.5	7.3 ± 0.3	NS
Immediate post-thaw survival (%)	159/238 (67%)	281/349 (81%)	<0.001
4-hour post-thaw survival (%)	155/238 (65%)	260/349 (75%)	<0.01
Fertilization (%)	104/155 (67%)	200/260 (77%)	<0.03
Cleavage from Day 1 to Day 2	74/104 (71%)	168/200 (84%)	<0.01
Biochemical pregnancies per cycle (%)	5/30 (17%)	22/48 (46%)	<0.01
Clinical pregnancies per cycle (%)	4/30 (13%)	18/48 (38%)	<0.02
Clinical pregnancy per oocytes thawed (%)	4/238 (1.7%)	18/349 (5.2%)	<0.03

NA = not applicable; NS = not significant. SE = standard error. *Adapted and reproduced with permission from Smith et al., 2010.*

Table 1. Oocyte survival and function following slow-freeze or vitrification for cryopreservation

5.4 Ovarian tissue cryopreservation

Ovarian tissue cryopreservation (oophoropexy) and transplantation can also be considered for female children who will survive childhood cancers but have potentially sterilizing chemotherapy and/or radiation. Ovarian tissue cryopreservation was first described using a sheep model (Gosden et al., 1994). After oophorectomy, strips of ovarian cortex were cryopreserved using a slow-freeze protocol with DMSO. Ovarian tissue was cooled to -140°C before being plunged into liquid nitrogen and stored for 3 weeks. Tissue was thawed and grafted back into the same animal after removal of the remaining ovary, after which animals were returned to the pasture and normal husbandry conditions. This protocol has been followed in human studies of ovarian tissue cryopreservation (Donnez et al., 2011). After thawing, decortication of the patient's atrophic ovaries occurs before transplantation of cryopreserved tissue (Donnez & Dolmans, 2009). Return of ovarian function appears to occur between 3.5-6.5 months after transplantation, as evidenced by an increase in E2 and decreased basal FSH levels. In a small case study, the duration of ovarian activity after transplantation appears to be about 2-5 years (Donnez et al., 2011). Heterotopic transplantation of fresh ovarian tissue to the forearm has been successful in 2 cancer patients with return of ovarian function (Oktay et al., 2001). Forearm heterotopic transplantation of cryopreserved ovarian tissue has been successful in primates (Schnorr et al., 2002), and preliminary studies of this technique in humans are ongoing.

6. Is oocyte cryopreservation "experimental?"

The American Society for Reproductive Medicine (ASRM) published a committee opinion in 2008 which stated that "the experimental nature of oocyte cryopreservation suggests potential for clinical application [...] it might therefore be acceptable [...] with appropriate informed consent under the auspices of an IRB" (ASRM Practice Committee, 2008). This "experimental" label was first published by ASRM in 2006. Some studies have looked at provider compliance with this ASRM practice guideline and likelihood of referral for oocyte cryopreservation. In a retrospective study of 530 IVF centers in the United States, 69% of these centers (365/530) were found to offer oocyte cryopreservation. Of these centers, only 62% do so under IRB approval, while 15% reported having an IRB pending and 18% did not use an IRB at all for oocyte cryopreservation (Beck et al., 2009). Compliance with ASRM guidelines was highest in the northeast (71%) and the size of the program was inversely related to the likelihood that oocyte cryopreservation occurred in conjunction with IRB approval. Still, these numbers indicate relatively high compliance with ASRM guidelines. In a different survey of healthcare providers at 5 United States IVF centers, physician preferences and recommendations were analyzed for practice patterns regarding oocyte cryopreservation. More than half of providers considered the ideal age for oocyte cryopreservation to be less than 35 years and 50% found it acceptable for a woman to preserve fertility in this way with a day 3 FSH value of <13 IU/L. A large proportion of providers were less likely to recommend egg freezing to patients with a low BAFC. Additionally, 89% of physicians were more likely to offer oocyte cryopreservation to their patients if there was a medical indication for the procedure, instead of elective reasons for fertility preservation (Luna et al., 2008). Providers recognized the emerging role oocyte cryopreservation will have in the field of fertility preservation. Thus, despite current reservations regarding which patients to refer for oocyte cryopreservation and ASRM guidelines, physicians view oocyte cryopreservation as a technique that will continue to be used with increasing frequency.

Discussion about the safety and efficacy of oocyte cryopreservation has focused on potential concern about meiotic spindle interruption from freezing, hardening of the zona pellucida (which may decrease rates of fertilization), and the potential risk of anomalies and abnormalities that may arise in the setting of a new technique without much outcome data. The meiotic spindle is a dynamic structure that forms during mitosis and meiosis to facilitate chromosomal segregation. Disruption of the meiotic spindle increases the risk of aneuploidy. These concerns were studied by Rienzi et al. by slow freezing oocytes and looking at the meiotic spindle using computer-assisted polarization microscopy (Rienzi et al., 2004). This technique allowed visualization of the spindle in real time by evaluating living oocytes. Previous studies had used electron microscopy or immunocytochemistry, which requires cell fixation and does not permit evaluation of dynamic spindle activity (reviewed in Eichenlaub-Ritter et al., 2002). Though spindles disappeared in oocytes during the thawing process, all surviving post-thaw oocytes were noted to have intact, functional meiotic spindles. Thus, it appears that cryopreserved oocytes are capable of reforming the meiotic spindle apparatus after thawing (Noyes et al., 2010). Hardening of the zona pellucida (ZP, the transparent glycoprotein envelope that surrounds a mature mammalian oocyte) is thought to occur due to premature cortical granule release during cryopreservation (Jain & Paulson, 2006). This release leads to early hardening of the ZP, which impedes penetration and fertilization by sperm. The advent of ICSI in 1992 introduced a solution to ZP hardening, in which the zona is bypassed by direct injection of the sperm into the oocyte. Additionally, vitrification of oocytes in calcium-free media appears to reduce zona pellucida hardening and leads to increased fertilization. Embryos obtained from cryopreserved oocytes have a similar incidence of chromosomal abnormalities when compared to control embryos using fluorescence in situ hybridization (FISH) (Cobo et al., 2001). Multiple recent studies have evaluated pregnancy, live birth, and early childhood outcomes in children born after mature oocyte cryopreservation (Borini et al., 2007; Chian et al., 2008a, 2008b; Noyes et al., 2009; Oktay et al., 2006; Wennerholm et al., 2009); these have not documented an increased rate of congenital anomalies among children born after oocyte cryopreservation. These studies are discussed in more detail in the next section.

In light of these concerns about cryopreservation, the ASRM maintains that oocyte cryopreservation should be considered an experimental procedure. The ASRM specifically states that assisted reproductive technology (ART) procedures should be considered "experimental" until "the published medical evidence regarding their [...] overall safety and efficacy is sufficient to regard them as standard medical practice. [This] medical evidence can derive only from appropriately designed, peer-reviewed, published studies performed by multiple independent investigators" (ASRM Practice Committee, 2009). Other authors have supported this statement, by noting that "because the largest demand for oocyte cryopreservation most probably is going to come from women who wish to delay childbearing electively, it is quite likely that several years will be required before sufficient births have occurred to determine the true safety of cryopreserved oocytes" (Jain & Paulson, 2006). Some authors, however, argue against this labeling of oocyte cryopreservation, stating that a variety of commonly used assisted reproductive technologies have never been studied "under the auspices of an IRB" before implementation into standard practice (Noyes et al., 2010).

The safety of oocyte cryopreservation has been evaluated through studies of pregnancy, perinatal, and childhood outcomes, in which over 900 infants have been evaluated (Chian et al., 2000b; Noyes et al., 2009). There does not seem to be an increased risk for adverse pregnancy outcomes or congenital anomalies in pregnancies conceived after oocyte cryopreservation, thaw, fertilization and embryo transfer. Additionally, cryopreservation may introduce an extra safety measure with regard to quarantine for infectious disease, similar to protocols in place for cryobanking of donor sperm. By freezing donated oocytes, additional infectious disease testing can be done months after oocyte retrieval to ensure optimal embryo transfer and pregnancy outcome. In an early 2007 paper by Barritt et al., 4 oocyte donors underwent synchronous ovarian stimulation with 4 recipient patients with impaired ovarian reserve, elevated basal FSH, and prior unsuccessful IVF treatments (Barritt et al., 2007). The donors were given a complete medical examination in accordance with ASRM guidelines for oocyte donors, which included a full history, physical exam, BAFC, and cervical cultures. In addition, these women had serological testing for infectious diseases, including HIV, hepatitis B and C, syphilis, gonorrhea, cytomegalovirus, and a urine drug screen. While this initial workup for oocyte donors seems exhaustive, additional checkpoints for infectious disease testing after an extended period of cryopreservation will further prevent the spread of communicable disease and improve pregnancy outcomes by preventing congenital infections. Data continue to emerge supporting the safety and efficacy of oocyte cryopreservation and it is likely that the "experimental" label will soon be removed from this technique.

Multiple studies from different investigators and institutions have compared the efficacy of oocyte cryopreservation to fresh oocyte cycles. An early IRB-approved prospective study of four donor-recipient oocyte cycles by Barritt et al. demonstrated high pregnancy and implantation rates following slow-freezing and overnight storage before thawing. After ICSI, the authors demonstrated an 89.7% fertilization rate and 91.8% of these fertilized oocytes cleaved normally. Of 23 transferred embryos, 26.1% implanted and 75% of implanted embryos led to clinical pregnancy (Barritt et al., 2007). Cobo et al. performed a study in which fresh oocytes from the same donor were either inseminated directly or vitrified for at least 1 hour before thaw and insemination (Cobo et al., 2008). In comparing embryo quality and clinical outcomes, they found that vitrified/thawed oocytes produced embryos capable of a 47.8% ongoing pregnancy rate, which was similar to fresh oocytes. In addition, Grifo and Noyes performed an age-matched control study of 23 oocyte cryopreservation cycles and fresh oocyte control cycles. Fertilization rates, blastocyst formation, and pregnancy rates were not significantly different between these two matched groups (Grifo & Noyes, 2010). This indicates that frozen/thawed oocytes perform as well as fresh oocytes in ART procedures. Finally, Nagy et al. demonstrated high efficiency of egg cryobanking, with a 55% implantation rate and delivery of 26 live infants. Furthermore, their study showed that twice-frozen gametes (i.e. oocyte cryopreservation followed by fertilization and supernumerary embryo vitrification) can lead to pregnancy after embryo thawing (Nagy et al., 2009). The efficacy of oocyte cryopreservation has thus been established by multiple independent groups in the literature, strengthening the argument to remove its experimental status.

For female cancer patients, treatment regimens of intensive chemotherapy, ionizing radiation, and bone marrow transplantation can lead to premature ovarian failure, with direct impact on the number and viability of remaining oocytes. Gonadotropin-releasing hormone (GnRH) analogues have been studied as a method for fertility preservation before cytotoxic treatments. By suppressing ovarian function and, essentially, rendering the ovary quiescent, it is thought that chemotherapeutics and radiation would not be able to affect post-treatment ovarian function. Unfortunately, this strategy does not have well-documented efficacy in the literature (Maltaris et al., 2009). Additionally, studies are lacking that have documented resultant oocyte and embryo quality following a course of chemotherapy (ASRM Practice Committee, 2008). Consequently, oocyte or ovarian tissue cryopreservation may be more reliable methods of fertility preservation for female cancer patients.

Finally, it is important to consider the ethical dilemmas of embryo cryopreservation that are bypassed by using oocyte cryopreservation. These issues, while not directly related to safety and efficacy of oocyte cryopreservation, provide additional support for arguments about the importance of this method to avoid the moral impasses generated by embryo cryopreservation and storage. Embryo cryopreservation has legal implications worldwide. Ovarian stimulation cycles and IVF procedures frequently lead to supernumerary cryopreserved embryos. Over 400,000 embryos are currently stored in the United States alone (Hoffman et al., 2003), leading to high rates of embryo abandonment in IVF clinics. The issue of embryo disposal versus continued cryopreservation is one which IVF clinics deal with daily.

7. Long-term pregnancy and health outcomes after oocyte cryopreservation

Given the evolving nature of the technology and the heterogeneity of patient-population and cryopreservation techniques, the actual "success rate" of egg freezing is unknown. Review of the literature, however, suggests that the efficiency of cryopreservation appears to be improving. One analysis of slow-freezing demonstrated improvement in live birth rates from 21.6% per transfer from 1996 to 2004 to 32.4% from 2002 to 2004 (Oktay et al., 2006). Vitrification data shows a similar trend of improvement: 29.4% live birth rate before 2005 versus 39% after 2005. In a study out of McGill University Health Center, 38 women underwent ovarian stimulation and vitrification of retrieved oocytes. After cryopreservation for one full menstrual cycle, there was an 81% thaw survival rate, 75.6% of oocytes were successfully fertilized, and a 50% pregnancy rate per cycle started was achieved (Chian et al., 2008a). Ultimately, 39.5% of women who initiated ovarian stimulation and cryopreservation cycles gave birth to live infants. Nine of these births were singleton, while the remaining six deliveries were multiples (five twins and one triplet). While initial high rates of spontaneous abortion were documented after oocyte cryopreservation (Borini et al., 2004), these rates have declined with a corresponding increase in live birth rates.

Since the early 2000s, studies have begun reporting pregnancy and neonatal outcomes following oocyte cryopreservation. Borini et al. reported 13 children born after slow freeze cryopreservation in 2004. All babies born were found to have a normal karyotype and no malformations were seen in their study group. They did note, however, a 20% spontaneous abortion rate in their cohort of patients who had undergone oocyte cryopreservation cycles (Borini et al., 2004). In a later study out of Italy, 149 pregnancies occurred after using a slow freeze protocol for oocyte cryopreservation. This group, again, had a relatively high spontaneous abortion rate of 23.5% (Borini et al., 2007). Reports of live births following oocyte cryopreservation have also emerged from groups in China. Chian et al. found that neonates born after ovarian stimulation and oocyte vitrification were all appropriate birthweights, none weighed <2500g. Additionally, all singletons in their cohort were born at term, with a mean gestational age at delivery of 39 1/7 weeks (Chian et al., 2008a). The same group analyzed 165 pregnancies resulting in 200 babies born after vitrification of oocytes at three centers. In their study, multiple gestations were more likely to deliver in the late preterm period (between 34-37 weeks' gestation) – 57% vs. 22% of singleton pregnancies. This is consistent with current expectations for multiple gestations in the general population. Additionally, 74% of multiples in their study were low birth weight (LBW, <2500g), with 5% of the cohort being very low birth weight (VLBW, <1500g). This was in comparison to singleton neonates born after vitrification, only 17% of which were LBW and 0.7% were VLBW. These birth weights are not significantly different when compared to women who spontaneously conceived or had fresh IVF (Chian et al., 2008b). Birthweight was also analyzed in a systematic review of pregnancy outcome data after oocyte cryopreservation and found to be consistently within normal limits (Wennerholm et al., 2009).

Some concerns have been raised about the rate of malformation or congenital anomalies seen in babies born after any assisted reproductive technology. Epigenetic syndromes (such as Beckwith-Weidemann Syndrome and Angelman Syndrome) have been reported as more common, specifically after ICSI (Noyes et al., 2009). With regard to egg freezing, of 105 babies studied by Borini et al. in 2007, only 2 malformations were seen; one infant was born with choanal atresia and the other with Rubenstein-Taybi syndrome (Borini et al., 2007). Chian et al. analyzed rates of malformations in their 2008 cohort of 200 babies born after vitrification. Overall, only 5 birth defects were noted, for a malformation rate of 2.5% (Chian et al., 2008b). This rate is consistent with that seen in spontaneously conceived pregnancies and those following fresh IVF (Tan et al., 1992). In the Chian study, 2 ventricular septal defects (VSD), 1 case of biliary atresia, 1 club foot and 1 skin hemangioma were described in neonates. In their systematic review of the literature, Wennerholm et al. found that children who underwent karyotype analysis after oocyte cryopreservation were all within normal limits (Wennerholm et al., 2009).

The largest study to date of congenital anomalies following oocyte cryopreservation was published in 2009 by Noyes, Porcu and Borini. In this literature review, the authors identified 936 infants born after oocyte cryopreservation. In this worldwide population of infants, only 12 of 936 had either a major or minor congenital anomaly, for a malformation rate of 1.3% (Noyes et al., 2009). Defects seen included 3 VSD, 3 clubfoot, 1 choanal atresia, 1 biliary atresia, 1 Rubenstein-Taybi syndrome, 1 Arnold-Chiari syndrome, 1 cleft palate, and 1 skin hemangioma; some of these defects have already been discussed from earlier studies (Borini et al., 2007; Chian et al., 2008b). No difference in rates of major or minor congenital anomalies was found when compared to the United States birth outcome data from the Centers for Disease Control and Prevention (CDC). The CDC reports major structural or genetic birth defects occurring in 3% of live births (CDC, 2011); the number of malformations seen after oocyte cryopreservation is, in fact, lower than this national average. Importantly, the birth defects amassed in this group mirror those seen most commonly in the general population. Additionally, the authors stratified the infants between those born after slow freeze versus vitrification protocols. There was no major difference in the rate of anomalies found after these methods of oocyte cryopreservation (1.1% versus 1.5%, respectively). No epigenetic syndromes were found in this international group of infants born after oocyte cryopreservation, though these have been reported for other types of ART.

Ovarian tissue cryopreservation, which has been less studied and is not as widely used as oocyte cryopreservation, has also resulted in successful pregnancies. The first birth after ovarian tissue cryopreservation and autotransplantation was documented in 2004 (Donnez et al., 2004). To date, there have been 13 infants born to 10 women after ovarian tissue cryopreservation (Donnez et al., 2011). Two of these women conceived and delivered two healthy infants in subsequent pregnancies from thawed, transplanted ovarian tissue. These 10 case-reports suggest that ovarian function may be restored anywhere from 2 to 5 years post-transplant of cryopreserved tissue. Women who received chemotherapy before taking measures to preserve ovarian tissue all had significantly decreased length of graft function, compared to those who cryopreserved ovarian tissue before initiating a chemotherapy regimen. All singleton gestations delivered at term, after 37 weeks' gestational age. Additionally, all of the infants born after this method of fertility preservation are alive and healthy, without any known congenital anomalies or perinatal morbidity (Donnez et al., 2011).

Studies of pregnancy outcome and neonatal well-being are extremely important with any new reproductive technology. Perhaps more crucial, however, is the ability to track and register pregnancies that arise out of oocyte cryopreservation cycles. The Human Oocyte

Preservation Experience (HOPE) is a phase IV, multicenter, observational registry in the United States that has been created to prospectively collect data on oocyte cryopreservation and subsequent outcomes (Ezcurra et al., 2009). The goals of this project are twofold: first, to evaluate the safety and efficacy of different oocyte cryopreservation techniques, and second, to assess the safety of these methods in relation to the babies resulting from cryopreserved oocytes. This initiative will follow 400 women over three years who are undergoing oocyte cryopreservation, thawing, and subsequent embryo transfer. Standardized data will be collected for all subjects, including demographic information, laboratory studies, and pregnancy outcomes. Additionally, all babies will be followed for the first year of life to evaluate perinatal and infant outcomes after oocyte cryopreservation. Studies of this nature are crucial for the validation of oocyte cryopreservation as a valuable method for fertility preservation in the United States and removal of its "experimental" categorization by ASRM.

8. Conclusion

A variety of ART strategies have been introduced over the past few decades without being deemed "experimental" or requiring IRB approval. Moreover, new procedures in ART have not historically been required to demonstrate improved efficacy over established protocols before being introduced into clinical practice (Noyes et al., 2010). One example is the introduction of ICSI in the 1990s (Palermo et al., 1995). Though ICSI is more invasive than conventional IVF, it was quickly embraced in the field and used widely for couples with severe male factor infertility after extensive informed consent. Other ART techniques, such as frozen embryo storage, prenatal genetic diagnosis (PGD), laser assisted hatching, and even human chorionic gonadotropin (hCG) agonist triggering of ovulation have not required implementation under the "auspices of an IRB." Instead, informed consent documents highlight risks and benefits of these procedures and infertility centers are expected to honestly present data regarding success rates and outcomes. In light of these inconsistencies, it seems incongruous to require such stringent, IRB-approved regulations for oocyte cryopreservation, which has been shown to produce high survival rates. Clinics should be transparent about their experience, their site-specific pregnancy rates, and the associated perinatal outcomes.

Though oocyte cryopreservation was first introduced more than three decades ago, the past several years have yielded significant enhancement of techniques and documentation of efficacy. Current and future advancements have the potential to preserve reproductive potential for young women with cancer prior to gonadotoxic treatments as well as for those seeking elective preservation of their fertility. As stimulation techniques are simplified, costs are contained, safety and efficacy are documented, and more wide-spread awareness of the reproductive aging process is achieved, it is likely that the number of women who are able to benefit from this new technology will continue to increase.

9. References

Ali J, Shelton JN. Design of vitrification solutions for the cryopreservation of embryos. *Journal of Reproduction and Fertility* 1993; 99(2):471-7.

Ashwood-Smith MJ, Morris GW, Fowler R, Appleton TC, Ashorn R. Physical factors are involved in the destruction of embryos and oocytes during freezing and thawing procedures. *Human Reproduction* 1988; 3(6):795–802.

ASRM Practice Committee. Definition of "experimental procedures." *Fertility and Sterility* 2009; 92(5):1517.

ASRM Practice Committee. Ovarian tissue and oocyte cryopreservation. *Fertility and Sterility* 2008; 90(Suppl 3):S241-6.

Barritt J, Luna M, Duke M, Valluzzo L, Howard M, Copperman AB. Prophylactic oocyte cryopreservation for fertility preservation in cancer patients. *Fertility and Sterility* 2008; 89(4 Supp 1):S19.

Barritt J, Luna M, Sandler B, Mukherjee T, Duke M, Copperman AB. Elective oocyte freezing for the preservation of fertility: an IRB approved pilot study of over 200 clients. *Fertility and Sterility* 2010; 94(4 Supp 1):S107.

Barritt J, Luna M, Duke M, Grunfeld L, Mukherjee T, Sandler B et al. Report of four donor-recipient oocyte cryopreservation cycles resulting in high pregnancy and implantation rates. *Fertility and Sterility* 2007; 87(1):189.e13-e17.

Beck LN, Exley A, Fischer C, Copperman K, Ezcurra D, Copperman AB. How Compliant Are We With ASRM Guidelines on Oocyte Cryopresevation? *Fertility and Sterility* 2009; 92(3 Supp 1):S194.

Beck LN, Sileo M, Copperman AB. The Average Cost of Fertility Preservation for Female Cancer Patients. *Fertility and Sterility* 2010; 94(4 Supp 1):S105.

Bielanski A, Nadin-Davis, S, Sapp T, Lutze-Wallace C. Viral contamination of embryos cryopreserved in liquid nitrogen. *Cryobiology* 2000; 40(2):110-6.

Bielanski A, Bergeron H, Lau PC, Devenish J. Microbial contamination of embryos and semen during long term banking in liquid nitrogen. *Cryobiology* 2003; 46(2):146-52.

Borini A, Bonu MA, Coticchio G, Bianchi V, Cattoli M, Flamigni C. Pregnancies and births after oocyte cryopreservation. *Fertility and Sterility* 2004; 82(3):601-5.

Borini A, Cattoli M, Mazzone S, Trevisi MR, Nalon M, Iadarola I, et al. Survey of 105 babies born after slow-cooling oocyte cryopreservation. *Fertility and Sterility* 2007; 88 (Suppl. 1):S13–S14.

Broekmans FJ, Visser JA, Laven JSE, Broer SL, Themmen APN, Fauser BC. Anti-Müllerian hormone and ovarian dysfunction. *Trends in Endocrinology and Metabolism* 2008; 19(9):340-7.

Byskov AG, Faddy MJ, Lemmen JG, Andersen CY. Eggs forever? *Differentiation* 2005; 73:438-46.

Center for Disease Control Report: http://www.cdc.gov/ncbddd/birthdefects/facts.html. Accessed August 2011.

Chen C. Pregnancy after human oocyte cryopreservation. *Lancet* 1986; 1:884 – 6.

Chian R, Huang JYJ, Gilbert L, Son W, Holzer H, Cui SJ, et al. Obstetric outcomes following vitrification of in vitro and in vivo matured oocytes. *Fertility and Sterility* 2008; 91(6):2391-8.

Chian R, Huang JYJ, Tan SL, Lucena E, Saa A, Rojas A, et al. Obstetric and perinatal outcome of 200 infants conceived from vitrified oocytes. *Reproductive BioMedicine Online* 2008; 16(5):608-10.

Cobo A, Kuwayama M, Perez S, Ruiz A, Pellicer A, Remohi J. Comparison of concomitant outcome achieved with fresh and cryopreserved donor oocytes vitrified by the CryoTop method. *Fertility and Sterility* 2008; 89(6):1657-64.

Cobo A, Rubio C, Gerli S, Ruiz A, Pellicer A, Remohi J. Use of fluorescence in situ hybridization to assess the chromosomal status of embryos obtained from cryopreserved oocytes. *Fertility and Sterility* 2001; 75(2):354-60.

Dolmans MM, Marinescu C, Saussoy P, Van Langendonckt A, Amorim C, Donnez J. Reimplantation of cryopreserved ovarian tissue from patients with acute lymphoblastic leukemia is potentially unsafe. *Blood*. 2010; 116(16):2908 – 14.

Donnez J, Dolmans MM. Cryopreservation of ovarian tissue: an overview. *Minerva Medica* 2009; 100:401 – 13.

Donnez J, Dolmans MM, Demylle D, Jadoul P, Pirard C, Squifflet J, et al. Livebirth after orthotopic transplantation of cryopreserved ovarian tissue. *Lancet* 2004; 364(9443):1405-10.

Donnez J, Silber S, Andersen CY, Demeestere I, Piver P, Meirow D, et al. Children born after autotransplantation of cryopreserved ovarian tissue. A review of 13 live births. *Annals of Medicine* 2011; 43:437-50.

Dunson DB, Baird DD, Colombo B. Increased infertility with age in men and women. *Obstetrics and Gynecology* 2004 Jan;103(1):51-6.

Eichenlaub-Ritter U, Shen Y, Tinneberg HR. Manipulation of the oocyte: possible damage to the spindle apparatus. *Reproductive BioMedicine Online* 2002; 5(2):117-24.

Eroglu A, Russo MJ, Bieganski R, Fowler A, Cheley S, Bayley H, et al. Intracellular trehalose improves the survival of cryopreserved mammalian cells. *Nature Biotechnology* 2000; 18:163-7.

Ezcurra D, Rangnow J, Craig M, Schertz J. The Human Oocyte Preservation Experience (HOPE) a phase IV, prospective, multicenter, observational oocyte cryopreservation registry. *Reproductive Biology and Endocrinology* 2009; 7:53.

Fabbri R, Porcu E, Marsella T, Rocchetta G, Venturoli S, Flamigni C. Human oocyte cryopreservation: new perspectives regarding oocyte survival. *Human Reproduction* 2001; 16(3):411-6.

Faddy MJ, Gosden RG, Gougeon A, Richardson SJ, Nelson JF. Accelerated disappearance of ovarian follicles in mid-life: implications for forecasting menopause. *Human Reproduction* 1992 Nov;7(10):1342-6.

Fahy GM. Vitrification: a new approach to organ cryopreservation. *Progress in Clinical and Biological Research* 1986; 224:305–35.

Flisser E, Sage CF, Uberti LM, Copperman K, Ezcurra D, Copperman AB. Trends in Oocyte Cryopreservation – Local and National Economy on Elective Medical Procedures. *Fertility and Sterility* 2009; 92(3 Supp 1):S134.

Frank Sage CF, Kolb BA, Treiser SL, Silverberg KM, Barritt J, Copperman AB. Oocyte Cryopreservation in Women Seeking Elective Fertility Preservation: A Multi-Center Analysis. *Obstetrics and Gynecology* 2008; 114(Supp 4):20S.

Gibreel A, Maheshwari A, Bhattacharya S, et al. Ultrasound tests of ovarian reserve; a systematic review of accuracy in predicting fertility outcomes. *Human Fertility* 2009; 12:95–106.

Gold E, Copperman K, Witkin G, Jones C, Copperman AB. A Motivational Assessment of Women Undergoing Elective Egg Freezing for Fertility Preservation. *Fertility and Sterility* 2006; 86(3 Supp 1):S201.

Gook DA, Edgar DH. Human oocyte cryopreservation. *Human Reproduction Update* 2007; 13(6):591-605.

Gosden RG, Baird DT, Wade JC, Webb R. Restoration of fertility to oophorectomized sheep by ovarian autografts stored at -196°C. *Human Reproduction* 1994; 9(4):597-603.

Grifo JA, Noyes N. Delivery rate using cryopreserved oocytes is comparable to conventional in vitro fertilization using fresh oocytes: potential fertility preservation for female cancer patients. *Fertility and Sterility* 2010; 93(2):391-6.

Gruijters MJG, Visser JA, Durlinger ALL, Themmen APN. Anti-Müllerian hormone and its role in ovarian function. *Molecular and Cellular Endocrinology* 2003; 211:85-90.

Herrero L, Martínez M, Garcia-Velasco JA. Current status of human oocyte and embryo cryopreservation. *Current Opinion in Obstetrics and Gynecology* 2011; 23(4):245-50.

Hoffman DI, Zellman GL, Fair CC, et al. Cryopreserved embryos in the United States and their availability for research. *Fertility and Sterility* 2003; 79(5):1063-9.

Hvidtfeldt UA, Gerster M, Knudsen L, Keiding N. Are low Danish fertility rates explained by changes in timing of births? *Scandinavian Journal of Public Health* 2010; 38(4):426-33.

Jain JK, Paulson RJ. Oocyte cryopreservation. *Fertility and Sterility* 2006; 86(Suppl 3):1037-46.

Johnson J, Canning J, Kaneko T, Pru JK, Tilly JL. Germline stem cells and follicular renewal in the postnatal mammalian ovary. *Nature* 2004; 428:145-50.

Jones KT. Meiosis in oocytes: predisposition to aneuploidy and its increased incidence with age. *Human Reproduction Update* 2008; 14(2):143-58.

Kolibianakis EM, Venetis CA, Tarlatzis BC. Cryopreservation of human embryos by vitrification or slow freezing: which one is better? *Current Opinion in Obstetrics and Gynecology* 2009; 21(3):270-4.

Kuleshova L, Gianaroli L, Magli C, Ferraretti A, Trounson A. Birth following vitrification of a small number of human oocytes: case report. *Human Reproduction* 1999; 14:3077-9.

Kuwayama M, Vajta G, Kato O, Leibo SP. Highly efficient vitrification method for cryopreservation of human oocytes. *Reproductive BioMedicine Online* 2005a; 11(3):300-8.

Kuwayama M, Vajta G, Leda S, Kato O. Comparison of open and closed methods for vitrification of human embryos and the elimination of potential contamination. *Reproductive BioMedicine Online* 2005b; 11:608-14.

La Marca A, Broekmans FJ, Volpe A, Fauser BC, Macklon NS. Anti-Müllerian hormone (AMH): what do we still need to know? *Human Reproduction* 2009; 24(9):2264-75.

Luna M, Barritt J, Sage CF, Jones C, Howard M, Copperman AB. Oocyte Cryopreservation Survey: Health Care Providers' Recommendations. *Fertility and Sterility* 2008; 89(Supp2):S21-22.

Maltaris T, Beckmann MW, Dittrich R. Fertility preservation for young female cancer patients. *In vivo* 2009; 23:123-30.

Nagy ZP, Chang CC, Shapiro DB, Bernal DP, Elsner CW, Mitchell-Leef D, et al. Clinical evaluation of the efficiency of an oocyte donation program using egg cryo-banking. *Fertility and Sterility* 2009; 92(2):520-6.

Noyes N, Boldt J, Nagy ZP. Oocyte cryopreservation: is it time to remove its experimental label? *Journal of Assisted Reproduction and Genetics* 2010; 27:69-74.

Noyes N, Knopman J, Labella P, McCaffrey C, Clark-Williams M, Grifo J. Oocyte cryopreservation outcomes including pre-cryopreservation and post-thaw meiotic spindle evaluation following slow cooling and vitrification of human oocytes. *Fertility and Sterility* 2010; 94(6):2078-82.

Noyes N, Porcu E, Borini A. Over 900 oocyte cryopreservation babies born with no apparent increase in congenital anomalies. *Reproductive BioMedicine Online* 2009; 18(6):769-76.

Oktay K. Evidence for limiting ovarian tissue harvesting for the purpose of transplantation to women younger than 40 years of age. *Journal of Clinical Endocrinology and Metabolism* 2002; 87(4):1907-8.

Oktay K, Cil AP, Bang H. Efficiency of oocyte cryopreservation: a meta-analysis. *Fertility and Sterility* 2006; 86:70 – 80.

Oktay K, Cil AP, Zhang J. Who is the best candidate for oocyte cryopreservation research? *Fertility and Sterility* 2010; 93(1):13-15.

Oktay K, Economos K, Kan M, Rucinski J, Veeck L, Rosenwaks Z. Endocrine function and oocyte retrieval after autologous transplantation of ovarian cortical strips to the forearm. *Journal of the American Medical Association* 2001; 286(12):1490-3.

Oktay K, Sönmezer M. Fertility preservation in gynecologic cancers. *Current Opinion in Oncology* 2007; 19:506-11.

Pacchiarotti J, Maki C, Ramos T, Marh J, Howerton K, Wong J, et al. Differentiation potential of germ line stem cells derived from the postnatal mouse ovary. *Differentiation* 2010; 79:159-70.

Palermo GD, Cohen J, Alikani M, Adler A, Rosenwaks Z. Intracytoplasmic sperm injection: a novel treatment for all forms of male factor infertility. *Fertility and Sterility* 1995; 63(6):1231–40.

Parte S, Bhartiya D, Telang J, Daithankar V, Salvi V, Zaveri K, et al. Detection, characterization, and spontaneous differentiation in vitro of very small embryonic-like putative stem cells in adult mammalian ovary. *Stem Cells and Development* 2011; Epub ahead of print.

Quintans CJ, Donaldson MJ, Bertolino MV, Pasqualini RS. Birth of two babies using oocytes that were cryopreserved in a choline-based freezing medium. *Human Reproduction* 2002; 17:3149-52.

Renard JP, Babinet C. High survival of mouse embryos after rapid freezing and thawing inside plastic straws with 1-2 propanediol as cryoprotectant. *Journal of Experimental Zoology* 1984; 230(3):443-8.

Rienzi L, Martinez F, Ubaldi F, Minasi MG, Iacobelli M, Tesarik J, et al. Polscope analysis of meiotic spindle changes in living metaphase II human oocytes during the freezing and thawing procedures. *Human Reproduction* 2004; 19(3):655-9.

Schnorr J, Oehninger S, Toner J, Hsiu J, Lanzendorf S, Williams R, et al. Functional studies of subcutaneous ovarian transplants in nonhuman primates: steroidogenesis, endometrial development, ovulation, menstrual patterns and gamete morphology. *Human Reproduction* 2002;17(3): 612–9.

Schorge JO, Schaffer JI, Halvorson LM, Hoffman BL, Bradshaw KD, Cunningham FG (Eds.). (2008). *Williams Gynecology*, The McGraw-Hill Companies, Inc., ISBN 978-0-07-147257-9, China.

Scott RT Jr, Hofmann GE. Prognostic assessment of ovarian reserve. *Fertility and Sterility* 1995; 63(1):1-11.

Shaw JM, Bowles S, Koopman P, Wood EC, Trounson OA. Fresh and cryopreserved ovarian tissue samples from donors with lymphoma transmit the cancer to graft recipients. *Human Reproduction* 1996; 11(8):1668 – 73.

Singer T, Barad DH, Weghofer A, Gleicher N. Correlation of antimüllerian hormone and baseline follicle-stimulating hormone levels. *Fertility and Sterility* 2009; 91(6):2616-9.

Smith GD, Serafini PC, Fioravanti J, Yadid I, Coslovsky M, Hassun P, et al. Prospective randomized comparison of human oocyte cryopreservation with slow-rate freezing or vitrification. *Fertility and Sterility* 2010; 94(6):2088-95.

Society for Assisted Reproductive Technology https://www.sartcorsonline.com/rptCSR_PublicMultYear.aspx?ClinicPKID=0. Accessed August 2011.

Sowers MR, Eyvazzadeh AD, McConnell D, Yosef M, Jannausch ML, Zhang D, et al. Anti-Mullerian hormone and inhibin-B in the definition of ovarian aging and the menopause transition. *Journal of Clinical Endocrinology and Metabolism* 2008; 93(9):3478-83.

Stachecki JJ, Cohen J, Willadsen SM. Cryopreservation of unfertilized mouse oocytes: the effect of replacing sodium with choline in the freezing medium. *Cryobiology* 1997; 37:346-54.

Tan SL, Doyle P, Campbell S, Beral V, Rizk B, Brinsden P, et al. Obstetric outcome of in vitro fertilization pregnancies compared with normally conceived pregnancies. *American Journal of Obstetrics and Gynecology* 1992; 167(3):778-84.

Trounson A, Mohr L. Human pregnancy following cryopreservation, thawing and transfer of an eight-cell embryo. *Nature* 1983; 305(5936):707-9.

Wallace WHB, Kelsey TW. Human ovarian reserve from conception to the menopause. *PLoS ONE* 2010; 5(1):e8772.

Wennerholm UB, Söderström-Anttila V, Bergh C, Aittomäki K, Hazekamp J, Nygren KG, et al. Children born after cryopreservation of embryos or oocytes: a systematic review of outcome data. *Human Reproduction* 2009; 24(9):2158-72.

Whittingham DG, Leibo SP, Mazur P. Survival of mouse embryos frozen to −196 degrees and −269 degrees C. *Science* 1972; 178(59):411–4.

Willett LL, Wellons MF, Hartig JR, Roenigk L, Panda M, Dearinger AT, et al. Do women residents delay childbearing due to perceived career threats? *Academic Medicine* 2010; 85(4):640-6.

Zou K, Yuan Z, Yang Z, Luo H, Sun K, Zhou L, et al. Production of offspring from a germline stem cell line derived from neonatal ovaries. *Nature Cell Biology* 2009; 11(5):631-6.

Part 4

Farm / Pet / Laboratory Animal ART

9

Cryopreservation of Embryos from Model Animals and Human

Wai Hung Tsang and King L. Chow

Division of Life Science, The Hong Kong University of Science and Technology
Hong Kong

1. Introduction

Diploidic germplasms such as embryos, compared to haploidic gametes, are theoretically a better choice for preservation of an animal species. However, there are significant challenges in cryopreservation of multicellular materials due to their size and physical complexity which affect the permeation of cryoprotectants and water, sensitivity to chilling and toxicity of cryoprotectants. While cryopreservation technologies are well developed and found feasible in embryos/larvae of some species, embryos of other species such as zebrafish failed to be cryopreserved. In addition, cryopreservation in many other emerging model organisms have not been developed at all. Hence, the limited cryopreservation technology has become a bottleneck in the development of various research areas, especially those relying on molecular genetics of emerging model organisms. Thorough understanding of the embryonic development and critical stages tolerant to cryopreservation needs to be identified so as to facilitate expansion of model systems available for specific biological and experimental interrogations.

1.2 Traditional and emerging animal models

Classical animal models, including species that represent major branches of the tree of life, are being used in biological studies. They include *Caenorhabditis elegans* (a nematode), *Drosophila melanogaster* (an arthropod), *Danio rerio* (a teleost fish), *Gallus gallus* (an avian), and *Mus musculus* (a mammal). They have been widely used in scientific research, primarily due to the ease of maintenance and specific features that facilitate experimental manipulations, genetic study and observation. As knowledge from these models has accumulated over the years, they offer important insights into the overall organization and functional composition of the general form of life. However, a comprehensive picture of variations of mechanistic innovation in the vast diversity of species in the Animal Kingdom is not available. Greater understanding of these organisms in different branches of the phylogenetic tree is in demand, in order to fill the gaps of existing findings. To meet this demand, more model organisms are emerging to provide unique perspectives of animal development and specific biological functions not yet uncovered in the study of other classical models. Emerging animal models include the brine shrimp *Artemia sinica*, starlet sea anemone *Nematostella vectensis*, non-parasitic flatworm *Planaria*, amphioxus *Branchiostoma floridae*, sea squirt *Ciona intestinalis*, sea lamprey *Petromyzon marinus*, Japanese quail *Coturnix japonica*, opossum *Monodelphis domestica* and marmoset *Callithrix jacchus*, etc. For example, sea lampreys are cyclostomes in a basal group of vertebrates. Comparative studies on lamprey and jawed fish, e.g., zebrafish, reveal key elements guiding jaw evolution. Characteristics shared between lamprey and other vertebrates but absent in non-vertebrate chordates include the presence of neural crest cells and a jaw. Comparative studies can direct us to the origins of these features. Therefore, study of these emerging model organisms does offer a unique approach to understand the relatedness of species in the living world. Another example illustrating the importance of new model systems in studying the evolution of body plan is shown in Figure 1 (Kosik, 2009).

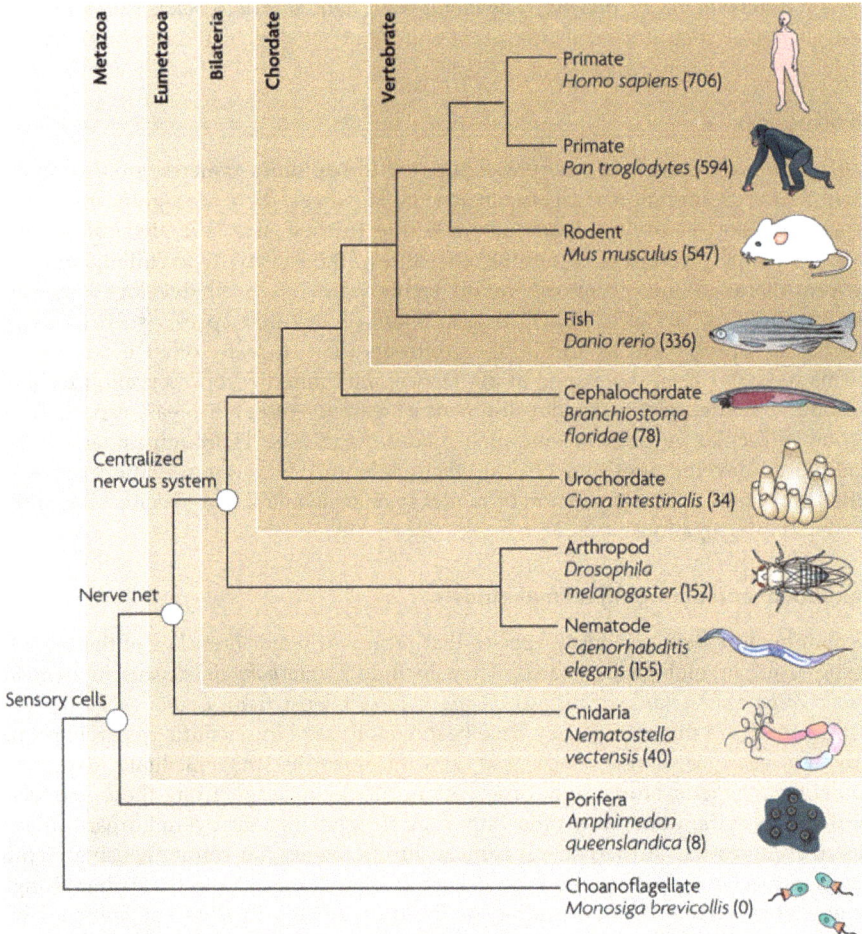

Fig. 1. The identification of miRNAs in different metazoan lineages revealed that the number of microRNAs (in brackets) is generally correlated with the complexity of body plan. Comparative studies involving "non-classical" model organisms further suggested a possible role of new microRNAs in evolutionary innovation. The figure is reproduced from Kosik (2009) with permission from the publisher.

1.3 Needs for cryopreservation

1.3.1 Archives of genetic resources

Modern evolutionary developmental biology and molecular genetic studies on model organisms largely rely on manipulation of tissues and genomes. The advancement of technologies for genomic modification of these model organisms, in turn, affects the popularity of usage of particular model organisms. Eventually, this leads to a rapid increase in the number of transgenic/mutant strains in each popular species.

For the *Caenorhabditis* Genetic Center, mutants had been deposited by individual research groups and various genome wide mutagenesis projects such as the National BioResource Project for The Nematode and *Caenorhabditis elegans* Gene Knockout Consortium. In the National BioResource Project, random mutagenesis was performed for nematodes with UV in the presence of trimethylpsoralen. Affected genes are identified by screening with a gene-specific primer set (Gengyo-Ando & Mitani, 2000). As of April 2010, about 4,400 mutants were available and mutants of some 2800 genes were being screened. Without convenient procedures for cryopreserving these species, maintenance of these strains is a heavy burden to these centers, and in other laboratories using these mutants extensively for their studies. On the other hand, passaging parasitic nematode models in plants and in donor animals, e.g., *Cooperia oncophora* in the cattle host (Borgsteede & Hendriks, 1979), is particularly labor intensive and costly and poses risk of cross contamination.

At Bloomington *Drosophila* Stock Center, more than 30,000 strains of *Drosophila* are currently present. Preparations are in progress to expand the facility to hold up to 70,000 stocks in order to meet the needs of stocking transgenic strains to be generated for a wide range of studies, including those made by tissue specific knocking-out of genes for the modeling of human diseases. In Flybase, 112,278 fly stocks were recorded for 2011 (Flybase FB2011_07 Release Notes).

In The Jackson Laboratory, over 4,000 mouse strains have been deposited and are available to the public. At the Medical Research Council Harwell and at The European Mouse Mutant Archive, over 1300 and 2200 mouse strains are stocked respectively (Eppig & Strivens, 1999). In various mouse stock centers in Japan, including BioResources Center (Riken) and Trans Genic Inc., about 8,000 mouse strains are stocked. More than 16,000 of the 24,954 protein coding genes in the mouse genome have been modified by the International Knockout Mouse Consortium (IKMC), as conditional knockout alleles in embryonic stem cells. So far, more than 1,000 mutant mice, each containing one of these conditional knockout alleles, are made available to the community. The ultimate goal is to generate different targeted alleles in embryonic stem cells (for targeted mice generation) or in targeted mice, to be available to the research community worldwide (Skarnes et al., 2011). On the other hand, about 1,200 zebrafish lines and about 100 *Xenopus* lines have been archived in the Zebrafish International Resource Center and European *Xenopus* Resource Centers, respectively.

Since keeping live animals is costly in terms of requirements of space, consumables and manpower, strains not being used need to be cryopreserved to reduce the running cost. It is, therefore, a very important technology that keeps various genome-wide knockout consortia affordable to average research laboratories.

1.3.2 Genetic stability control

The continuous passing of live animals in a small population may lead animal strains to accumulate spontaneous mutations and undergo genetic drift. In mouse, as an example of a lower mutation rate because of its relatively long life cycle, about 0.4 mutations are accumulated in each genome in each generation (Drake et al., 1998). Using this estimation, and assuming there are four generations per year, about ten mutations are accumulated in each descendent mouse diploid genome every 6.25 years (Tsang & Chow, 2010). After four years of sibling intercrosses, there is a 90% probability that more than one mutation can be fixed in a particular mouse line (Stevens et al., 2007). To circumvent this problem, the Jackson Laboratory (Bar Harbor, Maine, USA) adopted the Genetic Stability Program to refresh some mouse colonies with cryopreserved embryos once every five generations. This strategy aims at wiping out spontaneous mutations accumulated over time to ensure consistency of the mouse genome composition.

1.3.3 Genetic diversity maintenance

Maintaining genetic stability and diversity of wild parasitic nematodes collection is vital for research on parasite-host interactions, drug resistance and their applications. For example, prolonged passage of insecticidal nematodes (i.e. entomopathogenic nematodes) can cause a reduction of traits beneficial to pest control (Shapiro et al., 1996; Stuart & Gaugler, 1996; Wang & Grewal, 2002). Three continuous passages of *Galleria mellonella in vitro* resulted in a significant reduction in reproductive potential, and attenuated tolerance to heat, UV and desiccation (Wang & Grewal, 2002). These observations suggest that a selective pressure had been exerted on an isolated population that experienced continuous passages. The original genetic diversity in an isolated population will thus be largely reduced if cryopreservation is not practiced immediately after collection, before experimental analysis is performed.

1.3.4 Logistic advancement on assisted reproduction technologies

A traditional human *in vitro* fertilization cycle involves hormone induced ovarian stimulation for oocyte retrieval, *in vitro* fertilization, culture of embryos to blastocyst, selection of embryos and transfer of embryos into a recipient in an uninterrupted program. Surplus embryos are discarded. If the first pregnancy failed or another pregnancy is desired, the whole cycle has to be started again. Cryopreserving surplus embryos from the first round of the program can now be a backup for the second or more rounds of pregnancy. It can be done simply by transferring the thawed embryos to the recipients when the endometrial conditions are ready (Parriego et al., 2007).

In addition, for preimplantation genetic diagnosis programs, embryos must be transferred back to the mother at or younger than day-6 blastocysts, and so biopsies for prenatal diagnosis must be taken at the latest at day-6. More genetic materials (i.e. blastomeres or trophoblastic cells) can be obtained if the biopsy is performed later. However, accompanied shortcomings will be less time for genetic tests to be performed (Manipalviratn et al., 2009).

Cryopreservation of biopsied embryos can eliminate such conflicts between the quantity of biopsy material and the quality of genetic tests (as a function of available time; Figure 2). It has been demonstrated recently on mouse and then human embryos that biopsied embryos at various stages can survive cryopreservation well (Krzyminska & O'Neill, 1991; Wilton et al., 1989; Liu et al., 1993; Snabes et al., 1993 ; Zhang et al., 2009; Keskintepe et al., 2009).

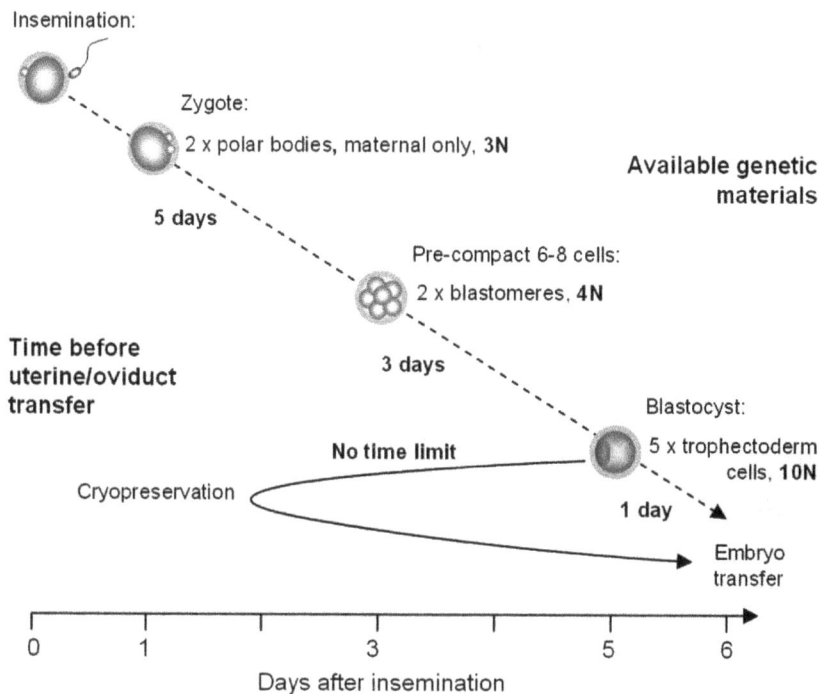

Fig. 2. Role of cryopreservation in assisted reproduction technology. Conflict between the time allowed for genetic diagnostic tests and the amount of genetic materials available (represented by the numbers of haploid genomes, N) from biopsies at different embryonic stages exists in a continuous pre-implantation genetic diagnosis program (dashed arrow). The release of the time constraint by a cryopreservation cycle after blastocyst biopsy is denoted by the solid arrow.

2. Cryopreservation of mammalian preimplantation embryos

2.1 Mammalian pre-implantation embryos at different stages

The mammalian zygote (Figure 3D) is formed by fertilization of the oocyte by spermatozoa, normally in the oviduct. With advancement of the embryo culture technology, fertilization can be initiated outside the body (i.e. *in vitro* fertilization). The embryonic development continues up to hatched blastocysts, i.e. maximum of 4 days in mouse and 6 days in human, without compromising the development of the embryos after they are transferred into the recipient's uterus/oviduct. The preimplantation stag eembryo is composed of a single cell or multiple blastomeres surrounded by an outer membrane called the zona pellucida, glycoprotein layer of a thickness of about 6μm in mouse and 8μm in human. The embryos have an outer diameter of about 0.10 mm in mouse and 0.12 mm in human. The volume enclosed by the mouse zona pellucida is limited to about 200 pl and the diameter of the cellular part of a mouse zygote is about 85μm (Zernicka-Goetz et al., 1997). In the presence of the perivitelline space, the cellular component is still in close contact with the zona pellucida, and is subjected to the immediate influence by the external medium due to the high permeability of the zona pellucida. The mouse embryo reaches two-cell stage and eight-cell stage at day 2 (dpc 1.5) and day 3 (dpc 2.5), respectively, after fertilization. Compaction usually occurs at day 3, causing a tight cell-cell association between the eight blastomeres to form a compact morula. At day 4 (dpc3.5), a blastocoel is evidenced as a cavity accumulated with fluid to form the blastocyst. The blastocoel expansion is limited by the non-growing zona pellucida. The blastocyst is composed of an embryonic inner cell mass and an extraembryonic trophectoderm which immediately surrounds the expanding blastocoel. At day 4, the blastocyst hatches from a breach in the zona pellucida and attaches onto the endometrium for further development.

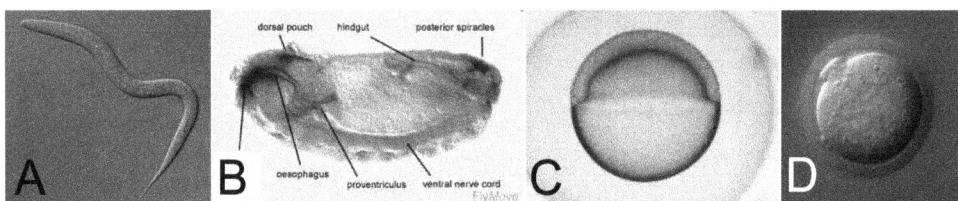

Fig. 3. Embryos and larva of model organisms to be cryopreserved. A, L1 stage larva of *C. elegans*; B, stage 15 (11-13hours) embryo of *Drosophila* (Weigmann et al., 2003); C; zebrafish at 50% epipoly; D, mouse embryo at zygote stage.

2.2 Cryoprotectant permeability of mammalian embryos

Prevention of ice formation by cryoprotectants is the key principle of protecting embryos from damage at cryogenic temperatures. To prevent intracellular ice formation, cryoprotectant molecules must penetrate the cells to exert intracellular cryoprotecting function. By measuring the changes in the volume of embryos immersed in cryoprotectant solutions (Emiliani et al., 2000), permeability of different cryoprotectant molecules at different developmental stages of preimplantation mouse embryos were compared (Table 1).

Embryos of different developmental stages show a differential permeating response to glycerol. The glycerol permeates compact morulae effectively and pre-compact 8-cell morulae moderately. One-cell embryos, 2-cell embryos and oocytes are virtually impermeable to glycerol. The highest permeability of embryos to cryoprotectant among all combinations is on compact morula to ethylene glycol. Acetamide also permeates at a relatively high degree in mouse 8-cell morulae but its permeability decreases dramatically in embryos at earlier developmental stages (Pedro et al., 2005). In sheep morula, ethylene glycol permeates faster than propylene glycol, dimethyl sulfoxide (DMSO) and glycerol (Songsasen et al., 1995). DMSO, on the other hand, shows moderate permeability on embryos at different preimplantation stages, with little difference. Being the most permeating cryoprotectants on 1-cell embryos and 2-cell embryos, propylene glycol permeates better than ethylene glycol. In general, there is a trend of increased permeability of the membrane towards various cryoprotectants when embryos develop (Mazur & Schneider, 1986; Pedro et al., 2005).

The findings indicate a dynamic change in permeability of cell membrane to different cryoprotectants during development. This permeability change does not correlate with the molecular size of the cryoprotectants. In addition, the dynamic changes in cryoprotectant permeability do not seem to be caused by the increase in the total surface area of the embryos. The mouse 8-cell embryos undergo compaction at late day 4, thus decreasing the total surface area drastically, but it is best penetrated at least by ethylene glycol, compared with 1-, 2- and pre-compacted 8-cell embryos (Pedro et al., 2005), which have a higher surface area to volume ratios. Altogether, these findings support the notion that permeability is a dynamic physiological change related to the cellular differentiation state, not a simple passive mechanism dictated by the physical size and surface area. Understanding the changes of permeability of embryos to cryoprotectants at the molecular level may help further develop the use of cryopreservation technologies on mammalian embryos and, more importantly, on other organisms that cannot be cryopreserved yet.

Cryoprotectants		Permeability to different mouse embryo			
Common name	Molecular property	1-cell	2-cell	Pre-compact morula (8-cell)	Compact morula (8-cell)
Acetamide	Amide, MW=59	Moderate	Moderate	High	High
Ethylene glycol	Polyhydric alcohol, MW=62.1	Moderate	Moderate	High	High
Propylene glycol	Polyhydric alcohol, MW=76.1	High	High	High	High
DMSO	Organosulfur, MW=78.1	High	High	High	High
Glycerol	Polyol, MW=92	Impermeable	Impermeable	Moderate	High

Table 1. Permeability of various common cryoprotectants to mouse preimplantation embryos (Information derived from Pedro et al., 2005)

2.3 Cryoprotectant toxicity to mammalian embryos

The higher the concentration of cryoprotecting agent is in a solution, the lower is the likelihood water crystals would be formed in the solution in a rapid-cooling process. However, most cryoprotecting molecules are toxic to embryos with toxicity positively correlating to their concentrations and the exposure time. When choosing the appropriate cryoprotectants, toxicity must be considered. Among five common permeating cryoprotecting agents to be tested, toxicity was determined to be dimethylformamide> erythritol > DMSO > glycerol and ethylene glycol on mouse morulae (Kasai et al., 1981). By electron probe microanalysis, Pogorelov et al. (2006 & 2007a) detected a dramatic decrease in intracellular potassium and sodium content in two-cell embryos treated with procedures mimicking vitrification in ethylene glycol, demonstrating a potential stress exerted on the embryos. When the mouse morulae were stored in 1.5M ethylene glycol or glycerol for 6 hours, the majority (>75%) of the embryos retained the capacity to develop into expanded blastocysts (Kasai et al., 1981). To minimize the toxic effect of the cryoprotecting solution to the embryos while retaining the cryoprotecting function, a mixture of two or more cryoprotecting agents could be used to decrease the relative concentration of each chemical. Macromolecules such as polyethylene glycol, ficoll and polyvinylpyrrolidone, which increase the viscosity of a solution, thus slowing down water molecules associating to form ice crystals when cooling, can also lower the concentration of cryoprotecting agents to be used in vitrification.

2.4 Osmotic flows in cryopreserving mammalian embryos

2.4.1 Slow-cooling

In a general slow-cooling procedure, embryos are immersed into permeating cryoprotectants. Intracellular water leaves the cells by osmosis and re-enters the cells together with the permeating cryoprotectants by diffusion. A temporal osmotic equilibrium state is acquired at the end. A 1-2M permeating cryoprotecting agent(s) is often used in slow-cooling. To cryopreserve the embryos, embryos and the surrounding freezing medium are loaded into a plastic straw and subjected to cooling to a temperature slightly lower than the freezing point, i.e. at about -7°C. Controlled ice nucleation is initiated by touching the straw with a cooler surface (e.g., a pair of forceps) to initiate the growth of ice inside the straw. The embryos themselves remain unfrozen but supercooled. The removal of water from the solution by the growing ice crystals increases the solute concentration of the extracellular medium. By osmosis, the intercellular water leaves the cells, resulting in an increase of cryoprotectant concentration in the cells. The subsequent slow cooling further dehydrates the embryos and concentrates the cryoprotectant in the cells to promote intracellular solidification, without intracellular ice formation, at a sufficiently low temperature. If the cooling is too fast, it leads to intracellular ice formation because the intracellular solute has not yet achieved a sufficiently high concentration. Too slow a cooling rate causes cells' death due to the prolonged exposure to hypertonic conditions. The cooling rate must be carefully controlled for each embryonic stage of each species because the permeability of cell membranes, and thus the hydrodynamics, of different samples, can be different.

2.4.2 Vitrification

In cryopreservation by vitrification (or rapid-cooling), equilibration of cryoprotectants in the embryos and the cryoprotecting medium are not required. Embryos are first permeated by cryoprotecting agents at a low concentration and then immersed in a moderately high concentration (4M or above) of the same cryoprotecting agent, sometimes together with non-permeating cryoprotectants such as 0.5M sucrose. In the presence of the non-permeating cryoprotectants, the embryos shrink osmotically, thus further increasing concentration of intracellular cryoprotectants. The high concentration of cryoprotecting agents in the medium prevents efflux of intracellular cryoprotecting agents by diffusion (Figure 4). The embryos are then loaded into a container and are rapidly cooled to solidify the embryos without the formation of ice crystals (Rall, 1987). Mathematical modelings such as the relativistic permeability approach is able to simulate the osmotic curve in these processes to facilitate the optimization of vitrification protocols in future (Katkov and Pogorelov, 2007).

2.5 The success of cryopreservation on mammalian embryos

The first mammalian embryo that survived cryopreservation was the mouse embryo 40 years ago by slow-cooling (Whittingham et al., 1972). More than 50% of preimplantation embryos survived after thawing and about 40% of the surviving embryos developed to full-term after being transferred to foster females. A similar protocol was applied successfully on cow embryos (Wilmut & Rowson, 1973), sheep embryos (Willadsen, 1977) and many other domestic mammals (Saragusty & Arav, 2011), suggesting that slow cooling is basically applicable to other mammalian embryos, provided the cooling rate can be optimized and well controlled. By using 1.5M DMSO as a cryoprotectant, sheep morulae and blastocysts were cooled at a rate of 0.1°C per minute. A survival rate of 80% can be achieved. With a similar protocol developed from mouse experiments, the first pregnancy after transfer of a cryopreserved human 8-cell embryo was recorded in 1983 (Trounson & Mohr, 1983). The pregnancy terminated at 24-weeks due to premature rupture of the membrane and *Streptomyces agalactiae* infection (Trounson & Mohr, 1983). A year later, Zeilmaker et al. (1984) described the first live birth after the transfer of cryopreserved human embryos. More reports of live births of cryopreserved human embryos were reported in the following year (Cohen et al., 1985; Downing et al., 1985) and in the years after that.

With the extensive characterization of embryo-cryoprotectant interactions, the feasibility of embryo cryopreservation by vitrification was demonstrated by Rall and Fahy (1985). The next challenges were to further lower the concentration of the cryoprotecting agent to be used and to increase the cooling rate. Increasing the cooling rate not only guarantees the absence of ice formation but also allows a further decrease in the amount of cryoprotecting agents being used, thus minimizing the potential toxic effects on the embryos. It can be illustrated by a recent report describing the use of a highly conductive micro-capillary to increase the cooling rate up to about 4,000°C per second. The cryoprotectant propylene glycol concentration can be reduced to 1.5M, compared to the normally used 4M or above (Lee et al., 2010). Over the past decade, a variety of holding devices have been developed to allow fast transmission of heat (thus a high cooling rate) from the sample to the coolants (reviewed in detail in Tsang and Chow, 2010). Some researchers have made use of containers with thinner walls, such as pulled-straw (Vajta et al., 1998). Others have made use of an open property of the devices to hold the sample on a surface to allow direct heat transfer between the samples and the coolant. These include the electronic microscopic metal grid (Martino et al., 1996), cryoloop (Lane et al., 1999), nylon mesh (Matsumoto et al., 2001), hemi-straw (Vanderzwalmen et al., 2003), cryotop (Kuwayama et al., 2005), vitrification spatula (Tsang & Chow, 2009) and plastic blade (Sugiyama et al., 2010). Using most of these open-systems, cryo-survival rates of above 80% are usually obtained. The remaining challenge is to select the right tool by considering the microbial surveillance requirement (a closed-storage system versus an open-storage system), the convenience factor and economic considerations in a routine facility operation.

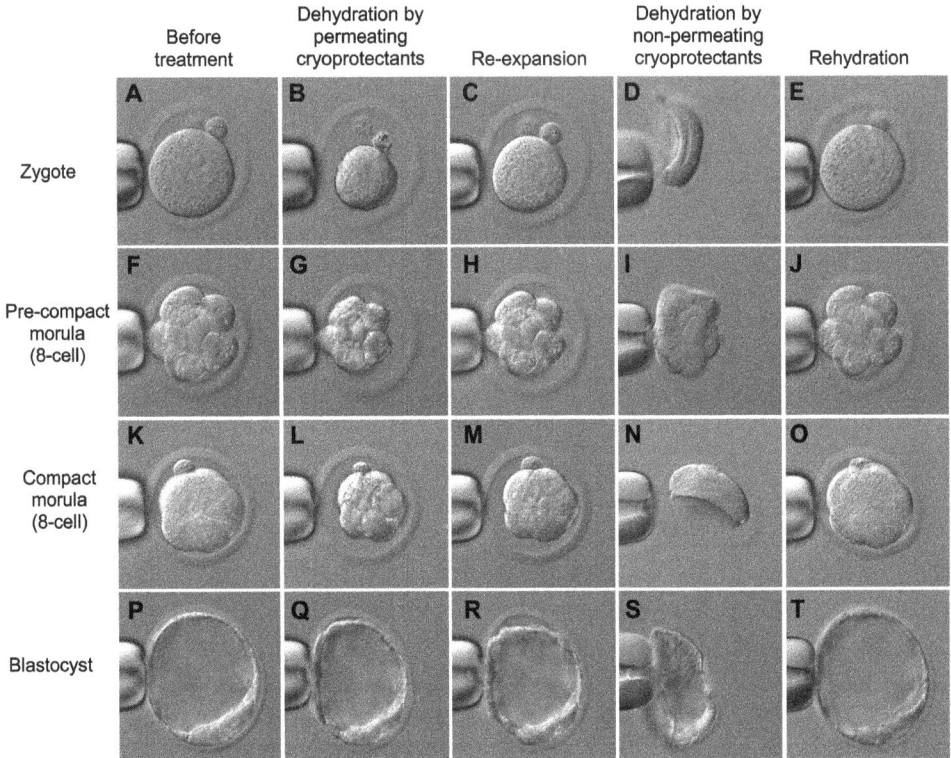

Fig. 4. Morphological changes of mammalian (mouse) preimplantation embryos in response to cryoprotectant treatments (for vitrification) and rehydration. Panel A-E: zygote (1-cell); panel F-J, pre-compact morula (8-cell); panel K-O, compact morula (8-cell); panel P-T, expanded blastocyst. Individual embryos were held by a glass micropipette by a slight suction under physiological isotonic medium (A, F, K and P). Vitrification solutions containing low concentration of permeating cryoprotectants and high concentration of permeating cryoprotectants plus non-permeating cryoprotectants were applied to the surroundings of the embryos, sequentially. About 30 seconds after the application of solution containing low concentration of permeating cryoprotectants, the embryos osmotically shrunk to minimal volumes (panel B, G, L and Q). The embryos (except the blastocyst) later re-expanded to a size closer to the original volumes, after an additional 30 seconds, when cryoprotectants and water re-entered the cells passively (panel C, H, M and R). After addition of the final vitrification solution containing high concentration of permeating cryoprotectants and non-permeating cryoprotectants, the embryos were further dehydrated by osmosis and shrunk without re-expansion. (panel D, I, N and S) High concentration of intracellular permeating cryoprotectants was achieved. Rapid cooling is normally done at this stage to vitrify the embryos but such cooling was not done in this demonstration. Step-wise rehydration of the embryos was done after dehydration to imitate the recovery steps after thawing the embryos from vitrification. After rehydration, the embryos re-expanded to their original size and re-gained normal morphologies (E, J, O and T).

Vitrification is now well accepted as a reliable means for cryopreserving mammalian embryos because of its simplicity; it does not require a controllable cooler. One factor responsible for the acceptance of this technology for cryopreservation of mammalian embryos has been the intensive studies on the interaction between different cryoprotectants and embryos, i.e. permeability and toxicity, at different developmental stages. Luckily, the most permeating cryoprotectants are not very toxic to the embryos. It allows the use of the cryoprotectants at a high concentration, yet below the lethal dose, to promote vitrification in response to a convenient cooling rate in most laboratories.

3. Cryopreservation of larvae of nematodes and platyhelminthes

3.1 Development of the nematode

Nematoda belong to the ecdysozoa, sharing the same clade with arthropoda. The members range from free living species to parasitic species in plants and animals. The life cycle of the nematode is generally divided into five morphological stages. Each successive larval stage is preceded by a molting process to remove the collagenous cuticle from the former larval stage. At hatching, the first-stage larva (Figure 3 A) consisting of 558 cells is under the protection by a cuticular layer. The animal grows in size after each hatching. In adulthood, a reproductive hermaphrodite is about 1mm long, 0.06mm in diameter, containing about 1,000 somatic cells. In comparison, a first-stage larva is about 0.37mm in length and 0.025mm in diameter.

Under favorable conditions, the development of the animal continues through the first- to the fourth-larval stage and finally to the reproductive adults. In many parasitic species such as the entomopathogenic species, the third-stage larvae are juveniles that are infective to their hosts. Under unfavorable conditions, i.e. outside the host body, a second-stage larva develops into the third-stage infective juvenile but retains the cuticle from the last larval stage to form a sheath. The entire animal is enclosed in the sheath until a suitable host is infected.

3.2 Cryopreservation of nematode without cryoprotectant additive

The simplest method for nematode cryopreservation was reported on ruminant nematodes. Infective juveniles were cooled directly in liquid nitrogen vapor after being unsheathed by sodium hypochlorite, and suspended in physiological saline (Campbell and Thomson, 1973; Van Wyk et al., 1977). For example, infective juveniles of sheep nematodes (*Haemonchus contortus, Ostertagia circumcincta, Trichostrongylus axei, Trichostrongylus colubriformis, Nematodirus spathiger and Oesophagostomum columbianum*) and the bovine nematodes (*Haemonchus placei, Ostertagia ostertagi, Nematodirus helvetianus, Oesophagostomum radiatum, Cooperia pectinata and Cooperia punctata*) survived these simple cryopreservation procedures.

James (1985) suggested that the presence of natural cryoprotectants plays a role in the cryo-survival of domestic animal parasitic nematodes. Bai et al. (2004) demonstrated in *Steinernema carpocapsae* and *Heterorhabditis bacteriophora* that the cryo-survival rates of infective juveniles are positively correlated with worm concentration during the cryoprotectant glycerol incubation step. The survival rates ranged from about 20% to 100%, which was proportional to worm concentration of 120-12,000 per ml (Bai et al., 2004). Infective juveniles indeed produce cryoprotecting molecules such as trehalose and glycerol, in response to thermal and other environmental stresses (Jagdale & Grew, 2003; Qiu & Bedding, 2002). The trehalose content in *Steinernema carpocapsae* increases from 4% to 8% after being incubated in 22% glycerol for 18 hours, before the animals are further processed for cryopreservation (Popiel & Vasquez, 1991). Production of natural cryoprotectants by the animal itself could, therefore, be the key to good animal survival in this cryopreservation procedure. Exploring an efficient way to induce the production of the natural cryoprotectants can improve cryo-survival. Identifying the molecular pathway responsible for cryoprotectant production may help make cryopreservation of these species simpler.

3.3 Slow freezing with DMSO and glycerol

Storage of live nematodes in liquid nitrogen was first demonstrated by Hwang (1970) who reported survival of animals in the genera *Aphelenchoides*, *Panagrellus*, *Turbatrix* and *Caenorhabditis*, after slow-cooling in heat-sealed glass ampoules, using DMSO as cryoprotectants. Later, glycerol was applied to slow cooling of *Caenorhabditis elegans* at the first-larval stage (Brenner, 1974) and since then this cryopreservation procedure has been routinely used for freezing this model organism in laboratories in the past 40 years. Up to 100% of the first-stage larvae survival can be achieved.

On the other hand, a slow-cooling method using DMSO as a cryoprotectant was used for cryopreserving another *Caenorhabditis* species popular in basic scientific research, i.e. *Caenorhabditis briggsae* (Haight et al., 1975). The worms were simply suspended in cooled 5% DMSO for 10 minutes and then slowly cooled at a rate of 0.2°C per minute to -100°C before storage in liquid nitrogen. About 75% of the animals at the second-larval stage and the third-larval stage, 50% of the forth-stage larvae and 3% of the adult animals survived the freezing/thawing cycle. In contrast to *Caenorhabditis elegans*, no animals at the first-larval stage survived the cryopreservation (Haight et al., 1975). Obviously, traits favoring the cryopreservation procedures are present in different larval stages in both species.

The dog parasitic nematode *Strongyloides stercoralis* can be cryopreserved by slow cooling after incubation for up to 60 minutes in a solution containing 10% DMSO. When thawed, third-stage larvae retain infectivity to dogs and the recovered first-stage larvae develop to the third-larval stage and regain infectivity (Nolan et al., 1988). The cryopreserved sheep nematodes *Haemonchus contortus*, *Trichostrongylus colubriformis* and *Ostertagia circumcincta*, at the first-larval stage, retained their infectivity as well as unfrozen worms, when cooled slowly in 10 % DMSO to -80°C before being transferred to liquid nitrogen for storage (Gill and Redwin, 1995).

3.4 Vitrification by an ethylene glycol two-step procedure

Before being used in mouse and fly embryo vitrification, ethylene glycol had been used for vitrifying the human platyhelminth *Schistosoma mansoni* and the farm animal nematode *Onchocerca microfilariae* in a similar two-stage procedure. *Schistosoma mansoni* were pre-incubated in 10% ethylene glycol at 37°C for 10 minutes, then cooled at 0°C for 5 minutes and finally incubated in 35% ethylene glycol for 10 minutes at 0°C. Before being rapidly cooled in liquid nitrogen, the worms were spread on a glass sliver (prepared from microscopic coverslips) which acted as an open carrier. About half of the thawed worms survived and remained infective in mice with an efficiency equivalent to half that of unfrozen worms (James, 1981). On the other hand, 70% of the *Onchocerca microfilariae* were viable and remained infective after vitrification and thawing (Ham et al., 1981).

A similar vitrification methodology was also applied to the plant parasitic nematode *Meloidogyne Graminicola*, a rice root-knot nematode (Bridge & Ham 1985). The second-stage larvae were pre-incubated in 10% ethylene glycol at 37°C for 15 minutes and then incubated in 40% ethylene glycol for 30-45 minutes. The worms were then rapidly cooled in liquid nitrogen in the same manner as James (1981) on *Schistosoma mansoni*. This two-stage vitrification procedure on plant nematodes was later modified by Triantaphyllou & McCabe (1989) who replaced the glass coverslip slivers with a small strip of chromatography paper as a carrier device. A survival rate of up to 90% was obtained. The author reported that the modified two-step method produced satisfactory results on other plant parasitic nematodes also, such as some *Meloidogyne* and *Heterodera* species (Triantaphyllou & McCabe, 1989).

3.5 Vitrification by a glycerol/methanol two-step procedure

An entirely different treatment methodology was developed primarily for vitrifying the entomopathogenic nematodes in the genera *Steinernema* and *Heterorhabditis*, which are effective as a biological control agent for insect pests in agriculture (Popiel & Vasquez, 1991). *Steinernema carpocapsae* were pre-incubated in 22% glycerol at room temperature for 24 hours and then in ice cold 70% methanol for 10 minutes. After removal of the majority of methanol by centrifugation, concentrated worms in methanol (i.e. 20µl) were spread on a small strip of filter paper before plunging into liquid nitrogen for rapid cooling. Up to 95% post-thaw survival rate could be obtained. On the other hand, a post-thaw survival rate of about 55% was obtained on *Heterorhabditis bacteriophora* when the optimal 14% glycerol was used (Popiel & Vasquez, 1991).

Curran et al. (1992) further optimized the protocol by replacing the centrifugation with a filtration step to remove glycerol prior to the methanol incubation. The optimal conditions for glycerol incubation for a number of entomopathogenic nematodes were also determined. Optimal glycerol pre-incubation conditions were determined to be 18% glycerol for 24 hours for *Steinernema carpocapsae*, 17% glycerol for 72 hours for *Heterorhabditis bacteriophora* and 13.8% glycerol for 72 hours for *Steinernema feltiae* and *Steinernema glaseri* (Curran et al., 1992). Other than that, 167 entomopathogenic nematodes were found to be able to survive the cryopreservation treatments, proving the feasibility of cryobanking of these worms. The mean survival rate of the *Steinernema* species is 58% (ranging from 25% to 97%) and that of *Heterorhabditis* species is 51% (ranging from 25% to 87%).

Based on the modification by Curran et al. (1992), Nugent et al. (1996) optimized cryopreservation on seven isolates of *Heterorhabditis*. Up to 8 days of pre-incubation in 11% or 15% glycerol is optimal for cryopreserving a couple of isolates. Nugent (1996) also found that glycerol can be replaced by DMSO in the pre-incubation step. For example, incubation of isolate HI82 in 8% DMSO for 3 days yielded a survival rate of about 80%, similar to those pre-incubated in 15% DMSO.

To the best of our knowledge, unlike mammals, there have been no studies of interactions between nematodes and cryoprotectants. Conversely, different protocols have been developed independently by different groups for specific worm species. Whether the different protocols are indeed applicable to other groups of nematodes or not requires further investigation. Nonetheless, we can interpret from the protocols that nematodes are generally resistant to cryoprotectant toxicity because most protocols involve a relatively long incubation time in cryoprotecting solutions. Surprisingly, most of the protocols do not involve removal of the cuticle, which is well known for its poor permeability, though successful cryoprotection requires the presence of enough intracellular concentration of cryoprotectant. It is possible that the cryoprotectants themselves can permeabilize the cuticle and then permeate into the cells beneath. Or the cryoprotectant enters the worm via the oral opening and the gut, into the rest of the body. Equally possible is that the "cryoprotectants" did not act as a cryoprotectant *per se* but acted as inducers to trigger the production of natural cryoprotectant in the worm. Therefore, thorough permeation of the chemicals was not required.

4. Cryopreservation of insect embryos

4.1 Development of insect embryos

Insecta is a class with the greatest interest among the *Arthropoda*. Pathogenic vectors associating mosquitoes, flies and bugs are of interest in medical and healthcare related research. Honeybee, silk moth, beetle and other moths are of economic interest to the beekeeping industry, silk industry and pest control in agriculture, respectively. The fruit fly *Drosophila melanogaster*, a species of the order *Diptera*, is chosen as an insect model in genetic studies because of its ease of keeping, handling and observing. It has a short life cycle and is fertile through out the year. A complete life cycle involves four distinct stages of development, which takes 8.5 days for a *Drosophila* egg to reach adulthood at 25°C.

A *Drosophila melanogaster* egg is about 0.18mm in width and 0.5mm in length, occupying a volume of about 9nl. The laid egg is a single cell with about 6,000 nuclei which later migrate to the plasma membrane to form a syncytial blastoderm at around 2 hours. Cell membranes form between the nuclei at around 2.5 hours. As gastrulation occurs, a ventral furrow and a cephalic furrow form. The midgut invaginates followed by a germ band extension (3-7 hours), stomodeal invagination (5-7 hours), germ band shortening, foregut and hindgut deeply invaginate (9-10 hours). The ectoderm closes dorsally and the head involutes at about 10-13 hours (Figure 3B). At around 13 hours, organogenesis has already begun in the 50,000-cell embryos. Organogenesis is completed and the gut regions are joined between 15 and 22 hours (i.e. at hatching) (Markow et al., 2009; Weigmann et al., 2003; Grumbling & Strelets, 2006). The hatched larva experiences three instar larval phases, before the skin of the third-instar larva hardens and encapsulates the animal to form a puparium. It takes four days for the third larva to transform into an adult by metamorphosis. At the end of metamorphosis, the adult of about 2.5mm length emerges from the puparium.

4.2 Permeabilization of insect embryos to cryoprotectants

The outermost layer of the eggshell of an insect embryo is a porous proteinacious chorion with a thickness ranging from 300-500nm in *Drosophila* to more than 40μm in saturniid moth (Magaritis et al., 1980; Fehrenbach, 1995). Underneath are: (1) the innermost crystalline chorion layer of about 40nm in *Drosophila* (Papassideri & Margaritis, 1996), (2) the waterproofing wax layer (Beament, 1946; Slifer, 1948) and (3) the vitelline layer (Papassideri et al., 1991). The wax layer is only 5nm in thickness in *Drosophila* and is intercalating with the outer crystalline chorion layer and the inner vitelline layer (Papassideri et al., 1991). The wax is mainly composed of n-alkanes and methyl-branched alkanes (Nelson & Leopold, 2003), making the eggshell impermeable even to water, thus protecting the embryos from desiccation. Extracting the dechorionated eggs with a wax removing solvent makes the vitelline membrane permeable to water, cytological stains and antibiotics (Schreuders et al., 1996), (Limbourg & Zalokar, 1973). It supports the notion that the wax component in the vitelline layer is the major factor blocking the cryoprotecting molecule from permeating into the embryos.

To facilitate permeation of cryoprotectants into insect embryos for further cryopreservation protocol development, attempts were made to permeabilize the eggshell but retain the viability of the embryos. Removal of the chorion can be done by exposing the eggs to about 2.5% sodium hypochlorite, without compromising the survival of the embryos (Lynch et al., 1989). Permeabilizing the inner layer of the eggshell with low injury can be done by a 2-step method. Dechorionated eggs of 12-13 hour embryos were rinsed with isopropanol and hexane. The embryos in permeabilized eggshell experienced minimal injury with a survival rate of 75% to 90% in culture medium (Lynch et al., 1989). This procedure resulted in 80%-95% of the treated eggs being permeabilized to water, ethylene glycol, propylene glycol, glycerol and DMSO. Mazur et al. (1992b) further optimized the two-step method to allow permeation of common cryoprotectants such as ethylene glycol and glycerol. The best result was obtained by exposing the dechorionated 12-14 hour embryos to 0.3%-0.4% 1-butanol in n-heptane for 90 seconds. At least 90% permeabilization and 80% survival can be obtained. Older embryos between 14-16 hours are much less sensitive to the above procedures, which are lethal to 3 hour embryos. Using procedures similar to Mazur (1992b), with decreased concentration of sodium hypochlorite in dechorionation and replacing isopropanol with air drying, all mosquito (*Anopheles gambiae*) embryos at 15-19 hours can be permeabilized in ethylene glycol with an acceptable survival rate of 30% (Valencia et al., 1996). On the other hand, heptane treatment seemed to be detrimental to the greater wax moth (*Galleria mellonella*) embryos. Replacing the heptane treatment with incubation in 1.25% sodium hypochlorite with 0.08% Tween-80 for 2 minutes permeabilized the moth embryos with 68% survival (Cosi et al., 2010). Determination of the optimal permeabilization procedures and the optimal embryonic stage to be permeabilized in different insecta species has opened the door for efficient cryopreservation of this largest class of animals on land.

4.3 Chilling sensitivity of insect embryos

Drosophila embryos are highly sensitive to chilling. When 15 hour eggs were incubated at -15°C, 50% of the embryos died within an hour even in the absence of ice formation. When younger embryos at 3 hours and 6 hours were cooled to the same temperature, the chilling injury increased dramatically (Mazur et al., 1992c). All 12 hour embryos died at -25°C when cooled at 1°C per minute. A similar phenomenon was found in case of honey bee (*Apis mellifera*) embryos (Collins & Mazur, 2006). The slow cooling approach that requires time for efflux of intracellular water osmotically to avoid intracellular ice formation is, therefore, theoretically impractical for handling insect embryo cryoprotection. It was estimated that cooling the embryos faster than 300°C per second can circumvent the chilling injury by shortening the time the embryo stays at such a low temperature. However, lethal intracellular ice forms at such a cooling rate when using the standard concentration of cryoprotectants (Mazur et al., 1992c). Vitrification is, therefore, the only possible way to cryopreserve insect embryos.

4.4 The success of insect embryo vitrification

Adopting and modifying the protocol for vitrifying mouse embryos with ethylene glycol (Rall & Fathy, 1985), Steponkus et al. (1990) first demonstrated successful vitrification of 13-14 hour *Drosophila* embryos. Instead of permeabilizing the vitelline layer with the currently developed method using an alkane, the eggs were "permeabilized" with a medium containing 2.125M ethylene glycol for 20 minutes. The intracellular concentration of ethylene glycol was further increased by dehydrating the embryos in 8.5M ethylene at 0°C for 8 minutes before plunging the embryos into nitrogen slush (-204°C), using a copper electronic microscopic grid as an open carrier (to achieve a cooling rate of about 400°C per second). After thawing, 18% of the eggs hatched and 3% developed into fertile adults. On the contrary, there were no embryos surviving at a lower cooling rate of 15°C per second when using a polypropylene straw as a carrier. The surrounding cryoprotecting solution vitrified *per se* since no crystals were detected by differential scanning calorimetry. The high lethality is probably due to the suboptimal permeabilization of the vitelline layer by ethylene glycol, leading to a lower concentration of cryoprotectants in the inner part of the multicellular/highly differentiated insect embryos. This may result in crystallization of water in the area with a lower concentration of cryoprotectant which would require a higher cooling rate to induce vitrification. A higher cooling rate was achieved by using a metal grid allowing vitrification to occur at such low concentrations of cryoprotectants, thus partially circumventing this potential permeating defect. Using a similar protocol, other *Dipteral* species such as blowfly (*Lucilia cuprina*, a parasite in sheep) (Leopold & Atkinson, 1999 20), midge (*Culicoides sonorensis*) (Nunamaker & Lockwood, 2001) and screw-worm (Leopold et al., 2001) were reported to be cryopreserved.

On the other hand, Mazur et al. (1992a) made use of the accumulated experience of wax removal by butanol-heptane based procedures to permeabilize the vitelline membrane prior to vitrification. In the optimized procedure, dechorionated *Drosophila* embryos were first permeabilized in 0.3% 1-butanol in n-heptane for 90 seconds. The embryos were then preincubated in 2M ethylene glycol for 30 minutes and then 8.5M ethylene glycol solution containing 10% polyvinylpyrrolidone for 5 minutes at 5°C before rapid cooling, using a filter membrane as an open carrier. The developmental stage of the embryos was also found to be critical for cryo-survival determination. Vitrifying precisely staged 14.5 hour embryos using the above mentioned method resulted in 60% of the cryopreserved embryos hatching and more than 40% of the hatched larvae developed into fertile adults (Mazur et al., 1992a).

The development of cryopreservation on *Drosophila* embryos suggests that permeabilization of the sample to cryoprotectant is the key to success even though the embryos are structurally complex. In nematode larvae, which are also susceptible to being cryopreserved, the organogenesis is even more advanced. This indicates that the body complexity brought about by organogenesis is not associated with the susceptibility of an embryo/larva to be cryopreserved, at least by vitrification. On the other hand, chilling injury can be circumvented by vitrification practically.

5. Cryopreservation of teleost embryos

5.1 Development of teleost embryos

The zygote of zebrafish (*Brachydanio rerio*) is about 0.7 mm in diameter when fertilization occurs. A few minutes later, the chorion swells to increase the diameter to about 1.2 mm, without much alteration in thickness, generating a significant vitelline space. The cytoplasm segregates to form the animal pole and the vegetal yolk with an approximate total volume of 128nl, not including the vitelline space and the chorion (Leung et al., 1998). The first cell cleavage occurs in the animal pole at about 45 minutes after fertilization. The blastomere gets divided five more times synchronously, each at about 15 minutes interval, producing a blastoderm with 64 cells in 2 hours. The daughter cells increase in number with a decrease in cell size. The blastomeres arrange themselves in a single cell layer before the fifth cleavage. Afterwards, newly formed daughter cells overlap with each other in the blastoderm. The multi-cell layered blastoderm spread over the yolk, reaching 30% epiboly at 4.7 hours and 50% epiboly at 5.25 hours (Figure 3C). At the gastrula period, epiboly continues at 5.3 hours. Two germ layers, i.e. epiblast and hypoblast, are formed by morphogenetic movement of involution, convergence and extension. Epiboly reaches 90% at 9 hours. At the end of this period, the tail bud and neural plate starts to form. The volume of the epiboly remains constant from 40% epiboly to 100% epiboly. Entering the segmentation period at 10 hours, segmentation processes such as formation of neuromeres, somites and the pharyngeal arch primordia occur. The embryo volume increases to 0.23 mm^3 at the six-somite stage at 12 hours when organogenesis starts (Hagedorn et al., 1997c). At the end of the 14 hour-period, the yolk largely reduces and tail movement can be seen. Pigment can be identified after 36 hours. At the third day, the primary organogenesis completes. Cartilage in the head and pectoral fin develops while hatching occurs anytime in the third day (Kimmel et al., 1995).

5.2 Complexity of teleost embryos

Fish embryos are composed of several components with distinct physical properties. They include the highly dynamic cellular part on the animal pole which contributes entirely to the future animal body, the yolk surrounded by the yolk syncytial layer occupying the majority of the early stage embryo and the chorion as the outermost mechanical protective shield. From 40-100% epiboly, the blastoderm and the yolk occupy 18% and 82% of the dechorionated embryo, respectively, in zebrafish (*Brachydanio rerio*). At the six-somite stage, the volume of the blastoderm increases to about 40%, leaving the yolk occupying 60% of the embryo (Hagedorn et al., 1997c). Between the chorion and the embryo is the perivitelline space filled with liquid with a chemical composition virtually identical to the surrounding medium (Rawson et al., 2000). The complexity of the teleost embryos is further increased by the unbalanced partial density of water in different compartments. At the six-somite stage, the blastoderm occupies about 40% of the dechorionated embryos but water constitutes 82% of its volume. Conversely, the yolk occupies about 60% of the volume of the dechorionated embryos but only 42% of this is constituted by water (Hagedorn et al., 1997b). It was estimated that the osmotically inactive volume in the one-somite stage and six-somite stage embryos are 72.9% and 82.6%, respectively (Zhang & Rawson, 1998).

5.3 Permeability of teleost embryos

The low permeability of the chorion and the perivitelline is evident from retarded permeation of radio-labeled DMSO from the external medium into the embryos by several folds (Harvey et al., 1983). Even after removal of the outermost barrier, the embryos were poorly permeated by cryoprotectants. By chemical shift selective magnetic resonance microscopy and magnetic resonance spectroscopy, kinetics of permeation of cryoprotectants methanol, DMSO and propylene were measured. While methanol can permeate the entire six-somite zebrafish (*Brachydanio rerio*) embryos in 15 minutes, DMSO and propylene are relatively poor in permeating into the embryos when applied to the medium (Hagedorn et at., 1996). Similar findings were obtained by osmometric measurements of volume changes in the embryos tested (Hagedorn et al., 1997c). Also, magnetic resonance imaging on the distribution of cryoprotectants, delivered to the external medium or injected into the yolk, in the three-somite stage *Brachydanio rerio* embryo revealed that the yolk is far less permeable than the blastoderm and the yolk syncytial layer is the major barrier to the cryoprotectants (Hagedorn et at., 1996; Hagedorn et al., 1997a).

To artificially promote permeation of cryoprotectants into fish embryos, ultrasound of 175V was used to increase the permeability by methanol to *Danio rerio* 50% epiboly (Wang et al., 2008). A high-intensity femtosecond laser was also used to introduce transient pores on blastomeres and the blastoderm-yolk boundary. Successful delivery of large molecules such as fluorescein isothiocyanate, streptavidin-conjugated quantum dots and DNA plasmid was detected on pec-fin stage (Kohli et al., 2007). Whether the physically induced permeation method can help cryopreservation of the whole teleost embryos requires experimental verification.

5.4 Chilling sensitivity of teleost embryos

Medaka (*Oryzias latipes*) embryos at early cleavage stage, i.e. 2-4-cell stage, are very sensitive to cooling at 0°C for 40 minutes. Only 38% of the embryos survived the chilling treatment. However, the same chilling treatment did not affect the survival of embryos in early gastrula stage (Valdez et al., 2005). Similarly, zebrafish (*Brachydanio rerio*) at cleavage stage are more sensitive to chilling than embryos at epiboly and at three-somite stage (Hagedorn et al., 1997c). This indicates that embryos at later developmental stages are more resistant to chilling. A similar phenomenon was found on other teleost species such as red sea bream, olive flounder and multicolorfin rainbowfish. In the same study, it was found that cleavage stage embryos responded to chilling by obstructing mitotic division and early gastrula stage embryos responded by delayed development at epiboly (Sasaki et al., 1998).

5.5 Cryoprotectant toxicity on teleost embryos

Incubation of the three-somite stage zebrafish (*Brachydanio rerio*) embryos in 1.5M DMSO and methanol for 30 minutes at room temperature did not adversely affect their survival. While propylene glycol is moderately toxic to the embryos, similar treatment with ethylene glycol or glycerol is lethal to all treated embryos. Of the cryoprotecting agents tested, ethylene glycol solution specifically led to the blastoderm being dissociated from the yolk (Hagedorn et al., 1997c). A similar phenomenon was observed on 14 to 20-somite stage embryos of *Danio rerio* (Higaki et al., 2010b). Treating the embryos with glycerol at a concentration of 2M for 30 minutes did not yield any viable cells in the embryos. The rest of the cryoprotectants tested (including methanol, ethylene glycol, DMSO, propylene glycol and 1,3-butylene glycol) gave a survival rate of between 90 to 100%. Treating the embryos with cryoprotectants at a higher concentration revealed that ethylene glycol is the next most toxic cryoprotectant, after glycerol; it kills all cells in the embryos at a concentration of 3M. In comparison, methanol and DMSO are moderately toxic. Propylene glycol and 1,3-butylene glycol are mildly toxic, killing only 58-78% of cells even at a concentration of 5M (Higaki et al., 2010b).

5.6 Attempts in vitrifying teleost embryos

So far, there have been no successful examples of live fish recovery after cryopreservation. The difficulty in controlling the dynamics of cryoprotecting agents and water in the highly structurally complex embryos may be the cause. However, studies have been conducted to assess the degree of protection provided by the cryoprotectant in vitrification. In a study, five-somite stages of turbot and zebrafish embryos were treated for 5 minutes with incremental concentrations of DMSO and then for a total time of 4 minutes in mixtures containing 5M DMSO, 2M methanol and 1M ethylene glycol, before being loaded into plastic straws and plunged into liquid nitrogen for vitrification. Although 50% of the overall glucose-6-phosphate dehydrogenase activity was retained, no embryo hatched after thawing (Robles et al., 2004).

The yolk and the surrounding syncytial layer were suggested to be a major reservoir of osmotically inactive water and a barrier to permeation of cryoprotectant to the blastoderm. After vitrifying yolk-removed zebrafish (*Danio rerio*) embryos at 14 to 20-somite stage in 20% ethylene glycol, 20% DMSO and 0.5M sucrose, no living embryos were obtained, but 87% of the cells survived after vitrification and up to 90% of the primordial germ cells were viable (Higaki et al., 2010a). Removal of yolk is deleterious to the development of the embryos. Eliminating the solute and water barrier by yolk removal is not the ultimate solution for cryopreserving fish embryos unless an artificial replacement of yolk is made feasible.

5.7 Alternatives for whole embryo cryopreservation

Due to the lack of progress in development of cryopreservation of fish embryos, isolated somatic cells are being explored as a means to preserve diploid genetic materials. The blastomere becomes one of the attractive candidates because of its abundance in embryos and its pluripotent property in the chimeric animals generated by blastomere implantation. The blastomere from genetically pigmented zebrafish embryos at mid-blastula stage were transplanted into an albino recipient embryo of the same developmental stage. In five out of the twenty-eight chimeric fish produced, blastomeres from the donor contributed to the germline, transmitting the pigmented phenotype to the next generation at a frequency of 1% to 40% (Lin et al., 1992). Slowly cooled zebrafish (*Danio rerio*) blastomeres, isolated from 50% epiboly, were cryopreserved with 1.5M DMSO and 0.1M sucrose in 0.25 ml straws by a programmable freezer. A survival rate of 70% was obtained after thawing (Lin et al., 2009). Combining these technologies, the germline transmission of the cryopreserved genetic materials through blastomeres-embryo chimera seems to be possible. More optimization, e.g., the stages from which the blastomeres are to be isolated, is needed to maximize germline transmission and to minimize operations to be conducted in a recovery procedure.

Another alternative diploid material often sought to be cryopreserved is the primordial germ cells. Compared with the blastomere, primordial germ cells are developmentally closer to the cell type to be differentiated *in vivo*, i.e. the germ cells. The first success in transplantation of primordial germ cells was demonstrated on rainbow trout (*Oncorhynchus mykiss*), a model with a relatively larger body size. Green fluorescent protein expressing primordial germ cells isolated from the genital ridge of hatchlings were injected into the peritoneal cavities of a wild type hatchling. The maker-labeled primordial germ cells were able to colonize the genital ridge of the recipient animal and transmit the donor characteristic to the next generation through sperm and eggs at a rate of up to about 4% (Takeuchi et al., 2003).

A similar operation in the smaller teleost species such as zebrafish is more challenging. A single primordial germ cell isolated from the pearl *Danio* (*Danio albolineatus*) at ten- to fifteen-somite stage was transplanted into the marginal region of each zebrafish (*Danio rerio*) embryo at the blastula stage and *vice versa*. The development of host germ cells was prevented in advance by injection of an antisense *dead end* morpholino oligonucleotide at an earlier embryonic stage (Slanchev et al., 2005). In the host, the transplanted primordial germ cell developed into a single gonad, making the animal regain fertility and transmit the donor genotype to the progenies. This complete germline replacement procedure can be applied to both goldfish (*Carassius auratus*) and loach (*Misgurnus anguillicaudatus*) (Saito et al., 2008). The success of these cases suggests that cryopreservation of primordial germ cells is a feasible approach to preserve the diploid germplasm. As the reservoirs of primordial germ cells, genital ridges from Rainbow trout (*Oncorhynchus mykiss*) embryos were cryopreserved by cooling in dry ice and then liquid nitrogen after treating with 1.8M ethylene glycol. About 51% of primordial germ cells survived. Fifteen to twenty surviving primordial germ cells were transplanted to the peritoneal cavity of each newly hatched animal. Germline transmission of the donor genotype could be found in 7.8% of the hosts and the germline transmission frequency was from 0.1 to 13.5%. (Kobayashi et al., 2007).

Later, Higaki et al. (2010b) vitrified whole zebrafish (*Danio rerio*) embryos at 14- to 20-somite stage with an optimized vitrification solution to cryopreserve primordial germ cells. With the use of 3M ethylene glycol and 0.5M sucrose, about 4 primordial germ cells, about 40% of all, survived in each embryo, after thawing. To increase cryo-survival, yolk-removed zebrafish (*Danio rerio*) embryos were vitrified in 20% ethylene glycol, 20% DMSO and 0.5M sucrose. Up to 90% live primordial germ cells were obtained. Half of the primordial germ cells retained pseudopodial movement. After transplanting the motile primordial germ cells into sterilized golden-type zebrafish blastulae, about 2.8% of the recipients developed normally and produced progenies with the donor's genotype (Higaki et al., 2010a).

Unless there is a breakthrough in cryopreserving and recovering whole fish embryos, cryopreservation of blastomeres or primordial germ cells seem to be the only methods for cryopreserving the fish diploid germplasm. Blastomeres may have advantages over primordial germ cells in generating germline transmitting chimera. Firstly, identification and isolation of primordial germ cells relies on a readily observable transgenic marker (Higaki et al., 2010b; Kobayashi et al., 2007). Breeding of a strain to a marker transgenic strain or freshly injecting DNA constructs is required before cryopreservation procedures, making the procedures more complicated. Removal of the marker from the recovered animals may also be required in some applications. Secondly, for germ-line replacement, the recipient blastulae have to be sterilized, e.g., by injecting a *dead end* antisense morpholino (Ciruna et al., 2002). This demands additional procedures in the entire cryopreservation/recovery cycle, making the primordial germ cell-base approach less attractive than the existing sperm cryopreservation. Delivery of antifreezing proteins to directly minimize water crystal formation or aquaporins to increase permeability to cryoprotectants and movement of water (Chauvigne et al., 2011) through transgenesis similarly complicate the cryopreservation procedures.

6. Concluding remarks

Successful cryopreservation relies on a number of conditions and properties of the embryos or larvae to be fulfilled. The conditions, which may be interdependent on each other, are (1) the chilling sensitivity of the embryos/larvae; (2) the permeability of the embryos/larvae to cryoprotectant and water; and (3) the sensitivity of the embryos/larvae to the cryoprotectant toxicity. The permeability of the embryos/larvae can be a function of size and structural heterogeneity. The toxicity of the cryoprotectant to the embryos/larvae can be a function of permeability at a particular developmental stage. Although a cryopreservation protocol can be as simple as slow freezing *Caenorhabditis elegans* in 15% glycerol, most of the other organisms require extensive optimization before being cryopreserved efficiently. Understanding the behavior of the interacting conditions can help initiate the development of cryopreservation of other model animals.

Chilling injury We learned from classical model organisms that chilling sensitivity coupled with a slow cooling procedure could be detrimental and vitrification can be a shortcut or even a better starting point to achieve the same goal. Vitrification of highly chilling-sensitive insect embryos is an excellent example. On the other hand, we have to keep in mind that vitrification requires a relatively high concentration of permeating cryoprotectant(s). If a new model organism to be cryopreserved is highly sensitive to the cryoprotectant(s) and has relatively low permeability, vitrification may not be feasible. Slow cooling, which requires a lower concentration of cryoprotectant, thus also allowing longer time for permeation, may be considered.

Permeability The permeability of a sample towards cryoprotectants is the major barrier to cryopreservation of *Drosophila melanogaster, Danio rerio* and probably some other model organisms. Understanding the complexity and structural properties of the embryo/larvae can make cryopreservation possible by developing a corresponding strategy to manage the flow of cryoprotecting agents and water at will. Although the studies on *Danio rerio* embryo complexity and development did not bring about successful cryopreservation of the whole embryo, they helped development of alternatives for cryopreserving diploidic germplasms. Cryopreservation of blastomeres and primordial germ cells using the optimized conditions leads to generation of germline-transmitting chimera after transplantation of cells.

Toxicity of cryoprotectants Knowing the toxicity of cryoprotectants at different developmental stages of an organism is critical in determining the combination of cryoprotectants with embryonic/larval stages to be chosen for effective cryopreservation during protocol development. For example, glycerol and dimethylformamide are very toxic to fish embryos and mammalian morulae, respectively (Higaki et al., 2010b; Kasai et al., 1981).

Model organisms offer a platform to address biological issues of a broad range of interests with ease. An ideal platform must have specific traits allowing convenient manipulations in a manner beneficial to specific fields of study. Use of classical model organisms such as the house mouse *Mus musculus*, zebrafish *Danio rerio*, fruit fly *Drosophila melanogaster* and nematode *Caenorhabditis elegans* for studying physiology, genetics, genomics, behavior, human diseases and their treatments, etc is well established. These are attractive model organisms from their representative evolutionary position. They are relatively more readily available, tractable, small in body size, rapid in development and have short reproductive cycles. They are still popular models because of the establishment of transgenic technologies related to these animals (Fire, 1986; Gordon et al., 1980; Rubin & Spradling, 1982; Zelenin et al., 1991). Such genome manipulation technologies make reverse genetics possible, allowing studies to be amendable for these model organisms.

Knowledge gained on the above-mentioned classical models and other less popular model organisms has been expanding for the past few decades. Thorough comparative studies in various fields of research will benefit translational research and our understanding of the evolutionary tree of life. More interdisciplinary studies on model organisms representing animals in various branches of the phylogenetic tree will enhance our comparative study. Promising transgenic techniques have been recently established in respect of some of these animals, e.g., planaria *Girardia tigrina*, brine shrimp *Artemia sinica*, amphipod crustacean *Parhyale hawaiensis*, red flour beetle *Tribolium castaneum*, sea anemone *Nematostella vectensis*, Mollusk dwarf surfclam *Mulinia lateralis*, sea squirt *Ciona intestinalis*, channel catfish *Ictalurus punctatus*, frog *Xenopus laevis* and *Xenopus tropicalis*, chicken *Gallus gallus*, Japanese quail *Coturnix japonica*, goat *Capra hircus*, dog *Canis familiaris* and marmoset *Callithrix jacchus*, (Berghammer et al., 1999; Chang et al., 2011; Dunham et al., 2002; Gonzalez-Estevez et al., 2003; Hong et al., 2009; Houdebine, 2009; Huss et al., 2008; Lu et al. 1996; Macha et al., 1997; Mozdziak & Petitte 2004 ; Pavlopoulos and Averof, 2005; Renfer et al., 2010; Sasaki et al., 2009; Sasakura et al., 2007;   Wheeler, 2003). The development of genome manipulation techniques for these emerging animal models will open the door to unlimited possibilities of *in vivo* investigations.

The cost of the knowledge explosion and scientific advancement will be the handling of an enormous number of transgenic strains generated. The cost of managing these invaluable resources can be a substantial burden on research laboratories or institutions world wide, which may impede further development. So far cryopreservation has been developed for embryos/larvae from non-classical model animals, including oyster *Crassostrea gigas*, hard clam *Meretrix lusoria*, sea urchin *Loxechinus albus*, amphioxus *Branchiotoma belcheri*, brine shrimp *Artimia franciscana*, euryhaline rotifer *Brachionus plicatilis* and marmoset *Callithrix jacchus*, etc. (Barros et al., 1997; Chao et al., 1997; Summers et al., 1987; Sun et al., 2007; Toleda & Kurokura, 1990; Yoshida et al., 2011).

The experience in cryopreserving embryos from such a broad evolutionary range will benefit the development of cryopreservation techniques in other emerging model organisms. The parameters highlighted in this review represent some keys for developing an effective cryopreservation protocol for any organisms for experimental use. The thorough understanding of these parameters in different model systems, the optimization therein, and improved procedures to store transgenic strains will not only release the management stress caused by the need for keeping the live animals but also eliminate the risk of their being affected by disease outbreaks and genetic drifts. It has great practical value for short term research purposes and daily operations. It is also beneficial for a longer term establishment of these models as alternative platforms for biomedical investigations.

7. Acknowledgement

We would like to thank Miss Mandy Chan and Prof. Andrew L. Miller (HKUST) for contribution to the photograph of a zebrafish embryo in Figure 3C. FlyMove, an internet educational resource (Weigmann 2003), is acknowledged for the permission in reproducing an image of a fly embryo in Figure 3B. WHT is a postdoctoral fellow supported by HKUST postdoctoral fellowship. This work was conducted and supported by Research Grants Council (HKUST 660407 and 660508).

8. References

Bai, C., Shapiro-Ilan, D.I., Gaugler, R. & Yi, S. (2004). Effect of entomopathogenic nematode concentration on survival during cryopreservation in liquid nitrogen. *Journal of Nematology*, Vol. 36, pp. (281-284).

Ballweber, P., Markl, J. & Burmester, T. (2002). Complete hemocyanin subunit sequences of the hunting spider cupiennius salei: Recent hemocyanin remodeling in entelegyne spiders. *The Journal of Biological Chemistry*, Vol. 277, pp. (14451-14457).

Barros, C., Muller, A. & Wood, M.J. (1997). High survival of spermatozoa and pluteus larvae of sea urchins frozen in Me2SO. *Cryobiology*, Vol. 35, pp. (341).

Beament, J.W.L. (1946). The waterproofing process in eggs of *Rhadnius prolius* Stahl. *Proceesdings of the Royal Society B: Biological Science*, Vol. 133, pp. (407-418).

Berghammer, A.J., Klingler, M. & Wimmer, E.A. (1999). A universal marker for transgenic insects. *Nature*, Vol. 402, pp. (370-371).

Borgsteede, F.H. & Hendriks, J. (1979). Experimental infections with *Cooperia oncophora* (Railliet, 1918) in calves. Results of single infections with two graded dose levels of larvae. *Parasitology*, Vol. 78, pp. (331-342).

Brenner, S. (1974). The genetics of *Caenorhabditis elegans*. *Genetics*, Vol. 77, pp. (71-94).

Bridge, J. & Ham, P.J. (1985). A technique for the cryopreservation of viable juveniles of *Meloidogyne graminicola*. *Nematologica*, Vol. 31, pp. (185-189).

Campbell, W.C. & Thomson, B.M. (1973). Survival of nematode larvae after freezing over liquid nitrogen. *Australian Veterinary Journal*, Vol. 49, pp. (110-111).

Chang, S.H., Lee, B.C., Chen, Y.D., Lee, Y.C. & Tsai, H.J. (2011). Development of transgenic zooplankton *Artemia* as a bioreactor to produce exogenous protein. *Transgenic Research*, (Jan2011). 1573-9368.

Chao, N.H., Lin, T.T., Chen, Y.J., Hsu, W.H. & Liao, I.C. (1997). Cryopreservation of the late embryos and early larvae in the oyster and hard clam. *Aquaculture*, Vol. 155, pp. (31-44).

Chauvigne, F., Lubzens, E. & Cerda, J. (2011). Design and characterization of genetically engineered zebrafish *aquaporin-3* mutants highly permeable to the cryoprotectant ethylene glycol. *BMC Biotechnology*, Vol. 11, pp. (34).

Ciruna, B., Weidinger, G., Knaut, H., Thisse, B., Thisse, C., Raz, E. & Schier, A.F. (2002). Production of maternal-zygotic mutant zebrafish by germ-line replacement.

Proceedings of the National Academy of Sciences of the United States of America, Vol. 99, pp. (14919-14924).

Cohen, J., Simons, R.F., Fehilly, C.B., Fishel, S.B., Edwards, R.G., Hewitt, J., Rowlant, G.F., Steptoe, P.C. & Webster, J.M. (1985). Birth after replacement of hatching blastocyst cryopreserved at expanded blastocyst stage. *Lancet,* Vol. 1, pp. (647).

Collins, A.M. & Mazur, P. (2006). Chill sensitivity of honey bee, *Apis mellifera,* embryos. *Cryobiology,* Vol. 53, pp. (22-27).

Cosi, E., Abidalla, M.T. & Roversi, P.F. (2010). The effect of Tween 80 on eggshell permeabilization in *Galleria mellonella* (L.) (Lepidoptera, pyralidae). *CryoLetters,* Vol. 31, pp. (291-300).

Curran, J., Gilbert, C. & Butler, K. (1992). Routine cryopreservation of isolates of *Steinernema* and *Heterorhabditis* spp. *Journal of Nematology,* Vol. 24, pp. (269-270).

Downing, B.G., Mohr, L.R., Trounson, A.O., Freemann, L.E. & Wood, C. (1985). Birth after transfer of cryopreserved embryos. *The Medical Journal of Australia,* Vol. 142, pp. (409-411).

Drake, J.W., Charlesworth, B., Charlesworth, D. & Crow, J.F. (1998). Rates of spontaneous mutation. *Genetics,* Vol. 148, pp. (1667-1686).

Dunham, R.A., Warr, G.W., Nichols, A., Duncan, P.L., Argue, B., Middleton, D. & Kucuktas, H. (2002). Enhanced bacterial disease resistance of transgenic channel catfish *Ictalurus punctatus* possessing cecropin genes. *Marine Biotechnology,* Vol. 4, pp. (338-344).

Emiliani, S., Van den Bergh, M., Vannin, A.S., Biramane, J. & Englert, Y. (2000). Comparison of ethylene glycol, 1,2-propanediol and glycerol for cryopreservation of slow-cooled mouse zygotes, 4-cell embryos and blastocysts. *Human Reproduction,* Vol. 15, pp. (905-910).

Eppig, J.T. & Strivens, M. (1999). Finding a mouse: The international mouse strain resource (IMSR). *Trends in Genetics,* Vol. 15, pp. (81-82).

Fehrenbach, H. (1995). Egg shells of *Lepidoptera* - fine structure and phylogenetic implications. *Zoologischer Anzeiger,* Vol. 234, pp. (19-41).

Fire, A. (1986). Integrative transformation of *Caenorhabditis elegans. The EMBO Journal,* Vol. 5, pp. (2673-2680).

Gengyo-Ando, K. & Mitani, S. (2000). Characterization of mutations induced by ethyl methanesulfonate, UV, and trimethylpsoralen in the nematode *Caenorhabditis elegans. Biochemical and Biophysical Research Communications,* Vol. 269, pp. (64-69).

Gill, J.H. & Redwin, J.M. (1995). Cryopreservation of the first-stage larvae of trichostrongylid nematode parasites. *International Journal for Parasitology,* Vol. 25, pp. (1421-1426).

Gonzalez-Estevez, C., Momose, T., Gehring, W.J. & Salo, E. (2003). Transgenic planarian lines obtained by electroporation using transposon-derived vectors and an eye-specific GFP marker. *Proceedings of the National Academy of Sciences of the United States of America,* Vol. 100, pp. (14046-14051).

Gordon, J.W., Scangos, G.A., Plotkin, D.J., Barbosa, J.A. & Ruddle, F.H. (1980). Genetic transformation of mouse embryos by microinjection of purified DNA. *Proceedings of the National Academy of Sciences of the United States of America,* Vol. 77, pp. (7380-7384).

Grumbling, G. & Strelets, V. (2006). FlyBase: Anatomical data, images and queries. *Nucleic Acids Research,* Vol. 34, pp. (D484-8).

Hagedorn, M., Hsu, E.W., Pilatus, U., Wildt, D.E., Rall, W.R. & Blackband, S.J. (1996). Magnetic resonance microscopy and spectroscopy reveal kinetics of cryoprotectant permeation in a multicompartmental biological system. *Proceedings of the National Academy of Sciences of the United States of America,* Vol. 93, pp. (7454-7459).

Hagedorn, M., Hsu, E., Kleinhans, F.W. & Wildt, D.E. (1997a). New approaches for studying the permeability of fish embryos: Toward successful cryopreservation. *Cryobiology,* Vol. 34, pp. (335-347).

Hagedorn, M., Kleinhans, F.W., Freitas, R., Liu, J., Hsu, E.W., Wildt, D.E. & Rall, W.F. (1997b). Water distribution and permeability of zebrafish embryos, *Brachydanio rerio.* The *Journal of Experimental Zoology,* Vol. 278, pp. (356-371).

Hagedorn, M., Kleinhans, F.W., Wildt, D.E. & Rall, W.F. (1997c). Chill sensitivity and cryoprotectant permeability of dechorionated zebrafish embryos, *Brachydanio rerio. Cryobiology,* Vol. 34, pp. (251-263).

Haight, M., Frim, J., Pasternak, J. & Frey, H. (1975). Freeze-thaw survival of the free-living nematode *Caenorhabditis briggsae. Cryobiology,* Vol. 12, pp. (497-505).

Ham, P.J., Townson, S., James, E.R. & Bianco, A.E. (1981). An improved technique for the cryopreservation of onchocerca microfilariae. *Parasitology,* Vol. 83, pp. (139-146).

Harvey, B., Kelley, R.N. & Ashwood-Smith, M.J. (1983). Permeability of intact and dechorionated zebra fish embryos to glycerol and dimethyl sulfoxide. *Cryobiology,* Vol. 20, pp. (432-439).

Higaki, S., Eto, Y., Kawakami, Y., Yamaha, E., Kagawa, N., Kuwayama, M., Nagano, M., Katagiri, S. & Takahashi, Y. (2010a). Production of fertile zebrafish (*Danio rerio*) possessing germ cells (gametes) originated from primordial germ cells recovered from vitrified embryos. *Reproduction,* Vol. 139, pp. (733-740).

Higaki, S., Mochizuki, K., Akashi, Y., Yamaha, E., Katagiri, S. & Takahashi, Y. (2010b). Cryopreservation of primordial germ cells by rapid cooling of whole zebrafish (*Danio rerio*) embryos. *The Journal of Reproduction and Development,* Vol. 56, pp. (212-218).

Hong, S.G., Kim, M.K., Jang, G., Oh, H.J., Park, J.E., Kang, J.T., Koo, O.J. *et al.* (2009). Generation of red fluorescent protein transgenic dogs. *Genesis,* Vol. 47, pp. (314-322).

Houdebine, L.M. (2009). Production of pharmaceutical proteins by transgenic animals. *Comparative Immunology, Microbiology and Infectious Diseases,* Vol. 32, pp. (107-121).

Huss, D., Poynter, G. & Lansford, R. (2008). Japanese quail (*Coturnix japonica*) as a laboratory animal model. *Lab Animal,* Vol. 37, pp. (513-519).

Hwang, S.W. (1970). Freezing and storage of nematodes in liquid nitrogen. *Nematologica,* Vol. 16, pp. (305-308).

Jagdale, G.B. & Grewal, P.S. (2003). Acclimation of entomopathogenic nematodes to novel temperatures: Trehalose accumulation and the acquisition of thermotolerance. *International Journal for Parasitology,* Vol. 33, pp. (145-152).

James, E.R. (1981). *Schistosoma mansoni:* Cryopreservation of schistosomula by two-step addition of ethanediol and rapid cooling. *Experimental Parasitology,* Vol. 52, pp. (105-116).

James, E.R. (1985). Cryopreservation of helminths. *Parasitology Today,* Vol. 1, pp. (134-139).

Kasai, M., Niwa, K. & Iritani, A. (1981). Effects of various cryoprotective agents on the survival of unfrozen and frozen mouse embryos. *Journal of Reproduction and Fertility,* Vol. 63, pp. (175-180).

Katkov, I.I. & Pogorelov, A.G. (2007). Influence of exposure to vitrification solutions on 2-cell mouse embryos: II. osmotic effects or chemical toxicity? *CryoLetters,* Vol. 28, pp. (409-427).

Keskintepe, L., Sher, G., Machnicka, A., Tortoriello, D., Bayrak, A., Fisch, J. & Agca, Y. (2009). Vitrification of human embryos subjected to blastomere biopsy for pre-implantation genetic screening produces higher survival and pregnancy rates than slow freezing. *Journal of Assisted Reproduction and Genetics,* Vol. 26, pp. (629-635).

Kimmel, C.B., Ballard, W.W., Kimmel, S.R., Ullmann, B. & Schilling, T.F. (1995). Stages of embryonic development of the zebrafish. *Developmental Dynamics : An Official Publication of the American Association of Anatomists,* Vol. 203, pp. (253-310).

Kobayashi, T., Takeuchi, Y., Takeuchi, T. & Yoshizaki, G. (2007). Generation of viable fish from cryopreserved primordial germ cells. *Molecular Reproduction and Development,* Vol. 74, pp. (207-213).

Kohli, V., Robles, V., Cancela, M.L., Acker, J.P., Waskiewicz, A.J. & Elezzabi, A.Y. (2007). An alternative method for delivering exogenous material into developing zebrafish embryos. *Biotechnology and Bioengineering,* Vol. 98, pp. (1230-1241).

Kosik, K.S. (2009). MicroRNAs tell an evo-devo story. *Nature Reviews.Neuroscience,* Vol. 10, No. 10, (Oct), pp. (754-759).

Krzyminska, U. & O'Neill, C. (1991). The effects of cryopreservation and thawing on the development in vitro and in vivo of biopsied 8-cell mouse embryos. *Human Reproduction,* Vol. 6, pp. (832-835).

Kuwayama, M., Vajta, G., Kato, O. & Leibo, S.P. (2005). Highly efficient vitrification method for cryopreservation of human oocytes. *Reproductive Biomedicine Online,* Vol. 11, pp. (300-308).

Lane, M., Schoolcraft, W.B. & Gardner, D.K. (1999). Vitrification of mouse and human blastocysts using a novel cryoloop container-less technique. *Fertility and Sterility,* Vol. 72, pp. (1073-1078).

Lee, H.J., Elmoazzen, H., Wright, D., Biggers, J., Rueda, B.R., Heo, Y.S., Toner, M. & Toth, T.L. (2010). Ultra-rapid vitrification of mouse oocytes in low cryoprotectant concentrations. *Reproductive Biomedicine Online,* Vol. 20, pp. (201-208).

Leung, C.F., Webb, S.E. & Miller, A.L. (1998). Calcium transients accompany ooplasmic segregation in zebrafish embryos. *Development, Growth & Differentiation,* Vol. 40, pp. (313-326).

Limbourg, B. & Zalokar, M. (1973). Permeabilization of *Drosophila* eggs. *Developmental Biology,* Vol. 35, pp. (382-387).

Lin, S., Long, W., Chen, J. & Hopkins, N. (1992). Production of germ-line chimeras in zebrafish by cell transplants from genetically pigmented to albino embryos. *Proceedings of the National Academy of Sciences of the United States of America,* Vol. 89, pp. (4519-4523).

Lin, C., Zhang, T. & Rawson, D.M. (2009). Cryopreservation of zebrafish (*Danio rerio*) blastomeres by controlled slow cooling. *CryoLetters,* Vol. 30, pp. (132-141).

Liu, J., Van den Abbeel, E. & Van Steirteghem, A. (1993). The in-vitro and in-vivo developmental potential of frozen and non-frozen biopsied 8-cell mouse embryos. *Human Reproduction,* Vol. 8, pp. (1481-1486).

Leopold, R.A. & Atkinson, P.W.(1999). Cryopreservation of sheep blow fly embryos, *Lucilia cuprina* (Diptera: Calliphoridae). *CryoLetters,* Vol. 20, pp. (37-44).

Leopold, R.A., Wang, W.B., Berkebile, D.R. & Freeman T.P.(2001). Cryopreservation of embryos of the new world screwworm *Cochliomyia hominivorax* (diptera: calliphoridae). *Annals of the Entomological Society of America,* Vol. 94, pp. (695-701).

Lu, J.K., Chen, T.T., Allen, S.K., Matsubara, T. & Burns, J.C. (1996). Production of transgenic dwarf surfclams, *Mulinia lateralis,* with pantropic retroviral vectors. *Proceedings of the National Academy of Sciences of the United States of America,* Vol. 93, pp. (3482-3486).

Lynch, D.V., Lin, T.T., Myers, S.P., Leibo, S.P., Macintyre, R.J., Pitt, R.E. & Steponkus, P.L. (1989). A two-step method for permeabilization of *Drosophila* eggs. *Cryobiology,* Vol. 26, pp. (445-452).

Macha, J., Stursova, D., Takac, M., Habrova, V. & Jonak, J. (1997). Uptake of plasmid RSV DNA by frog and mouse spermatozoa. *Folia Biologica,* Vol. 43, pp. (123-127).

Manipalviratn, S., DeCherney, A. & Segars, J. (2009). Imprinting disorders and assisted reproductive technology. *Fertility and Sterility,* Vol. 91, pp. (305-315).

Margaritis, L.H., Kafatos, F.C. & Petri, W.H. (1980). The eggshell of *Drosophila melanogaster.* I. fine structure of the layers and regions of the wild-type eggshell. *Journal of Cell Science,* Vol. 43, pp. (1-35).

Markow, T.A., Beall, S. & Matzkin, L.M. (2009). Egg size, embryonic development time and ovoviviparity in *Drosophila* species. *Journal of Evolutionary Biology,* Vol. 22, pp. (430-434).

Martino, A., Songsasen, N. & Leibo, S.P. (1996). Development into blastocysts of bovine oocytes cryopreserved by ultra-rapid cooling. *Biology of Reproduction,* Vol. 54, pp. (1059-1069).

Matsumoto, H., Jiang, J.Y., Tanaka, T., Sasada, H. & Sato, E. (2001). Vitrification of large quantities of immature bovine oocytes using nylon mesh. *Cryobiology,* Vol. 42, pp. (139-144).

Mazur, P. & Schneider, U. (1986). Osmotic responses of preimplantation mouse and bovine embryos and their cryobiological implications. *Cell Biophysics,* Vol. 8, pp. (259-285).

Mazur, P., Cole, K.W., Hall, J.W., Schreuders, P.D. & Mahowald, A.P. (1992a). Cryobiological preservation of *Drosophila* embryos. *Science (New York, N.Y.),* Vol. 258, pp. (1932-1935).

Mazur, P., Cole, K.W. & Mahowald, A.P. (1992b). Critical factors affecting the permeabilization of *Drosophila* embryos by alkanes. *Cryobiology,* Vol. 29, pp. (210-239).

Mazur, P., Schneider, U. & Mahowald, A.P. (1992c). Characteristics and kinetics of subzero chilling injury in *Drosophila* embryos. *Cryobiology,* Vol. 29, pp. (39-68).

Mozdziak, P.E. & Petitte, J.N. (2004). Status of transgenic chicken models for developmental biology. *Developmental Dynamics,* Vol. 229, pp. (414-421).

Nelson, D.R. & Leopold, R.A. (2003). Composition of the surface hydrocarbons from the vitelline membranes of dipteran embryos. *Comparative Biochemistry and Physiology.Part B, Biochemistry & Molecular Biology,* Vol. 136, pp. (295-308).

Nolan, T.J., Aikens, L.M. & Schad, G.A. (1988). Cryopreservation of first-stage and infective third-stage larvae of *Strongyloides stercoralis. The Journal of Parasitology,* Vol. 74, pp. (387-391).

Nugent, M.J., O'Leary, S.A. & Burnell, A.M. (1996). Optimised procedures for the cryopreservation of different species of *Heterorhabditis. Fundamental and Applied Nematology,* Vol. 19, pp. (1-6).

Nunamaker, R.A. & Lockwood, J.A. (2001). Cryopreservation of embryos of *Culicoides sonorensis* (Diptera: Ceratopogonidae). *Journal of Medical Entomology,* Vol. 38, pp. (55-58).

Papassideri, I., Margaritis, L.H. & Gulik-Krzywicki, T. (1991). The egg-shell of *Drosophila melanogaster.* VI, structural analysis of the wax layer in laid eggs. *Tissue & Cell,* Vol. 23, pp. (567-575).

Papassideri, I.S. & Margaritis, L.H. (1996). The eggshell of *Drosophila melanogaster:* IX. synthesis and morphogenesis of the innermost chorionic layer. *Tissue & Cell,* Vol. 28, pp. (401-409).

Parriego, M., Sole, M., Aurell, R., Barri, P.N. & Veiga, A. (2007). Birth after transfer of frozen-thawed vitrified biopsied blastocysts. *Journal of Assisted Reproduction and Genetics,* Vol. 24, pp. (147-149).

Pavlopoulos, A. & Averof, M. (2005). Establishing genetic transformation for comparative developmental studies in the crustacean *Parhyale hawaiensis. Proceedings of the National Academy of Sciences of the United States of America,* Vol. 102, pp. (7888-7893).

Pedro, P.B., Yokoyama, E., Zhu, S.E., Yoshida, N., Valdez, D.M.,Jr, Tanaka, M., Edashige, K. & Kasai, M. (2005). Permeability of mouse oocytes and embryos at various developmental stages to five cryoprotectants. *The Journal of Reproduction and Development,* Vol. 51, pp. (235-246).

Pogorelov, A.G., Katkov, I.I. & Pogorelova, V.N. (2007). Influence of exposure to vitrification solutions on 2-cell mouse embryos: I. intracellular potassium and sodium content. *CryoLetters,* Vol. 28, pp. (403-408).

Pogorelov, A.G., Katkov, I.I., Smolyaninova, E.I. & Goldshtein, D.V. (2006). Changes in intracellular potassium and sodium content of 2-cell mouse embryos induced by exposition to vitrification concentrations of ethylene glycol. *CryoLetters,* Vol. 27, pp. (87-98).

Popiel, I. & Vasquez, E.M. (1991). Cryopreservation of *Steinernema carpocapsae* and *Heterorhabditis bacteriophora. Journal of Nematology,* Vol. 23, pp. (432-437).

Qiu, L. & Bedding, R.A. (2002). Characteristics of protectant synthesis of infective juveniles of *Steinernema carpocapsae* and importance of glycerol as a protectant for survival of the nematodes during osmotic dehydration. *Comparative Biochemistry and Physiology.Part B, Biochemistry & Molecular Biology,* Vol. 131, pp. (757-765).

Rall, W.F. & Fahy, G.M. (1985). Ice-free cryopreservation of mouse embryos at -196 degrees C by vitrification. *Nature,* Vol. 313, pp. (573-575).

Rall, W.F. (1987). Factors affecting the survival of mouse embryos cryopreserved by vitrification. *Cryobiology,* Vol. 24, pp. (387-402).

Rawson, D.M., Zhang, T., Kalicharan, D. & Jongebloed, W.L. (2000). FESEM and TEM studies of the chorion, plasma membrane and syncytial layers of the gastrula stage embryo of the zebrafish (*Brachydanio rerio*): A consideration of the structural and functional relationship with respect to cryoprotectant penetration. *Aquaculture Research*, Vol. 31, pp. (325-336).

Renfer, E., Amon-Hassenzahl, A., Steinmetz, P.R. & Technau, U. (2010). A muscle-specific transgenic reporter line of the sea anemone, *Nematostella vectensis*. *Proceedings of the National Academy of Sciences of the United States of America*, Vol. 107, pp. (104-108).

Robles, V., Cabrita, E., de Paz, P., Cunado, S., Anel, L. & Herraez, M.P. (2004). Effect of a vitrification protocol on the lactate dehydrogenase and glucose-6-phosphate dehydrogenase activities and the hatching rates of zebrafish (*Danio rerio*) and turbot (*Scophthalmus maximus*) embryos. *Theriogenology*, Vol. 61, pp. (1367-1379).

Rubin, G.M. & Spradling, A.C. (1982). Genetic transformation of *Drosophila* with transposable element vectors. *Science*, Vol. 218, pp. (348-353).

Saito, T., Goto-Kazeto, R., Arai, K. & Yamaha, E. (2008). Xenogenesis in teleost fish through generation of germ-line chimeras by single primordial germ cell transplantation. *Biology of Reproduction*, Vol. 78, pp. (159-166).

Saragusty, J. & Arav, A. (2011). Current progress in oocyte and embryo cryopreservation by slow freezing and vitrification. *Reproduction*, Vol. 141, pp. (1-19).

Sasaki, K., Kurokura, H. & Kasahara, S. (1988). Changes in low temperature tolerance of the eggs of certain marine fish during embryonic development. *Comparative Biochemistry and Physiology*, Vol. 91A, pp. (183-187).

Sasaki, E., Suemizu, H., Shimada, A., Hanazawa, K., Oiwa, R., Kamioka, M., Tomioka, I. *et al.* (2009). Generation of transgenic non-human primates with germline transmission. *Nature*, Vol. 459, pp. (523-527).

Sasakura, Y., Oogai, Y., Matsuoka, T., Satoh, N. & Awazu, S. (2007). Transposon mediated transgenesis in a marine invertebrate chordate: *Ciona intestinalis*. *Genome Biology*, Vol. 8 Suppl 1, pp. (S3).

Scavarda, N.J. & Hartl, D.L. (1984). Interspecific DNA transformation in *Drosophila*. *Proceedings of the National Academy of Sciences of the United States of America*, Vol. 81, pp. (7515-7519).

Schreuders, P.D., Kassis, J., Cole, K.W., Schneider, U., Mahowald, A.P. & Mazur, P. (1996). Kinetics of embryo drying in *Drosophila melanogaster* as a function of the steps of permeabilization: Experimental. *Journal of Insect Physiology*, Vol. 42, pp. (501-516).

Shapiro, D.I., Glazer, I. & Segal, D. (1996). Trait stability and fitness of the heat tolerant entomopathogenic nematode *Heterorhabditis bacteriophora* IS5 strain. *Biological Control*, Vol. 6, pp. (238-244).

Skarnes, W.C., Rosen, B., West, A.P., Koutsourakis, M., Bushell, W., Iyer, V., Mujica, A.O. *et al.* (2011). A conditional knockout resource for the genome-wide study of mouse gene function. *Nature*, Vol. 474, pp. (337-342).

Slanchev, K., Stebler, J., de la Cueva-Mendez, G. & Raz, E. (2005). Development without germ cells: The role of the germ line in zebrafish sex differentiation. *Proceedings of the National Academy of Sciences of the United States of America*, Vol. 102, pp. (4074-4079).

Slifer, E.H. (1948). Isolation of a wax-like material from the shell of the grasshopper egg. *Discussions of the Faraday Society,* Vol. 3, pp. (182-187).

Snabes, M.C., Cota, J. & Hughes, M.R. (1993). Cryopreserved mouse embryos can successfully survive biopsy and refreezing. *Journal of Assisted Reproduction and Genetics,* Vol. 10, pp. (513-516).

Songsasen, N., Buckrell, B.C., Plante, C. & Leibo, S.P. (1995). In vitro and in vivo survival of cryopreserved sheep embryos. *Cryobiology,* Vol. 32, No. 1, (Feb), pp. (78-91).

Steponkus, P.L., Myers, S.P., Lynch, D.V., Gardner, L., Bronshteyn, V., Leibo, S.P., Rall, W.F., Pitt, R.E., Lin, T.T. & MacIntyre, R.J. (1990). Cryopreservation of *Drosophila melanogaster* embryos. *Nature,* Vol. 345, pp. (170-172).

Stevens, J.C., Banks, G.T., Festing, M.F. & Fisher, E.M. (2007). Quiet mutations in inbred strains of mice. *Trends in Molecular Medicine,* Vol. 13, pp. (512-519).

Stuart, R.J. & Gaugler, R. (1996). Genetic adaptation and founder effect in laboratory populations of the entomopathogenic nematode *Steinernema glaseri. Canadian Journal of Zoology,* Vol. 74, pp. (164-170).

Sugiyama, R., Nakagawa, K., Shirai, A., Sugiyama, R., Nishi, Y., Kuribayashi, Y. & Inoue, M. (2010). Clinical outcomes resulting from the transfer of vitrified human embryos using a new device for cryopreservation (plastic blade). *Journal of Assisted Reproduction and Genetics,* Vol. 27, pp. (161-167).

Summers, P.M., Shephard, A.M., Taylor, C.T. & Hearn, J.P. (1987). The effects of cryopreservation and transfer on embryonic development in the common marmoset monkey, *Callithrix jacchus. Journal of Reproduction and Fertility,* Vol. 79, pp. (241-250).

Sun, Y., Zhang, Q.J. & Wang, Y.Q. (2007). Programmed cryopreservation of the amphioxus *Branchiotoma belcheri* embryos. *Current Zoology,* Vol. 53, pp. (524-530).

Takeuchi, Y., Yoshizaki, G. & Takeuchi, T. (2003). Generation of live fry from intraperitoneally transplanted primordial germ cells in rainbow trout. *Biology of Reproduction,* Vol. 69, pp. (1142-1149).

Toledo, J.D. & Kurokura, H. (1990). Cryopreservation of the euryhaline rotifer *Brachionus plicatilis. Aquaculture,* Vol. 91, pp. (385-394).

Triantaphyllou, A.C. & McCabe, E. (1989). Efficient preservation of root-knot and cyst nematodes in liquid nitrogen. *Journal of Nematology,* Vol. 21, pp. (423-426).

Trounson, A. & Mohr, L. (1983). Human pregnancy following cryopreservation, thawing and transfer of an eight-cell embryo. *Nature,* Vol. 305, pp. (707-709).

Tsang, W.H. & Chow, K.L. (2009). Mouse embryo cryopreservation utilizing a novel high-capacity vitrification spatula. *BioTechniques,* Vol. 46, pp. (550-552).

Tsang, W.H. & Chow, K.L. (2010). Cryopreservation of mammalian embryos: Advancement of putting life on hold. *Birth Defects Research.Part C, Embryo Today : Reviews,* Vol. 90, , pp. (163-175).

Vajta, G., Holm, P., Kuwayama, M., Booth, P.J., Jacobsen, H., Greve, T. & Callesen, H. (1998). Open pulled straw (OPS) vitrification: A new way to reduce cryoinjuries of bovine ova and embryos. *Molecular Reproduction and Development,* Vol. 51, pp. (53-58).

Valdez, D.M.,Jr, Miyamoto, A., Hara, T., Edashige, K. & Kasai, M. (2005). Sensitivity to chilling of medaka (*Oryzias latipes*) embryos at various developmental stages. *Theriogenology,* Vol. 64, pp. (112-122).

Valencia, M.D., Miller, L.H. & Mazur, P. (1996). Permeabilization of eggs of the malaria mosquito *Anopheles gambiae. Cryobiology,* Vol. 33, pp. (149-162).

Van Wyk, J.A., Gerber, H.M. & Van Aardt, W.P. (1977). Cryopreservation of the infective larvae of the common nematodes of ruminants. *The Onderstepoort Journal of Veterinary Research,* Vol. 44, pp. (173-194).

Vanderzwalmen, P., Bertin, G., Debauche, C., Standaert, V., Bollen, N., van Roosendaal, E., Vandervorst, M., Schoysman, R. & Zech, N. (2003). Vitrification of human blastocysts with the hemi-straw carrier: Application of assisted hatching after thawing. *Human Reproduction,* Vol. 18, pp. (1504-1511).

Wang, X. & Grewal, P.S. (2002). Rapid genetic deterioration of environmental tolerance and reproductive potential of an entomopathogenic nematode during laboratory maintenance. *Biological Control,* Vol. 23, pp. (78).

Wang, R.Y., Guan, M., Rawson, D.M. & Zhang, T. (2008). Ultrasound enhanced methanol penetration of zebrafish (*Danio rerio*) embryos measured by permittivity changes using impedance spectroscopy. *European Biophysics Journal,* Vol. 37, pp. (1039-1044).

Weigmann, K., Klapper, R., Strasser, T., Rickert, C., Technau, G., Jackle, H., Janning, W. & Klambt, C. (2003). FlyMove--a new way to look at development of *Drosophila. Trends in Genetics,,* Vol. 19, pp. (310-311).

Wheeler, M.B. (2003). Production of transgenic livestock: Promise fulfilled. *Journal of Animal Science,* Vol. 81 Suppl 3, pp. (32-37).

Whittingham, D.G., Leibo, S.P. & Mazur, P. (1972). Survival of mouse embryos frozen to -196 degrees and -269 degrees C. *Science,* Vol. 178, pp. (411-414).

Willadsen, S.M. (1977). Factors affecting the survival of sheep embryos during-freezing and thawing. *Ciba Foundation Symposium,* Vol. (52), pp. (175-201).

Wilmut, I. & Rowson, L.E. (1973). Experiments on the low-temperature preservation of cow embryos. *The Veterinary Record,* Vol. 92, pp. (686-690).

Wilton, L.J., Shaw, J.M. & Trounson, A.O. (1989). Successful single-cell biopsy and cryopreservation of preimplantation mouse embryos. *Fertility and Sterility,* Vol. 51, pp. (513-517).

Yoshida, T., Arii, Y., Hino, K., Sawatani, I., Tanaka, M., Takahashi, R., Bando, T., Mukai, K. & Fukuo, K. (2011). High hatching rates after cryopreservation of hydrated cysts of the brine shrimp *A. franciscana. CryoLetters,* Vol. 32, pp. (206-215).

Zeilmaker, G.H., Alberda, A.T., van Gent, I., Rijkmans, C.M. & Drogendijk, A.C. (1984). Two pregnancies following transfer of intact frozen-thawed embryos. *Fertility and Sterility,* Vol. 42, pp. (293-296).

Zelenin, A.V., Alimov, A.A., Barmintzev, V.A., Beniumov, A.O., Zelenina, I.A., Krasnov, A.M. & Kolesnikov, V.A. (1991). The delivery of foreign genes into fertilized fish eggs using high-velocity microprojectiles. *FEBS Letters,* Vol. 287, pp. (118-120).

Zernicka-Goetz, M., Pines, J., McLean Hunter, S., Dixon, J.P., Siemering, K.R., Haseloff, J. & Evans, M.J. (1997). Following cell fate in the living mouse embryo. *Development,* Vol. 124, pp. (1133-1137).

Zhang, T. & Rawson, D.M. (1998). Permeability of dechorionated one-cell and six-somite stage zebrafish (*Brachydanio rerio*) embryos to water and methanol. *Cryobiology,* Vol. 37, pp. (13-21).

Zhang, X., Trokoudes, K.M. & Pavlides, C. (2009). Vitrification of biopsied embryos at cleavage, morula and blastocyst stage. *Reproductive Biomedicine Online*, Vol. 19, pp. (526-531).

10

Cryopreservation of Porcine Gametes, Embryos and Genital Tissues: State of the Art

Heriberto Rodriguez-Martinez

Department of Clinical and Experimental Medicine, Faculty of Health Sciences,
Linköping University, Linköping
Sweden

1. Introduction

Preservation of germplasm (e.g. a term hereby applied to collectively gather spermatozoa, oocytes or early embryos whose use would –eventually- lead to offspring) for research, repository building and propagation of genetic material using Assisted Reproductive Techniques (ART) has a long lasting priority (Mazur et al 2008). The first approaches, besides those historically anecdotic (see Flowers 1999) were directed to the application of artificial insemination (AI) of domestic species (Foote 1999) pertaining dissemination of genetics to a general population of, particularly, production animals. Positive effects for simple cryo-protectant agents (CPA) such as glycerol on animal sperm cryoprotection were demonstrated already by the end of the 1930`s (Bernschtein & Petropavloski 1937) and a decade later it became apparent that spermatozoa could be cooled, frozen and thawed in solutions containing egg yolk and glycerol (Polge et al 1949). For some species, such as bovine, the fact that bull semen could be easily frozen with an acceptable sperm survival post-thaw and accompanied by acceptable fertility after intra-uterine AI led to the rapid development of such primary reproductive biotechnology (Rodriguez-Martinez & Barth 2007). Attempts in other species of domesticated animals followed, and it was soon realised that the success seen with bovine could not be reached, primarily due to low sperm survival, difficulties in attaining an optimal deposition or proper timing towards spontaneous ovulation. Differences in survival and fertility varied not only among species but also between individuals of a given species or even ejaculates within sires (Holt 2000).

Porcine male germplasm freezing started already by the 1950´s (Polge 1956) but their post-thaw fertility was not reassured using cervical AI until a decade later (Crabo and Einarsson 1971, Graham et al 1971, Pursel & Johnson 1971), which revealed major constrains when applying cryopreservation on boar spermatozoa. Today, despite documented efforts to reach acceptable fertility and prolificacy after AI (Eriksson et al 2002, Roca et al 2011), the cryosurvival of boar spermatozoa is still consistently low in comparison to other species, owing to damage during a processing that is time-consuming, costly and yields few doses per ejaculate (see Rath et al 2009, Rodriguez-Martinez & Wallgren 2011). Number of piglets born is lower than for cooled or neat semen implying that sperm lifespan, deposition site and closeness to ovulation are yet significant hurdles to be overcome (Roca et al 2006b, Wongtawan et al 2006). Preservation of male genetics can also be performed by freezing of epidydimal spermatozoa (retrieved by biopsy of the cauda) or by tissue sampling through testicular biopsies (Keros et al 2005, Curaba et al 2011). However, these approaches are not relevant for porcine breeding. Cauda epididymal spermatozoa are easier to slow freeze than ejaculated spermatozoa (Rath & Niemann 1997) and testicular biopsies are not advisable in boars owing to their highly vascularized testicular capsule (Ohanian et al 1979).

Preservation of female genetics can be done either by freezing of germplam (e.g. oocytes or embryos) or of ovarian tissue (slices or whole ovary), from which oocytes can thereafter be harvested for ART. Germplasm freezing in pigs has also followed a tortuous road, with deceiving results for decades, particularly related to the high sensitivity of pig oocytes and early embryos to chilling, similarly to other species containing large deposits of intracellular lipids (Zhou & Li 2009), in contrast to blastocysts where the lipid amounts were lower. Delipation (or side-dislocation of lipid depots by ultracentrifugation) was soon shown to increase the survival of oocytes/embryos subjected to freezing (Nagashima et al 1995), survival that could be enhanced if the cytoskeleton could be preserved from damage using exogenous chemicals (Shi et al 2006). Use of alternative methods such as vitrification instead of slow cooling led to better survival (see Massip 2001) including the birth of offspring (Berthelot et al 2000). However, large variation was seen among methods, sources and laboratories (Holm et al 1999, Cuello et al 2007, Somfai et al 2008, Ogawa et al 2010), including the method used for intrauterine deposition (Rodriguez-Martinez 2007b, Roca et al 2011).

Cryopreservation of ovarian tissue (or even of whole ovaries) has been tested in several species including human (Isachenko et al 2009) pertaining the recovery of follicular oocytes for ART or ultimate autographs (Kim et al 2010). Procedures for porcine ovarian samples have followed methods tested in other species (Imhof et al 2004, Borges et al 2009) with promising results, albeit yet at an academic level, pertaining the advancement of xenografting (Moniruzzaman et al 2009). As well, experimental models using the porcine species have been developed for cryopreservation of genital tissues, particularly the uterus (Dittrich et al 2004, 2006) paving the way for human transplantation procedures (Diaz-Garcia et al 2011).

Thus, interest has been large to attempt routine cryopreservation of porcine gametes, embryos and genital tissues, yet with various degrees of success. Therefore, the present review aims to summarize the state-of-the-art regarding established and emerging methods for the cryopreservation of porcine gametes and embryos as germplasm, intending a critical revision of the underlying problems that still constrain their application for establishing repositories, their use in reproductive biotechnologies and, ultimately, for breeding. As well, it intends to describe our level of knowledge when attempting cryogenics of gonads and other genital tissues for comparative research, particularly on human regenerative medicine. The review is not exhaustive and focus on methodological aspects of procedures.

2. What happens during cryopreservation?

Independently of the cell or tissue above mentioned being considered, the current methods for their cryopreservation fall into one of the two following categories: (i) slow equilibrium freezing or, (ii) rapid non-equilibrium vitrification, and variations within. In either case, the entire process basically concerns the way we handle the presence of water in and around the cells and whether its freezing is allowed (conventional cryopreservation, slow equilibrium freezing) or totally prevented (vitrification).

In the first method, which is the one traditionally used in biomedicine, particularly for sperm preservation, cells are subjected to slow cooling to temperatures below zero, with freezing rates of 0.5-100 ∘C/min). The method allows ice to form and solute to concentrate alongside the change in water phase. Both ice and high solute concentrations can cause direct (either initial or eutectic, Han & Bischof 2004a), respectively secondary damage, jeopardizing cell survival or handicapping vital cell functions post-thaw. At some moment during the process, water freezes to form ice, primarily extracellular, but even intracellular. Ice grows and becomes over time surrounded by an increasing amount of solutes which move to the areas where water did not yet changed phase. Cells balance ion concentrations at either side of the plasma membrane thus keeping proper osmotic pressure. Depending on the relative amounts of free and bound water, such a change of phase (either formation or dissolution of ice and de/rehydration phenomena) implies changes in ionic concentration caused by directional movement of water across the membrane, disturbing the homeostatic osmotic pressure of the cell/s (Pegg 2007). Cells respond by allowing water to leave the intra-cellular compartment, to compensate the increasing hyper-osmotic extra-cellular compartment caused by the progressive formation of ice. Those water movements lead both to cell dehydration and to a toxic hyper-concentration of solutes intracellular which, ultimately, affects cells viability (Watson & Fuller 2001). Freezing injury can then be related to high electrolyte concentration effects (solute effects), presence of intracellularly ice (formed direct or eutectic) and also the pressure of large extracellular crystals on the veins of concentrated (i.e. vitrified) extender and cells (Saragusty et al 2009). See **Figures 1and 2** for an illustration of these events.

Freezing injury during slow freezing can be minimized. Intracellular freezing is generally lethal but can be avoided by sufficiently slowing the rate of cooling. Solute-caused damage, which is the dominating feature under conventional slow freezing especially in cells in liquid suspension, can be minimized by the addition of CPA. Most CPA´s (as glycerol, dimethyl sulfoxide (DMSO), ethylenglycol (EG), propyleneglycol (PG)) are highly soluble, permeating compounds of low-to-medium toxicity, whose primary role is to reduce the amount of ice formed at any given sub-zero temperature, by simply increasing the total concentration of all solutes in the system, thus defining the concept of slow equilibrium freezing (Pegg 2002 2007). Introduction of sufficient CPA would eventually avoid freezing and a glassy of vitreous state could be produced instead. Such concept is the theoretical rationale for the second method listed above: rapid non-equilibrium vitrification. Vitrification is the physical process by which a highly concentrated cryoprotective solution supercools to very low temperatures (often to -120 to -130 ∘C) to finally solidify into a metastable glass, without undergoing crystallization at a practical, high speed cooling rate (i.e. dipping onto LN_2). Use of ultra-high speed voids the need of penetrating CPA, open for using non-penetrating CPA (such as sucrose, fructose, glucose), but demands the use of small (5-50µL) suspension droplets. The glassy state is defined by its viscosity reaching 10-13 poises, sufficient for the aqueous material to behave as a solid, without any water crystallization. Once again, this waives the above listed sources for cell injury: ice crystals and increased/ill distributed solute concentrations. The CPA used to vitrify cells include those used during conventional freesing but at very high concentrations (10-fold higher compared to slow freezing), near the maximum tolerated by the cells, thus becoming potentially harmful (Pegg 2005). Penetrating CPA-free vitrification was attempted already by the early 1940´s using rabbit spermatozoa plunged into LN$_2$ (Hoagland & Pincus 1942). Use of non-penetrating "CPA" (CPA-free concept) such as sucrose has proven feasible for the spermatozoa of some species, including human (Isachenko et al 2004, 2005, 2008, Hossain & Osuamkpe 2007), primates (Dong et al 2009), or canine (Sanchez et al 2011), where sperm suspensions were vitrified (either drop-wise, Isachenko et al (2004, 2005) or contained in 50µL-plastic capillaries (Isachenko et al 2011)) by plunging in LN$_2$, with a cooling rate of ˜ 10,000 ₒC/min. Basically, vitrification is therefore always determined by a relation between cooling rate, medium viscosity and sample size.

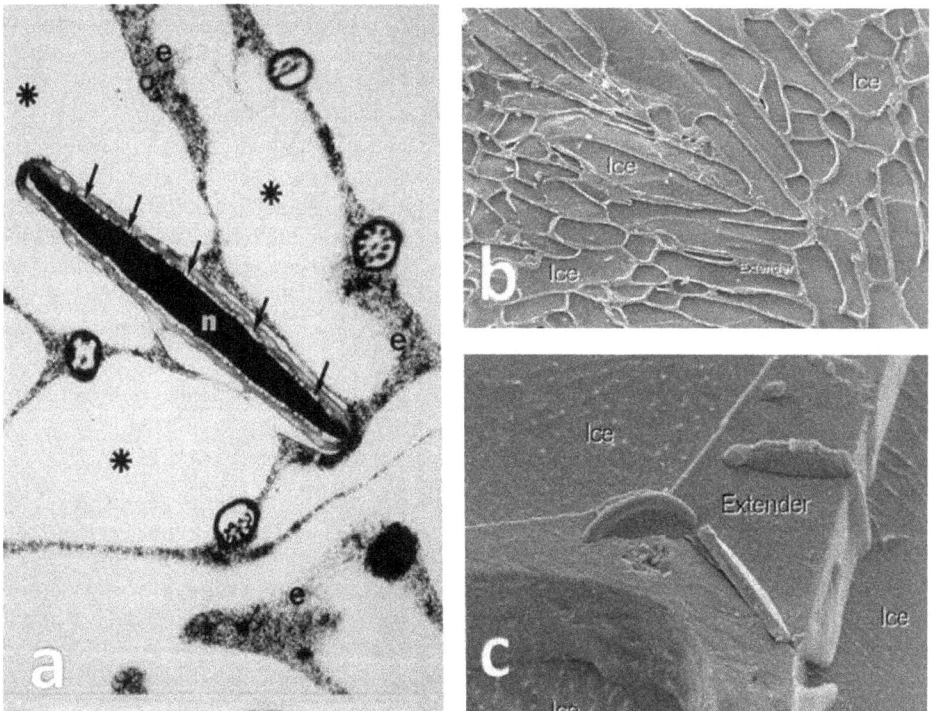

Fig. 1. Micrographs of frozen boar semen illustrated with (a) transmission electron microscopy or (b,c) Cryo-scanning electron microscopy. Spermatozoa were extended and conventionally frozen in maxi-straws (a, 5 mL) or FlatPack™ (b,c, 5 mL) and subjected to freeze-substitution (a) or partial sublimation (b,c) to depict extracellular ice lakes (* in a, marked with legend in b,c) and the veins of concentrated extender (e in a, legend in b,c). Note the presence of intracellular ice marks (arrows in a) and the dislocation of axoneme structures in the tails. Such marks are not seen in the FlatPack™ material (Photo: Dr Hans Ekwall, Uppsala, Sweden).

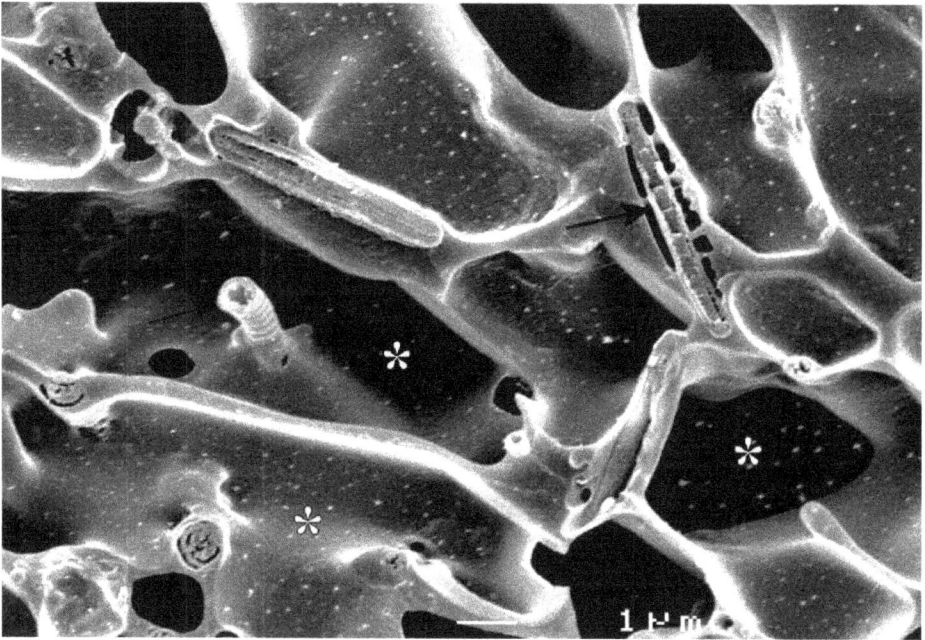

Fig. 2. Cryo-scanning electron microscopy (cryo-SEM) micrograph at higher magnification showing the contents of a maxi-straw frozen at the speed of 1,200°C/min (by direct plunging into LN$_2$ after initial cooling to +5°C). The ice lakes (*) are small, and surrounded by prominent veins. This fast cooling caused freezing of water both extra- and intra-cellularly, with clear evidence of sub-cellular distortion caused by the presence of intracellular ice crystals in the peri-nuclear and peri-axonemal areas, owing to a lack of sperm dehydration during the process. Note the fractured sperm head (large arrow) with marks of lethal intracellular ice, and the tail entrapped by extracellular ice (small arrow) with dislocation of the axoneme (Courtesy of Dr Hans Ekwall).

However, we should bear in mind that the physical phenomenon of vitrification (e.g. the process by which a liquid begins to behave as a solid during cooling without substantial changes in molecular arrangement or thermodynamic state variables) is as relevant to conventional freezing, where the cells survive in this glassy medium between ice crystals (see **Figure 1a-c**) as to vitrification *per se*, where the entire sample is vitrified (Wowk 2010). Therefore, seeding is quite relevant for supercooled vitrification solutions in conventional freezing, while it does not play any role during pure vitrification, provided that cooling rates are high. For instance, use of LN$_2$-slush (e.g. lowering the temperature to near the freezing point of LN$_2$, -205 to -210 °C by applying negative pressure to the LN$_2$, Yavin et al 2009) increases the cooling rate 2 to 7-fold compared to simple plunging in LN$_2$. Viscosity also plays a major role and must increase during cooling, until the glass transition (i.e. the change from liquid to solid) is reached. This concept opens for the freezing of highly concentrated semen samples, provided the size of the sample is small enough.

When thawing or re-warming occurs, the events above described basically reverse. Slow re-warming allows water to reflux to the areas where solutes are concentrated in cells treated by slow freezing, but the time elapsing is not short enough to avoid the toxicity that the solute concentration exerts on the cells, either leading to cell death or dysfunction. If the re-warming is too slow, ice (intracellular in particular) can damage organelles and the cytoskeleton. Rapid rewarming diminishes these risks since the toxic solutes or CPA are only momentarily present.

For either method listed, the CPA has to gain access to all areas of the cell/tissue/organ. Traditional cooling and re-warming rates affect the fluidity of the membranes of the cell and the organelles through the rearrangement of structural proteins and the dislocation of constituent lipids. If these changes affect diffusion and/or osmosis, they can jeopardize -by causing changes in the viscosity of fluids or inducing osmotic inbalance- the proper distribution of the CPA, its introduction and removal and ultimately, the freezing and the thawing process (Morris 2006, Morris et al 2007). Cooling can disrupt the integrity of the cytoskeleton and of the chromatin structure, including DNA damage (Watson & Fuller 2001, Fraser et al 2011). In cells in suspension, such as spermatozoa, both the form and volume of the sample to be cooled/re-warmed, and the concentration of the contained cells play major roles during the most damaging interval in the process, i.e. during the changes in phase of the extra-cellular water, when heat is either dissipated (during cooling) or incorporated (during re-warming) (Mazur & Cole 1989, Morris et al 1999). It is therefore obvious that samples (cells, tissues, organs) have to pass cooling and re-warming under conditions where cell injury can be minimized (Morris 2006, Morris et al 2007).

3. Cryopreservation of boar semen: State-of-the-art

Porcine AI with liquid-stored semen where either the entire ejaculate or only the sperm-rich fraction (SRF) of the ejaculate is collected, the spermatozoa are re-suspended at low concentrations in chemically defined extenders and stored at 16–20°C for several days before use, most often for up to 3 days) has increased exponentially in the past 25 years. Globally, AI is practiced in 75% of sows (range 10-99%) using >160 million semen doses, with countries in Europe and the Americas having basically all sows under AI (Riesenbeck 2011). Fertility rates are similar to those obtained after natural mating (for a review, see Rodriguez-Martinez 2007a). Liquid semen is therefore used both for production breeding and for genetic improvement at national or regional level, with some export countries having a major international trade. However, the limited shelf-life of liquid semen, its decline in fertility over transit time, and risks of damage due to temperature, pressure or handling changes, all call for alternatives, with a focus on frozen-thawed semen.

Boar spermatozoa are still being "best" cryopreserved (in terms of cryosurvival) using protocols originally devised in the mid 1970´s (Westendorf et al 1975) with modifications (most often empirically introduced). Most methods use standard lactose-egg yolk (or LDL)-based cooling and freezing media, following the removal of most of the seminal plasma by extension in chelate-containing (often EDTA) buffers and centrifugation. The freezing media most often include a surfactant (often laurylsulphate, Orvus es Paste-OEP) and glycerol as CPA (2-3% final concentration added at +5°C). Spermatozoa are further cooled beyond the eutectic temperature at 30 to 50°C/min. Thawing is done at 1,000-1,800 °C/min. The entire procedure takes most often 8-9 hours from semen collection to storage of the frozen doses in LN$_2$, being still tedious (many different steps) and, inconvenient, producing few AI-doses per ejaculate (5-8). For examples of current protocols see Eriksson & Rodriguez-Martinez (2000), Saravia et al (2005), Parrilla et al (2009) or Rath et al (2009) and methods cited therein.

This general current protocol fits most boars but considering the large variation between ejaculates and –particularly– among boars for their capacity to sustain cryopreservation (Roca et al 2006a), the protocol has to be modified to accommodate those with sub-optimal sperm freezability (the so-called bad freezers), particularly regarding glycerol concentration and warming rates (Hernandez et al 2007a). Those changes usually allow for minimum acceptable cryosurvival (i.e. around 40%). However, it clearly shows that the methodology is still sub-optimal. Current semen cryopreservation techniques are technically demanding and expensive, both in terms of labour- and laboratory equipment costs, as well as time-consuming (rev by Roca et al 2006b, 2011). Last but not least, there is a lack of reliable laboratory tests for the accurate assessment of semen quality in vitro, that limits our capacity to properly monitor the methods used to freeze-thaw boar semen and, particularly, its relationship to AI-fertility (Rodriguez-Martinez, 2007b). This is critical, since despite having acceptable post-thaw survival (even above 60%) this cryosurvival is not reflected in fertility after AI. Thus, boar spermatozoa are considered one of the most demanding cell types with respect to sustaining viability during freezing and thawing, with a large proportion of the spermatozoa not surviving these procedures (Penfold & Watson 2001). Moreover, those surviving spermatozoa are usually a mixture of cells, some of which survive well while others show modified motility and a shortened lifespan, factors which compromise their fertilising ability. Insemination with such spermatozoa leads, ultimately, to lowered pregnancy rates and fewer piglets born, compared with AI using liquid-stored semen (Knox 2011). In sum, although freezing methods are nowadays rather stable in many laboratories and yield above 50% of sperm survival post-thaw, fertility after AI is extremely variable (Parrilla et al 2009). The major constrain is not only the inherent difficulties to freeze spermatozoa from this species (Holt, 2000a,b), but -within the species- the sire-dependent cryosurvivability to the current procedures (Eriksson et al 2002, Holt et al 2005, Gil et al 2005, Waterhouse et al 2006, Hernandez et al 2006, 2007a, Roca et al., 2006a, Parrilla et al 2009, Roca et al 2011).

This variation is usually compensated by the AI of excessive sperm numbers (at least 5×10^9 spermatozoa per AI-dose), i.e. double the numbers of total spermatozoa present in liquid semen doses. Fertility post-AI is nowadays substantially better, closer to AI with liquid semen (Eriksson et al 2002). See **Table 1** for an overview of fertility after conventional (cervical) AI with frozen-thawed boar semen. Fertility with lower sperm numbers is also becoming acceptable when deep intrauterine AI is practiced, although data are still restricted in numbers (Bathgate et al 2006, Roca et al 2006b, 2011). But, even with these huge sperm numbers, overall fertility (as farrowing rates) and prolificacy (as litter size) are still lower than for liquid semen (around 10-30 % lower farrowing rates, and 1-3 less piglets), indicating that other factors are limiting, such as the timing of insemination respective to spontaneous ovulation (Bolarín et al 2006, Wongtawan et al 2006). This implies that we are far from reaching the goals set up by the industry for the use of frozen-thawed semen: 85% of conception rates and a litter size of 11 piglets (Knox 2011). So, frozen-thawed boar semen is still basically limited to research, genetic banking or the export of semen for selected nuclei lines, constituting barely above 1% of all AIs.

Package	AIs	AI-dose (x10^9)	Sows	Farrowed (%)	Litter size (n)	References
Pellet	2	6	334	40	6.4	Didion & Schoenbeck 1996
MS	2-4	2.5	392	48	10.4	Almlid & Hofmo 1996
MS	2-4	2.5	496	57	12.2	Almlid & Hofmo 1996
MS	2-4	2.5	350	50	9	Almlid & Hofmo 1996
MS	2	5	190	62.6	9	Hofmo & Grevle 2000
FlatPack™	2	5	352	72.2	10.7	Eriksson et al 2002

Table 1. Fertility after conventional (cervical) AI in field trials (>100 sows) with frozen-thawed boar semen (modified from Roca et al 2006).

Therefore, it seems -at first sight- unlikely that deep frozen semen will replace the use of fresh semen on an extensive basis even if the fertility levels were similar. It is too expensive considering that the current cryopreservation protocols barely yield half of the doses produceable per ejaculate. Since the amount of spermatozoa per dose is minimum twice that of liquid-stored semen, such equation is simply undependable from a commercial point of view both in production costs for AI-doses and the sub-optimal boar use. However, having a reliable cryopreservation method for boar semen would (a) allow selection of genetics from all over the world, (b) enable planned, essential AIs´ at the top of the breeding pyramid and so (c) facilitate preservation of top quality genetic lines for ongoing or future breeding programmes and/or (d) offer an extra health safeguard, by allowing completion of any health test specified by a country or breeding organization before use. The challenge is there, undoubtedly.

3.1 Improvements in boar semen freezing

Over the past decade, the cryobiology of boar semen has diminished its empiric approach towards a more experimental one. Major areas of research have involved: (i) the determination of *in vivo* features (particularly regarding seminal plasma (SP) and the presence of characteristic fractions of the ejaculate), (ii) the action of specific additives and different CPA, (iii) the use of automated freezers and of directional gradient freezing and, (iv) the use of novel containers adapted for the freezing of concentrated spermatozoa. Fertility post-AI is nowadays substantially better, closer to AI with liquid semen, even when using lower sperm numbers and alternative sites of sperm deposition, such as deeply intra-utero (Roca et al 2011).

Specific additives and different CPA´s: Glycerol, a small, poly-hydroxylated solute highly soluble in water since it interacts with it by hydrogen bonding, and able to permeate across the plasma membrane, at a low rate, is by far the mostly used CPA for boar semen conventional freezing. Since glycerol disturbs cell metabolism at body temperature, boar spermatozoa are usually exposed to this CPA at ~5°C, which –unfortunately- further slows its low rate permeation. Mixed with the other solutes of the extender in solution, it depresses their freezing point and ameliorates the rise in sodium chloride concentration during dehydration. Moreover, glycerol increases viscosity with lowering temperatures to more than 100,000 cP by -55°C (Morris et al 2006), leading to a retardation of both ice crystal growth and of dehydration speed on a kinetic basis. Moreover, glycerol eliminates eutectic phase changes of the extender (Han & Bischof 2004b), making it a very suitable CPA when added at 2-3% rates. While such interval does not affect cryosurvival in "good-freezer" boars, those considered moderate or bad freezers benefit from a minimum of 3% glycerol (Hernandez et al 2007a). A broad range of other solutes (mostly alcohols, sugars, diols and amides) have also been tested for CPA capacity (Fuller 2004, Buranaamnuay et al 2011), but boar spermatozoa react variably. Alcohols and diols can induce membrane blebbing. Sugars (such as the dissacharides sucrose, raffinose or trehalose which both increase viscosity and stabilise the membrane by interacting with phospholipids) are not better than glycerol, regarding cryosurvival (Hu et al 2008), but shows synergistic effects (Gutierrez-Perez et al 2009, Hu et al 2009). On the other hand, replacing glycerol with amides (formamide; methyl- or dimethylformamide, MF- DMF; acetamide; methyl- or dimethylacetamide (MA- DMA) at ~5% concentration, has proven beneficial for cryo-susceptible boars, probably because the amide permeates the plasma membrane more effectively than glycerol, thus causing less osmotic damage during thawing (Bianchi et al 2008). Other additives enhance cryosurvival of boar spermatozoa, such as L-glutamine (de Mercado et al 2009) or low rates (<0.1%) of N-acetyl-D-glucosamine (Yi et al 2002a), the latter possibly interacting with the surfactant OEP (Yi et al 2002b). Laurylsulphate, albeit its mode of action is yet unexplained in detail regarding interaction with egg yolk and the sperm plasma membrane, has repeatedly proven valuable (Karosas & Rodriguez-Martinez 1993, Buranaamnuay et al 2009). Use of low-density lipoproteins (LDLs), isolated from egg-yolk from different species (Jiang et al 2007), has proven beneficial for sperm function post-thaw, particularly for DNA-integrity. Similarly, sperm cryosurvival has been enhanced by the addition of antioxidants (Peña et al 2003, 2004a, Roca et al 2005, Jeong et al 2009, Kaeoket et al 2010), hyaluronan (Peña et al 2004b), or platelet-activating factor (PAF, Bathgate et al 2007), although the beneficial effects vary, particularly when different sperm sub-populations are used. Cryosurvival of several cold-shock susceptible species, of which the porcine is one, has been found to improve when cholesterol-loaded cyclodextrins (CLC) are used as additives before cooling (Zeng & Terada 2001, Mocé et al 2010). Cyclodextrins can encapsulate hydrophobic compounds, such as cholesterol, and transfer the cholesterol into membranes down a concentration gradient (Zidovetzki & Levitan 2007). However, it is yet to determine if the effects are substantial and not only individually-related (Waterhouse et al 2006).

Automated freezers and directional gradient freezing: Controlled freezing using programmed freezers improves cryosurvival by use of "optimal" cooling (and thawing) rates e.g. those that substantially diminish the period during which heat is released/absorbed in the sample when water changed phase (i.e. ice was formed/melt). Interestingly enough, experimentally-determined optimal rates of the range 30-50°C/min (Thurston et al 2003, Medrano et al 2009, Juarez et al 2011) have been theoretically predicted (Devireddy et al 2004, Woelders & Chaveiro 2004) and confirmed by use of novel procedures such as directional freezing where the thermal gradient is monitored by modifying the velocity at which the liquid-ice interface grows so that the size and shape of the ice crystals is maintained within optimal limits. In this methodology, derived from the principle of seeding, the biological material is moved through a linear temperature gradient, so that both the freezing rate and the ice front propagation are controlled (Arav et al 2002, Woelders et al 2005, Saragusty et al 2007). The method can be advantageously applied both to large samples, frozen in large containers moving along a rim of seeding or to highly concentrated samples, in smaller (i.e. mini-straws) containers.

Cryobiologically best-suited packaging containers: use of cryobiologically adequate packaging systems for the extended spermatozoa, showed a direct improvement of cryosurvival. Boar spermatozoa has been processed in plastic straws of different volumes (0.25 to 5 mL, Johnson et al 2000), in flattened 5 mL straws (Weitze et al 1987), in metal (Fraser & Strzezek 2007) or in plastic bags of various types and constitution (Bwanga et al 1991, Mwanza & Rodriguez-Martinez 1993, Ortman & Rodriguez-Martinez 1994; Eriksson & Rodriguez-Martinez 2000, Eriksson & Rodriguez-Martinez 2000, Saravia et al 2005). The latter developed, denominated "FlatPacks™" proven equally good or better than 0.25 mL straws in terms of sperm cryosurvival despite of the fact that they held 5 mL of semen (an entire dose for cervical AI, 5 billion spermatozoa), thus waiving the need of pooling innumerable straws when thawing. Fertility after conventional cervical AI of FlatPack™ frozen-thawed semen yield acceptable farrowing rates and litter sizes (Eriksson et al 2002). See **Figure 3** for a schematic description of the differences between containers. The FlatPack™ was considered as cryobiologically convenient (very thin and with a large surface) to dissipate heat during cooling and warm rapidly, as those small containers tested. It is important to remember that the freezing in these containers, with high heat dissipation, inflicts less damage to the cells by intra-container mechanical pressure (Saragusty et al 2009). **Figure 4** shows cryological differences in shape and size of frozen water lakes between mini-straws and FlatPack™.

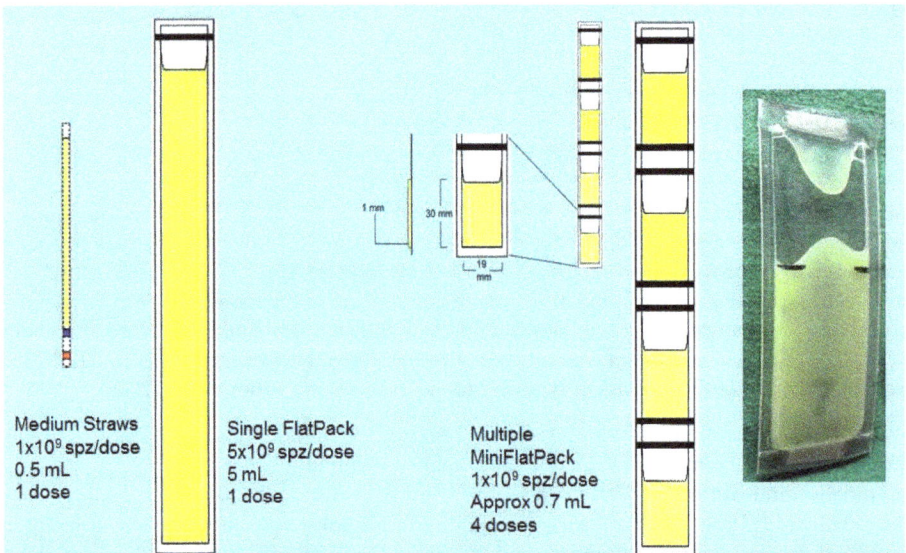

Fig. 3. Schematic representation of the major differences between plastic 0.25 mL mini-straws, with single and Multiple FlatPack™ (the latter also named "MiniFlatPack™, see adjacent photograph of a filled and sealed MiniFlatPack™)(Diagramme/photograph: courtesy of Dr Fernando Saravia).

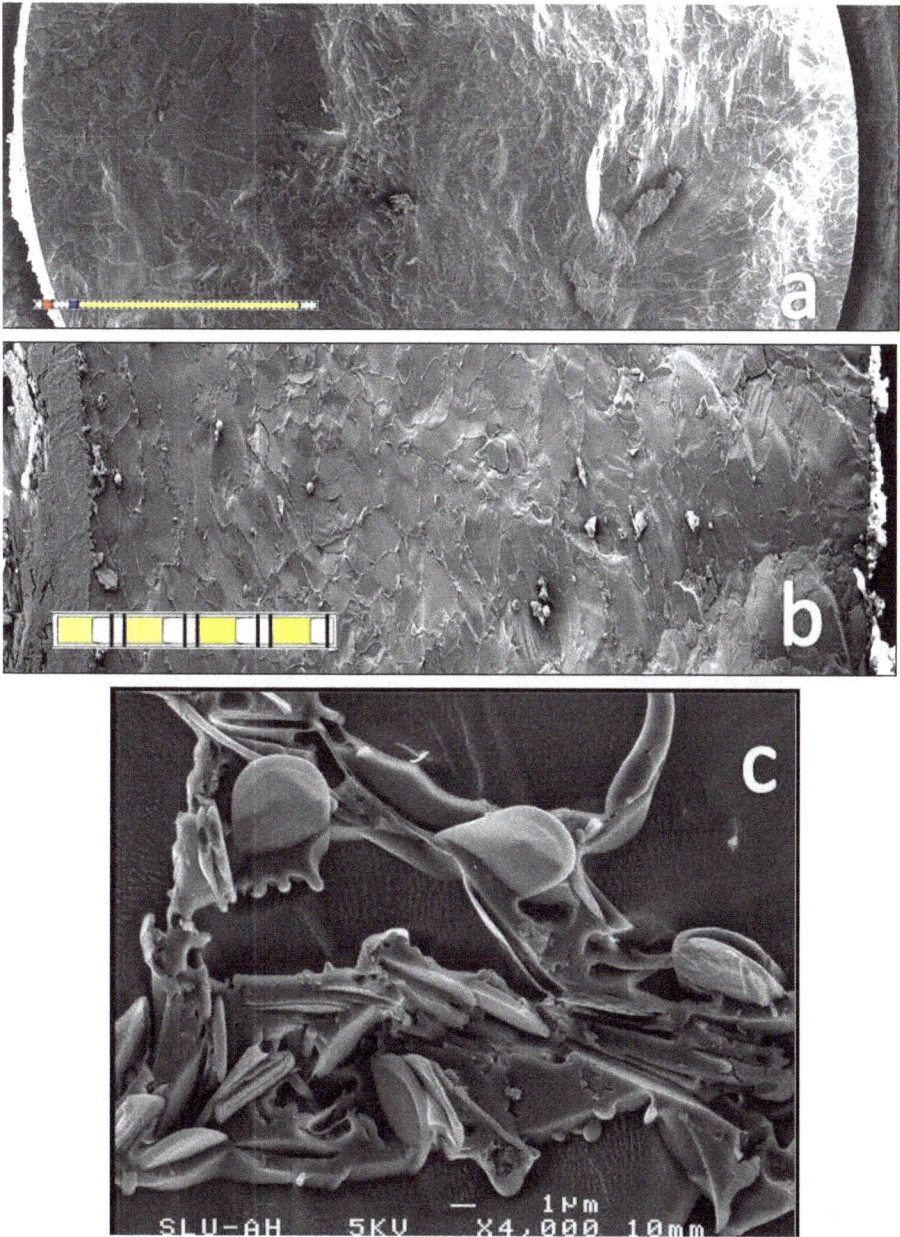

Fig. 4. Cryo-SEM micrographs at low magnification of cross sectioned frozen mini-straw (a) and a MiniFlatPack™ (b) depicting major differences in the orientation and size frozen water lakes/extender veins. In (c) a higher magnification of (b) showing morphologically well preserved boar spermatozoa from the sperm-peak portion of the ejaculate (Courtesy of Dr Hans Ekwall).

However, doses with such large sperm numbers conspire against the best use of the ejaculates and, with the introduction of intrauterine deposition of semen, it opened for the use of smaller containers with high numbers of spermatozoa to contain a single AI-dose. Recently, boar spermatozoa have been frozen, highly concentrated, in small volumes (0.5-0.7 mL) in novel containers, the so-called "MiniFlatPack™" (Saravia et al 2005, 2010, Pimenta-Siqueira et al 2011), as 1-2 billion spermatozoa/mL. Interestingly, cryosurvival (see **Table 2**) was equal or higher than for 0.5 mL plastic straws, suggesting the shape maintained the cryobiological advantages of the FlatPack™ (Ekwall et al 2007), including fertility when using deep-intrauterine AI (Wongtawan et al 2006).

Sperm vitrification: Boar spermatozoa packed in 0.12 mm thick film plastic bags were frozen ultra-rapidly at various stages of conventional freezing-thawing and besides survival, samples were explored ultrastructurally, for presence of ice damage. Survival was minimal and ice presence was detected, indicating that cooling rates, although high, were not enough to handle the volumes assayed (Bwanga et al 1991b). Non-penetrating sugars have either been used for vitrification (Meng et al 2010) and also for empirical improvement of slow freezing (Malo et al 2010). There is, *a priori*, nothing against the use of vitrification for freezing small suspensions of boar spermatozoa (for instance for intracytoplasmic sperm injection, ICSI, Meng et al 2010) but there is no practical use for breeding, since the amounts needed are too large to achieve ultra-rapid cooling and thawing rates.

Simplified freezing (SF, 3.5h)		Conventional freezing (CF, 8-9h)	
P1- sperm	SRF-sperm	P1-sperm	SRF-sperm
62.9 ± 3.13	54.2 ± 3.50	70.0 ± 4.40	64.0 ± 2.60

Table 2. Cryosurvival (Computer Assisted Sperm Analysis, CASA, mean±SEM), as percentages of motile spermatozoa, 30 min post-thaw at 38 °C) of ejaculated boar spermatozoa from the sperm-peak portion (P1, first 10mL of the sperm-rich fraction, SRF) or the entire SRF subjected to a simplified (SF, 3.5h) or a conventional freezing (CF, 8h) and an equal thawing (35°C for 20 seconds) (Modified from Saravia et al 2010).

Learning from the ejaculate: Boar SP is a composite, heterogeneous fluid composed by fractioned secretions of the epididymal caudae and the accessory sexual glands. *In vivo*, spermatozoa contact some of these fractions but not necessarily all, and different effects (sometimes deleterious, sometimes advantageous) have been recorded *in vitro* when removing (Fraser et al 2007) alternatively keeping boar spermatozoa in its own SP, depending on the fraction used (Guthrie & Welch 2005, Rodriguez-Martinez et al 2009, 2011). The SP or the sperm-rich fraction (SRF) might not be necessary for cryosurvival or fertility, since spermatozoa from boars that were semino-vesiculectomised were able to sustain freezing and thawing equally well as spermatozoa bathing in seminal vesicular proteins (Moore et al 1977). However, we have recently determined that boar spermatozoa contained in the first 10 mL of the SRF (also called sperm-peak portion or P1, where about ¼ of all spermatozoa in the SRF are) were more resilient to handling (from extension to cooling) and cryopreservation that the spermatozoa contained in the rest of the ejaculate (Peña et al 2003, Rodriguez-Martinez et al 2009, Saravia et al 2007, 2010, Rodriguez-Martinez et al 2008). It appeared that it was actually the SP in this sperm-peak P1 portion that was beneficial for spermatozoa, either because of its higher contents of cauda epididymal fluid and specific proteins, or its lower amounts of seminal plasma spermadhesins, bicarbonate or zinc levels (Rodriguez-Martinez et al 2011), compared to other fractions of the ejaculate (Saravia et al 2010).

In an attempt to simplify the freezing protocol, only the P1-spermatozoa were frozen in concentrated form for eventual use with deep-intrauterine AI. These spermatozoa were firstly kept in their SP for 30 min, and thereafter, without centrifugation (i.e. without removal of the SP) they were mixed with lactose-egg yolk (LEY) extender and cooled down to +5°C within 1.5 h, before being mixed with LEY+glycerol (3%) and OEP and packed into MiniFlatPack™'s for customary freezing using 50°C/min cooling rate. This "simplified" entire procedure (SF), lasted 3.5 h compared to the "conventional freezing" (CF) that was used as control procedure, which lasted 8 h. As controls, spermatozoa from the SRF were compared to P1-spermatozoa. Cryosurvival was, as seen in **Table 2**, equally good (above 60% of the processed cells (Saravia et al 2010, Pimenta-Siqueira et al 2011). Moreover, the spermatozoa in the sperm-peak-fraction of the boar ejaculate showed a maintained plasmalemmal intactness and fluidity and a lower flow of Ca^{2+} under capacitation conditions post-thaw, which might account for their higher membrane stability after cryopreservation (Hossain et al 2011).

There are several advantages of using this simplified, shorter protocol, namely the exclusion of the customary primary extension and the following removal of this conspicuously beneficial SP-aliquot by centrifugation. As well, it waives the need of expensive refrigerated centrifuges. Moreover, inter-boar variation was minimised by use of P1-spermatozoa which, not only were the "best" spermatozoa to be cryopreserved, but uses a portion of the sperm-rich fraction where the documented "fertility-associated" proteins are present (Rodriguez-Martinez et al 2011). Finally, the procedure frees the rest of the collected spermatozoa (75% of the total sperm count) for additional processing of liquid semen AI-doses. This simpler protocol ought thus to be an interesting alternative for AI-studs to –using the one and the same ejaculate- freeze boar semen (P1) for gene banking or for repopulation or commercial distribution, along with production of conventional semen doses for AI with liquid semen, using the rest of the ejaculate. Such procedures would not disturb routine handling of boars or their ejaculates. Inseminations in the field (deep intra-utero) have shown acceptable figures for farrowing and litter size (Wallgren, personal communication).

4. Cryopreservation of oocytes and embryos

The slow freezing technique developed for oocytes and embryos in the 1970´s (Willadsen et al 1978) has been thoroughly established by the increasing repertoire of CPA where they were gradually exposed to. Cultured cells/embryos are exposed to relatively low concentration of permeating CPA´s (glycerol, DMSO, EG or PG at 1-1.5 M (oocytes) or 1.3-1.5 M (embryos) alternatively non-permeating CPA in the culture medium, loaded into mini-straws and cooled at -5 to -7∘C, equilibrated for some minutes followed by seeding of extracellular ice nucleation, to be thereafter slowly cooled at~0.3-0.5∘C/minute to -40/-65∘C and final plunge in LN_2 for storage of the now carefully dehydrated and vitrified germplasm (for a comprehensive review see Saragusty & Arav 2011). However, pig oocytes, zygotes and cleavage embryos are rich in cytoplasmic lipids, and very sensitive to temperatures below 15°C (Wilmut 1972), a sensitivity that decreases -along with the amount of lipids- with development, towards peri-hatching blastocysts (Niimura and Ishida 1980). Offspring has been obtained after embryo transfer (ET) of slow-frozen and thawed 2-4 cell pig embryos where these cytoplasmic lipids were removed in vitro (de-lipation) before cooling (Hayashi et al 1989, Nagashima et al 1994, 1995, 1996) and thereafter the technique, albeit cumbersome, has been thoroughly applied (Yoneda et al 2004). The results enhanced when the cytoskeleton was preserved from damage using exogenous chemicals (Shi et al 2006).

Over the past years, vitrification (Rall and Fahy 1985) appeared as a better alternative for long-term storage of pig oocytes and embryos. One one hand, the small size of the material to process provided another dimension: vitrification could be modulated via size of sample (10 µL in most cases) so that neither cooling rate nor CPA-amounts ruled so that the method was more practical and less risky. Samples could be handled and carried/stored through either "surface" methods (e.g on liquid loops, mesh of different materials etc) or "tubing" carriers (thin straws, cyopipettes, ultrathin tubing etc). Both yield high cooling rates but while the surface type has the highest warming rates, the other is much easier to handle and, safer (Saragusty & Arav 2011).

On the other hand, vitrification of oocytes and embryos differ in degree of difficulty. As already mentioned, oocytes are more sensitive than embryos, particularly morulas or blastocysts since oocytes have a high cytoplasmic lipid contents (chilling sensitive). Moreover, oocytes have easily disrupted submembranous actin microtubules (which decreases plasmalemman robustness) and fragile meiotic spindle and cytoskeleton, which complicates the resumption of development. Lastly, the process of freezing and thawing can increase the risk for ROS-attack and the premature emptying of cortical granules, thus changing the structure of the zona pellucida (ZP) (Gajda 2009). Therefore, chemical stabilization of the cytoskeleton (Esaki et al 2004) and the use of increased pressure following vitrification (Du et al 2008) had been successfully applied, obtaining development post-rewarming towards the fetal stage (Ogawa et al 2010). Other measures, such as induction of osmotic stress (by exposure to NaCl) has shown to improve developmental competence after vitrification (Lin et al 2008). Centrifugation (lipid depot relocation) for vitrification appears detrimental for *in vitro*-matured oocytes, but not in zygotes or later stages (Somfai et al 2008)

Vitrification of in vivo-developed, ZP-intact pig embryos, where lipids were polarized by centrifugation of the blastomeres, by delipation and/or treatment with cytochalasin for cytoskeleton stabilization, has resulted after rewarming and ET, in piglets (Dobrinsky 1997, Dobrinsky et al 2000, 2001, Kobayashi et al 1998, Berthelot et al 2000, 2003, Cameron et al 2000). Blastocysts were also developed by *in vitro* fertilization (IVF) of follicular oocytes vitrified as cumulus-oocyte complexes from offal porcine follicles (Somfai et al 2010).

Recently, piglets were even obtained following vitrification of delipated 4-8 cell stages of in vitro produced (IVP), parthenogenetic embryos and ET (Nagashima et al 2007). Vitrification, usually done within 0.25 mL plastics-straws, yield better embryo survival post-warming when Open Pulled Straws (OPS; Vajta et al 1997), which increases the cooling rate achievable in 0.25 mL straws (2,500°C/min) by almost 8-fold (Cuello et al 2004a-b), were used, again resulting in piglets born (Berthelot et al 2000; 2001). Higher cooling-rates

(>20,000°C/min) can nowadays be reached using using cryo-loops (Lane et al 1999) or with straws with a smaller inner diameter and wall thickness (the Superfine Open Pulled Straws: SOPS; Isachenko et al 2003), and by applying immersion in LN$_2$-slush, which allowed for the use of lower concentrations of toxic cryoprotectant.

Vitrification of untreated morulae and blastocysts has resulted in high survival rates after warming (Berthelot et al 2003), especially when re-warming after SOPS is done in one stage (direct warming, a very practical solution for ET, Cuello et al 2004b), yielding live litters (Cuello et al 2005). For blastocysts (See **Figure 5**), use of the SOPS waived the need for centrifugation (dislocation of lipids) or microtubule stabilization, thus making the method a very practical one and indicating the procedure is now reaching maturity for commercial application (Cameron et al 2004, Beebe et al 2005, Martinat-Botté et al 2006, Cuello et al 2008, 2010, Sanchez-Osorio et al 2009, 2010). Cryopreservation of in IVP-pig embryos -owing to differences in the cytoskeleton and the distribution of the lipid deposits- has been, until recently (Esaki et al 2004), considered as more difficult than for *in vivo*-developed, but the birth of piglets resulting from ET of IVP, transgenic pig embryos, has modified this view opening for the commercialization of highly valuable, modified genetic material (Li et al 2006, Kawagami et al 2008). Despite peri-hatching blastocyst stage embryos are the ones best sustaining vitrification and warming with continued *in vitro* development (Dobrinsky 2001), this particular embryo stage can not be commercially used since there is no ZP.

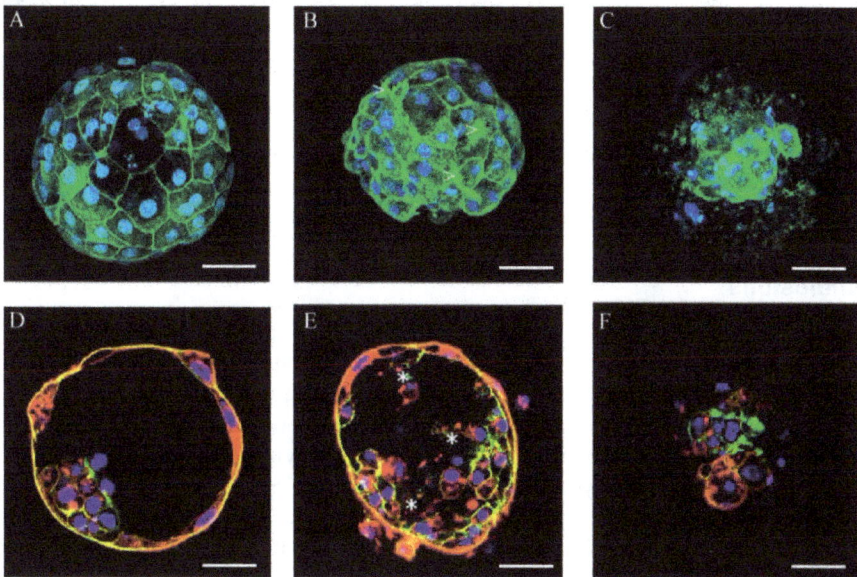

Fig. 5. Laser scanning confocal microphotographs of grade I (A and D), grade II (B and E) or grade III (C and F) in vivo-derived fresh (A-C) and superfine open pulled straws (SOPS)-vitrified (D-F) porcine blastocysts, following uploading of Hoescht H-33342 (blue, cell nuclei), phalloidin-Alexa Fluor 488 (green, actin filaments) and wheat germ agglutinin-Alexa fluor 594 (red, lectin reactive membrane elements). Note the high degree of morphological intactness even after rewarming compared to fresh controls (Reprinted from Cuello et al 2010, with permission).

5. Cryopreservation of genital tissues

Freezing of ovarian tissue in humans relate primarily, but not only, to a dramatic measure to warrant availability of oocytes in cases of oncotherapy, when sterility is foreseen, similar to the ongoing sperm banking prior to onco- or hormonal therapy. Rescue of oocytes from frozen samples of ovarian cortex is then feasible for ART (Shaw & Trounson 2002). Both slow freezing and directional freezing had been assayed with acceptable results (Arav & Natan 2009), opening possibilities for the cryopreservation of large samples and even of whole ovary for autografting purposes and possibly evolving in oocyte banking as an insurance against childlessnes. Adult testicular samples (aspiration or biopsy) are mainly issued during biopsy for recovery of spermatids for ICSI (Keros et al 2005, Curaba et al 2011). However, the strongly ongoing research in adult stem cells shall be based on the absolute need of properly cryopreserving pre-pubertal testicular tissues. Transplantation of other organs or tissues (uterine in particular) is also within the scope of not-far, albeit discussable, scenarios (Bredkjaer & Grudzinskas 2001).

Regarding the porcine species, although there is no obvious rationale for most of the above considerations in human, it provides an excellent animal model for experimental reproductive medicine, particularly considering transplantation surgery. Porcine whole uteri were arterially perfused with CPA (DMSO) prior to slow controlled freezing. Rewarmed tissues were able to present live cells 7 h post rewarming (Dittrich et al 2006) and even to demonstrate contractility *in vitro* 60 min post-rewarming (Dittrich et al 2010). As such, comparative analyses of equilibrium freezing and vitrification procedures have involved pig ovarian fragments (Gandolfi et al 2006, Borges et al 2009), or whole ovaries (Imhof et al 2004). These attempts were all done using slow freezing, but evidence is now provided that vitrification of thin slices of ovarian cortex is feasible and that rewarmed primordial follicles from these samples were able to develop (albeit slower than controls) in murine xenografts (Moniruzzaman et al 2009). Further development in this area is expected.

6. Conclusions

Vitrification as a method for cryopreservation in porcine applies thus far to small samples that can be managed at high cooling and rewarming rates without need of applying permeating CPA of potential toxicity. Therefore, the technique has developmental potential for oocytes, COCs and embryos for IVF and ET. Boar spermatozoa are yet to follow this path, and although there is a potential breach for vitrifying limited volumes of sperm suspensions, such approach is yet solely academic in nature. Semen for breeding ought to be frozen conventionally, albeit with a focus on increased cell lifespan, and managing concentrated semen doses for deep intrauterine AI. There is much yet to be learned from the ejaculate and the relationships between specific components of the seminal plasma and sperm function.

7. Acknowledgements

The studies of the author have been made possible by grants from the Swedish Research Council Formas, the Swedish Farmer´s Foundation for Agricultural Research (SLF), and The Swedish Research Council (VR), Stockholm, Sweden.

8. References

Arav A, Yavin S, Zeron Y, Natan D, Dekel I, Gacitua H (2002) New trends in gamete´s cryopreservation. Mol Cell Endocrinol 187: 77-81.

Arav A, Natan Y (2009) Directional freezing: a solution to the methodological challenges to preserve large organs. Semin Reprod Med 27: 438-442.

Bathgate R, Eriksson B, Maxwell WMC, Evans G (2005) Low dose deep intrauterine insemination of sows with fresh and frozen-thawed spermatozoa. Theriogenology 63: 553–554.

Bathgate R, Maxwell WMC, Evans G (2007) Effects of platelet-activating factor and platelet activating factor acetylhydrolase on in vitro post-thaw boar sperm parameters. Theriogenology 67 886-892.

Beebe LFS, Cameron RDA, Blackshaw AW, Keates HL (2005) Changes to procine blastocyst vitrification methods and improved litter size after transfer. Theriogenology 64: 879-890.

Bernschtein AD, Petropavloski VV (1937) Influence of non-electrolytes on viability of spermatozoa. Bull Exp Biol Med III: 21-25.

Berthelot F, Martinat-Botte F, Locatelli A, Perreau C, Terqui M (2000) Piglets born after vitrification of embryos using the Open Pulled Straw method. Cryobiology 41:116-124.

Berthelot F, Martinat-Botté F, Perreau C, Terqui M (2001) Birth of piglets after OPS vitrification and transfer of compacted morula stage with intact zona pellucida. Reprod Nutr Dev 41: 267-272.

Berthelot F, Martinat-Botte F, Vajta G, Terqui M (2003) Cryopreservation of porcine embryos: state of the art. Livest Prod Sci 83: 73-83.

Bianchi I, Calderam K, Maschio EF, Madeira EM, da Rosa Ulguim R, Corcini CD, Bongalhardo DC, Correa EK, Lucia T Jr, Deschamps JC, Correa MN (2008) Evaluation of amides and centrifugation temperature in boar semen cryopreservation. Theriogenology 69: 632-638.

Bolarín A, Roca J, Rodriguez-Martinez H, Hernandez M, Vazquez JM, Martínez EA (2006) Dissimilarities in sows' ovarian status at the insemination time could explain differences in fertility between farms when frozen-thawed semen is used. Theriogenology, 65, 669-680.

Borges EN, Silva RC, Futino DO, Rocha-Junior CM, Amorim CA, Báo SN, Lucci CM (2009) Cryopreservation of swine ovarian tissue: effect of different cryoprotectants on the structural preservation of preantral follicle oocytes. Cryobiology 59:195-200.

Bredkjaer HE, Grudzinskas JG (2001) Cryobiology in human assisted reproductive technology. Would Hyppocrates approve? Early Pregnancy 5: 211-213.

Buranaamnuay K, Grossfeld R, Struckmann C, Rath D (2011) Influence of cryoprotectans glycerol and maides, combined with antioxidants on quality of frozen-thawed boar semen. Anim Reprod Sci 127: 56-61.

Buranaamnuay K, Tummaruk P, Singlor J, Rodriguez-Martinez H, Techakumphu M (2009) Effects of straw volume and Equex-STM® on boar sperm quality after cryopreservation. Reprod Domest Anim 44: 69-73.

Bussiere JF, Bertaud G, Guillouet P (2000) Conservation of boar semen by freezing. Evaluation in vitro and after insemination. 32èmes Journées de la Recherche Porcine en France 32: 429–432.

Bwanga CO, Einarsson S, Rodriguez-Martinez H (1991a) Freezing of boar semen in plastic bags and straws. Reprod Domest Anim 26 117-125.

Bwanga CO, Ekwall H, Rodriguez-Martinez H (1991b) Cryopreservation of boar semen: III- Ultrastructure of boar spermatozoa frozen ultra-rapidly at various stages of conventional freezing/thawing. Acta Vet Scand 32 463-471.

Cameron RDA, Beebe LFS, Blackshaw AW, Higgins A, Nottle MB (2000) Piglets born from vitrified early blastocysts using a simple technique. Austr Vet J 78: 195-196.

Cameron RDA, Beebe LFS, Blackshaw AW, Keates HL (2004) Farrowing rates and litter size following transfer of vitrified porcine embryos into a commercial swine herd. Theriogenology, 61, 1533-1543.

Crabo B, Einarsson S (1971) Fertility of deep frozen boar spermatozoa. Acta Vet Scand 12: 125–127.

Cuello C, Berthelot F, Martinat-Botte F, Venturi E, Guillouet P, Vazquez JM, Roca J, Martinez EA (2005) Piglets born after non-surgical deep intrauterine transfer of vitrified blastocysts in gilts. Anim Reprod Sci 85: 275-286.

Cuello C, Gil MA, Parrilla I, Tornel J, Vazquez JM, Roca J, Berthelot F, Martinat-Botte F, Martinez EA (2004a) Vitrification of porcine embryos at various developmental stages using different ultra-rapid cooling procedures. Theriogenology 62: 353-361.

Cuello C, Gil MA, Parrilla I, Tornel J, Vazquez JM, Roca J, Berthelot F, Martinat-Botte F, Martinez EA (2004b) In vitro development following one-step dilution of OPS- vitrified porcine blastocysts. Theriogenology 62: 1144-1152.

Cuello C, Gil MA, Almiñana C, Sanchez-Osorio J, Parrilla I, Caballero I, Vazquez JM, Roca J, Rodriguez-Martinez H, Martinez EA (2007) Vitrification of in vitro cultured porcine two-to-four cell embryos. Theriogenology 68: 258-264.

Cuello C, Martinez E, Sanchez-Osorio J, Almiñana C, Gil MA, Caballero I, Parrilla I, Roca J, Vazquez JM, Rodriguez-Martinez H (2008) Super Open Pulled Straw Vitrification of Porcine Blastocysts: Effect of Centrifugation and Cytoskeletal Stabilization. Biol Reprod SI-2008: 114-115 (Proc 41st SSR Annual Meeting, May 27-30, 2008, Kailua- Kona, Hawaii, SSRABSTRACTS/2008/007187 Nr 262).

Cuello C, Sanchez-Osorio J, Almiñana C, Gil MA, Parrilla I, Roca J, Vazquez JM, Martinez EA, Rodriguez-Martinez H (2010) Superfine open pulled straws vitrification of porcine blastocysts does not require pretreatment with cytochalasin B and/or centrifugation. Reprod Fert Dev 22: 808-817.

Curaba M, Poels J, van Langendonckt A, Donnez J, Wyns C (2011) Can prepubertal human testicular tissue be cryopreservad by vitrification? Fertil Steril 95: 2123 e9-12.

De Mercado E, Hernandez M, Sanz E, Rodriguez A, Gomez E, Vazquez JM, Martinez EA, Roca J (2009) Evaluation of L-glutamine for cryopreservation of boar spermatozoa. Anim Reprod Sci 115: 149-157.

Devireddy RV, Fahrig B, Godke RA, Leibo SP (2004) Subzero water transport characteristics of boar spermatozoa confirm observed optimal cooling rates. Mol Reprod Dev 67: 446-457.

Díaz-García C, Johannesson L, Enskog A, Tzakis A, Olausson M, Brännström M (2011) Uterine transplantation research: laboratory protocols for clinical application. Mol Hum Reprod (in press)

Dittrich R, Maltaris T, Mueller A, Dimmler A, Hoffmann I, Kiesewetter F, Beckmann MW (2006) Successful uterus cryopreservation in an animal model. Horm Metab Res 38: 141-145.

Dittrich R, Beckmann MW, Mueller A, Binder H, Maltaris T (2010) Uterus cryopreservation: maintenance of uterine contractility by the use of different cryoprotocols. Reprod domest Anim 45: 86-91.

Dobrinsky JR (2001) Cryopreservation of pig embryos: adaptation of vitrification technology for embryo transfer. Reproduction 58: 325-333.

Dobrinsky JR, Nagashima H, Pursel VG, Schreier LL, Johnson LA (2001) Cryopreservation of morula and early blastocyst stage swine embryos: Birth of litters after transfers. Theriogenology 55: 303.

Dobrinsky JR, Pursel VG, Long CR, Johnson LA (2000) Birth of piglets after transfer of embryos cryopreserved by cytoskeletal stabilization and vitrification. Biol Reprod 62: 564-570.

Dobrinsky JR (1997) Cryopreservation of pig embryos. J Reprod Fertil 52: 301-312.

Dong Q, Correa LM, VandeVoort CA (2009) Rhesus monkey sperm cryopreservation with TEST-yolk extender in the absence of permeable cryoprotectant. Cryobiology 58: 20-27.

Du Y, Pribenszky CS, Molnar M, Zhang X, Yang H, Kuwayama M, Pedersen AM, Villemoes K, Bolund L, Vajta G (2008) High hydrostatic pressure: a new way to improve in vitro developmental competence of porcine matured oocytes after vitrification. Reproduction 135: 13-17.

Ekwall H (2009) Cryo-scanning electron microscopy discloses differences in dehydration of frozen boar semen stored in large containers. Reprod Domest Anim 44 62-68.

Ekwall H, Hernández M, Saravia F & Rodríguez-Martínez H (2007) Cryo-scanning electron microscopy (Cryo-SEM) of boar semen frozen in medium-straws and MiniFlatPacks. Theriogenology 67: 1463-1472.

Eriksson BM, Rodriguez-Martinez H (2000) Deep freezing of boar semen in plastic film "Cochettes". J Vet Med A 47: 89-97.

Eriksson BM, Rodriguez-Martinez H (2000) Effect of freezing and thawing rates on the post-thaw viability of boar spermatozoa frozen in large 5 ml packages (FlatPack). Anim Reprod Sci 63: 205-220.

Eriksson BM, Petersson H, Rodriguez-Martinez H (2002) Field fertility with exported boar semen frozen in the new FlatPack container. Theriogenology 58: 1065-1079.

Esaki R, Ueda H, Kurome M, Hirakawa K, Tomii R, Yoshioka H, Ushijima H, Kurayama M, Nagashima H (2004) Cryopreservation of porcine embryos derived from in vitro-matures oocytes. Biol Reprod 71: 432-437.

Flowers WL (1999) Artificial insemination, in animals. In: Knobil E, Neill JD (Eds) Encyclopedia of Reproduction. Vol I, Academic Press, San Diego: 291-302.

Foote RH (1999) Development of reproductive biotechnologies in domestic animals from artificial insemination to cloning: a perspective. Cloning 1: 133-142.

Fraser L, Dziekonska A, Strzezek R, Strzezek J (2007) Dialysis of boar semen prior to freezing-thawing: its effects on post-thaw characteristics. Theriogenology 67: 994-1003.

Fraser L, Strzezek J (2007) Effect of different procedures of ejaculate collection, extenders and packages on DNA integrity of boar spermatozoa following freezing-thawing. Anim Reprod Sci 99: 317-329.

Fraser L, Strzezek J, Kordan W (2011) Effect of freezing on sperm nuclear DNA. Reprod domest Anim 46 S2: 14-17.

Fuller BJ (2004) Cryoprotectants: The essential antifreezes to protect life in the frozen state. Cryo Letters 25: 375-388.

Gandolfi F, Paffni A, Papasso Brambilla E, Bonetti S, Brevini TAL, Ragni G (2006) Efficiency if equilibrium cooling and vitrification procedures for the cryopreservation of ovarian tissue: comparative analysis between human and animal models. Fertil Steril 85 S1: 1150-1156.

Gajda B (2009) Factors and methods of pig oocyte and embryo quality improvement and their application in reproductive biotechnology. Reprod Biol 9: 97-112.

Gil MA, Roca J, Cremades T, Hernandez M, Vazquez JM, Rodriguez-Martinez H, Martinez EA (2005) Does multivariate analysis of post-thaw sperm characteristics accurately estimate in vitro fertility of boar individual ejaculates? Theriogenology 64: 305–316.

Graham EF, Rajamannan AHJ, Schmehl MKL, Maki-Laurila M, Re B (1971) Preliminary report on procedure and rationale for freezing boar spermatozoa. AI Digest 19: 12–14.

Guthrie HD, Welch GR (2005) Impact of storage prior to cryopreservation on plama membrane function and fertility of boar sperm. Theriogenology 63: 396-410.

Gutierrez-Perez O, Juarez-Mosqueda Mde L, Carvajal SU, Ortega ME (2009) Boar spermatozoa cryopreservation in low glycerol/trehalose enriched freezing media improves cellular integrity. Cryobiology 58: 287-292.

Han B, Bischof JC (2004a) Direct cell injury associated with eutectic crystallization during freezing. Cryobiology 48: 8-21.

Han B & Bischof JC (2004b) Thermodynamic nonequilibrium phase change behaviour and thermal properties of biological solutions for cryobiology applications. J Biochem Eng 126: 196-203.

Hayashi S, Kobayashi K, Mizuno J, Saitoh S (1989) Birth of piglets from frozen embryos. Vet Record 125: 43-44.

Hernandez M, Ekwall H, Roca J, Vazquez JM, Martinez E, Rodriguez-Martinez H (2007b) Cryo-scanning electron microscopy (Cryo-SEM) of semen frozen in medium-straws from good and sub-standard freezer AI-boars. Cryobiology 54: 63-70.

Hernandez M, Roca J, Ballester J, Vázquez JM, Martinez EA, Johannisson A, Saravia F, Rodriguez-Martinez H (2006) Differences in SCSA outcome among boars with different sperm freezability. Int J Androl 29: 583–591.

Hernandez M, Roca J, Gil MA, Vázquez JM, Martínez EA (2007a) Adjustments on the cryopreservation conditions reduce the incidence of boar ejaculates with poor sperm freezability. Theriogenology 67: 1436–1445.

Holm P, Vatja G, Machaty Z, Schmidt M, Prather RS, Greve T, Callesen H (1999) Open pulled straw (ops) vitrification of porcine blastocysts: simple procedure yielding excellent in vitro survival, but so far no piglets following transfer. CryoLetters 20: 307-310.

Holt WV (2000a) Fundamental aspects of sperm cryobiology: the importance of species and individual differences. Theriogenology 53: 47–58.

Holt WV (2000b) Basic aspects of frozen storage of semen. Anim Reprod Sci 62: 3–22.

Holt WV, Medrano A, Thurston LM, Watson PF (2005) The significance of cooling rates and animal variability for boar sperm cryopreservation: insights from the cryomicroscope. Theriogenology 15: 370–382.

Hossain AM, Osuamkpe CO (2007) Sole use of sucrose in human sperm cryopreservation. Arch androl 53: 99-103.

Hossain MdS, Johannisson A, Pimenta Siqueira A, Wallgren M, Rodriguez-Martinez H (2011) Spermatozoa in the sperm-peak-fraction of the boar ejaculate show a lower flow of Ca2+ under capacitation conditions post-thaw which might account for their higher membrane stability after cryopreservation. Anim Reprod Sci 128: 37-44.

Hu JH, Li QW, Jiang ZL & Li WY (2008) Effects of different extenders on DNA integrity of boar spermatozoa following freezing-thawing. Cryobiology 57: 257-262.

Hu JH, Li QW, Jiang ZL, Yang H, Zhang SS, Zhao HW (2008) The cryoprotective effect of trehalose supplementation on boar spermatozoa quality. Reprod domest Anim 44: 571-575.

Imhof M, Hofstetter G, Bergmeister H, Rudas M, Kain R, Lipovac M, Huber J (2004) Cryopreservation of a whole ovary as a strategy for restoring ovarian function. J Assist Reprod Genet 21: 459-65.

Isachenko V, Folch J, Isachenko E, Nawroth F, Krivokharchenko D, Vajta G, Dattena M, Alabart JL (2003) Double vitrification of rat embryos at different developmental stages using an identical protocol. Theriogenology, 60, 445-452.

Isachenko E, Isachenko V, Katkov II, Rahimi G, Schondorf T, Mallmann P, Dessole S, Nawroth F (2004) DNA integrity and motility of human spermatozoa after standard slow freezing versus cryoprotectant-free vitrification. Human Reprod 19: 932-939.

Isachenko E, Isachenko V, Montag M, Zaeva V, Kriokharchenko A, Nawroth F, Dessole S, Katkov I, van der Ven H (2005) Clean technique for cryoprotectant-free vitrification of human spermatozoa. Reprod Biomed Online 10: 350-354.

Isachenko E, Isachenko V, Weiss JM, Kreienberg R, Katkov II, Schulz M, Lulat AGMI, Risopatron MJ, Sanchez R (2008) Acrosomal status and mitochondrial activity of human spermatozoa vitrified with sucrose. Reproduction 136: 167-173.

Isachenko E, Maettner R, Petrunkina AM, Sterzik K, Mallmann P, Rahimi G, Sanchez R, Risopatron J, Damjanoski I, Isachenko E (2011) Vitrification of human ICSI/IVF spermatozoa without cryoprotectants: new capillary technology. J Androl (in press).

Isachenko V, Lapidus I, Isachenko E, Krivokharchenko A, Kreienberg R, Woriedh M, Bader M, Weiss JM (2009) Human ovarian tissue vitrification versus conventional freezing: morphological, endocrinological, and molecular biological evaluation. Reproduction 138: 319-327.

Jeong YJ, Kin MK, Song HJ, Kang EJ, Ock SA, Kumar BM, Balasubramanian S, Rho GJ (2009) Effect of lpha-tocopherol supplementation during boar semen cryopreservation on sperm characteristics and expression of apoptosis related genes. Cryobiology 58: 181-189.

Jiang ZL, Li QW, Li WY, Hu JH, Zhao HW, Zhang SS (2007) Effect of low density lipoproteins on DNA integrity of freezing–thawing boar sperm by neutral comet assay. Anim Reprod Sci 99 401-407.

Johnson LA, Weitze KF, Fiser P, Maxwell WM (2000) Storage of boar semen. Anim Reprod Sci 18 143-172.

Juarez JD, Parrilla I, Vazquez JM, Martinez EA, Roca J (2011) Boar semen can tolerate rapid cooling rates prior to freezing. Reprod Fertil Dev 23: 681-690.

Kaeoket K, Chanapiwat P, ummaruk P, Techakumphu M (2010) Supplemental effect of varying L-cysteine concentrations on the quality of cryopreserved boar semen. Asian J Androl 12: 760-765.

Karosas J, Rodriguez-Martinez H (1993) Use of two detergents for freezing of boar semen in plastic bags. Biomed Res 4: 125-136.

Kawakami M, Kato Y, Tsunoda Y (2008) The effects of time of first cleavage, developmental stage, and delipidation of nuclear-transferred porcine blastocysts on survival following vitrification. AnimReprod Sci 106: 402-411.

Keros V, Rosenlund B, Hultenby K, Aghajanova L, Levkov L, Hovatta O (2005) Optimizing cryopreservation of human testicular tissue: comparison of protocols with glycerol, propanediol and dimethylsulphoxide as cryoprotectants. Hum Reprod 20: 1676-1787.

Kim SS (2010) Time to re-think: ovarian tissue transplantation versus whole ovary transplantation. Reprod Biomed Online 20: 171-174.

Knox RV (2011) The current value of frozen-thawed boar semen for commercial companies. Reprod Domest Anim 46 S2: 4-6.

Kobayashi S, Takei M, Kano M, Tomita M, Leibo SP (1998) Piglets produced by transfer of vitrified porcine embryos after stepwise dilution of cryoprotectants. Cryobiology 36: 20-31.

Kumar S, Millar JD, Watson PF (2003) The effect of cooling rate on the survival of cryopreserved bull, ram and boar spermatozoa: a comparison of two controlled-rate cooling machines. Cryobiology 46: 246–253.

Lane M, Schoolcraft WB, Gardner DK (1999) Vitrification of mouse and human blastocysts using a novel cryoloop container-less technique. Fertil Steril 72: 1073-1078.

Li N, Lai L, Wax D, Hao Y, Murphy CN, Rieke A, Samuel M, Linville ML, Korte SW, Evans RW, Turk JR, Kang JX, Witt WT, Dai Y, Prather RS (2006) Cloned transgenic swine via in vitro production and cryopreservation. Biol Reprod 75: 226-230.

Lin L, Kragh PM, Kuwayama M, Du Y, Zhang X, Yang H, Bolund L, Callesen H, Vajta G (2008) Osmotic stress induced by sodium chloride, sucrose or threhalose improves cryotolerance and developmetnalcompetence of porcine oocytes. Reprod Fertil Dev 21: 338-344.

Malo C, Gil L, Gonzalez N, Cano R, de Blas I, Espinosa E (2010) Comparing sugar type supplementation for cryopreservation of boar semen in egg yol-based extender. Cryobiology 61: 17-21.

Martinat-Botte F, Berthelot F, Plat M, Madec F (2006) Cryopreservation and transfer of pig in vivo embryos: state of the art. Gynecol Obstét Fert 34: 754-759.

Massip A (2001) Cryopreservation of embryos of farm animals. Reprod Domest Anim 36: 49-55.

Mazur P, Cole KW (1989) Roles of unfrozen fraction, salt concentration, and changes in cell volume in the survival of slowly frozen human erythrocytes. Cryobiology 26: 1–29.

Mazur P, Leibo SP, Seidel GEJr (2008) Cryopreservation of the germplasm of animals used in biological and medical research: importance, impact, status, and future directions. Biol Reprod 78: 2-12.

Medrano A, Holt WV, Watson PF (2009) Controlled freezing studies on boar sperm cryopreservation. Andrologia 41: 246-250.

Meng X, Gu X, Wu C, Dai J, Zhang , Xie Y, Wu Z, Liu L, Ma H, Zhang D (2010) Effect of trehalose on the free-dried boar spermatozoa. Sheng Wu Gong Cheng Xue Bao 26: 1143-1149.

Mocé E, Blanch E, Tomás C, Graham JK (2010) Use of cholesterol in sperm cryopreservation: present moment and perspectives to future. Reprod Domest Anim 45: 57-66.

Moniruzzaman M, Bao RM, Taketsuru H, Miyano T (2009) Development of vitrified porcine primordial follicles in xenografts. Theriogenology 72: 280-288.

Moore HDM, Hibbitt KG (1977) Fertility of boar spermatozoa after freezing in the absence of seminal vesicular proteins. J Reprod Fert 50: 349-352.

Morris GJ (2006) Rapidly cooled human sperm: no evidence of intracellular ice formation. Human Reprod 21: 2075–2083.

Morris GJ, Acton E, Avery S (1999) A novel approach to sperm cryopreservation. Human Reprod 14: 1013–1021.

Morris GJ, Goodrich M, Acton E, Fonseca F (2006) The high viscosity encountered during freezing in glycerol solutions: effects on cryopreservation. Cryobiology 52: 323–334.

Morris GJ, Faszer K, Green JE, Raper D, Grout BWW, Fonseca F (2007) Rapidly cooled horse spermatozoa: loss of viability is due to osmotic imbalance during thawing, not intracellular ice formation. Theriogenology 68: 804–812.

Mwanza A, Rodriguez-Martinez H (1993) Post-thaw motility, acrosome morphology and fertility of deep frozen boar semen packaged in plastic PVC-bags. Biomed Res 4 21-29.

Nagashima H, Kashiwazaki N, Ashman RJ, Grupen CG, Nottle MB (1995) Cryopreservation of porcine embryos. Nature 374: 416.

Nagashima H, Kashiwazaki N, Ashman RJ, Grupen CG, Seamark RF, Nottle MB (1994) Removal of cytoplasmic lipid enhances the tolerance of porcine embryos to chilling. Biol Reprod 51: 618-622.

Nagashima H, Kuwayama M, Grupen CG, Ashman RJ, Nottle MB (1996) Vitrification of porcine early cleavage stage embryos and oocytes after removal of cytoplasmic lipid droplets. Theriogenology 45: 180.

Nagashima H, Hiruma K, Saito H, Tomii R, Ueno S, Nakayama N, Matsunari H, Kurome M (2007) Production of live piglets following cryopreservation of embryos derived from in vitro-matured oocytes. Biol Reprod 76: 900-905.

Niimura S, Ishida K (1980) Histochemical observation of lipid droplets in mammalian eggs during the early development. Jap J Animal Reprod 26: 46-49.

Ogawa B, Ueno S, Nakayama N, Matsunari H, Nakano K, Fujiwara T, Ikezawa Y, Nagashima H (2010) Developmental ability of porcine in vitro matured oocytes at the meiosis II stage after vitrification. J Reprod Dev 56: 356-361.

Ohanian C, Rodriguez H, Piriz H, Martino I, rieppi G, Garofalo EG, Roca RA (1979) Studies on the contractile activity and ultrastructure of the boar testicular capsule. J Reprod Fertil 57: 79-85.

Ortman K, Rodriguez-Martinez H (1994) Membrane damage during dilution, cooling and freezing-thawing of boar spermatozoa packaged in plastic bags. J Vet Med A 41: 37-47.

Parrilla I, Vazquez JM, Caballero I, Gil MA, Hernandez M, Roca J, Lucas X, Martinez EA (2009) Optimal characteristics of spermatozoa for semen technologies in pigs. Soc Reprod Fertil S66: 37–50.

Pegg DE (2002) The history and principles of cryopreservation. Semin Reprod Med 20: 5-13.

Pegg DE (2005) The role of vitrification techniques of cryopreservation in reproductive medicine. Hum Fertil (Camb) 8: 231-239.

Pegg DE (2007) Principles of cryopreservation. Methods Mol Biol 368: 39-57.

Penfold LM, Watson PF (2001) The cryopreservation of gametes and embryos of cattle, sheep, goats and pigs. In: (Watson PF, Holt WV (Eds.) Cryobanking the Genetic Resource. New York, NY: Taylor & Francis: 279–316.

Peña FJ, Johannisson A, Wallgren M, Rodriguez-Martinez H (2003) Antioxidant supplementation in vitro improves boar sperm motility and mitochondrial membrane potential after cryopreservation of different fractions of the ejaculate. Anim Reprod Sci 78: 85-98.

Peña FJ, Johannisson A, Wallgren M, Rodriguez-Martinez H (2004a) Antioxidant supplementation of boar spermatozoa from different fractions of the ejaculate improves cryopreservation: changes in sperm membrane lipid architecture. Zygote 12: 117-124.

Peña FJ, Johannisson A, Wallgren M, Rodriguez Martinez H (2004b) Effect of hyaluronan supplementation on boar sperm motility and membrane lipid architecture status after cryopreservation. Theriogenology 61: 63-70.

Pimenta Siqueira A, Wallgren M, Johannisson A, Md Hossain, Rodriguez-Martinez H (2011) Quality of boar spermatozoa from the sperm-peak portion of the ejaculate after simplified freezing in MiniFlatpacks compared to the remaining spermatozoa of the sperm-rich fraction. Theriogenology 75:1175-1184.

Polge C (1956) Artificial Insemination in Pigs. Vet Record 68: 62–76.

Pursel VG, Johnson LA (1971) Procedure for the preservation of boar spermatozoa by freezing. USDA, ARS Bulletin Notes 44–227: 1–5.

Rall WF, Fahy GM (1985) Ice-free cryopreservation of mouse embryos at -196°C by vitrification. Nature 313: 573-575.

Rath D, Niemann H (1997) In vitro fertilization of porcine oocytes with fresh and frozen-thawed epididymal semen obtained from identical boars. Theriogenology 47: 785-793.

Rath D, Bathgate R, Rodriguez-Martinez H, Roca J, Strzezek J, Waberski D (2009) Recent advances in boar semen cryopreservation. In: "Control of Pig Reproduction VIII" Rodriguez-Martinez H, Vallet JL, Ziecik AJ (Eds). Nottingham University Press, UK: 51-66 (ISBN 978-1-904761-39-6).

Riesenbeck A (2011) Review on international trade with boar semen. Reprod Domest Anim 46 S2: 1-3.

Roca J, Gil MA, Hernandez M, Parrilla I, Vazquez JM, Martinez EA (2004) Survival and fertility of boar spermatozoa after freeze-thawing in extender supplemented with butylated hydroxytoluene. J Androl 25: 397-405.

Roca J, Rodríguez MJ, Gil MA, Carvajal G, García EM, Cuello C, Vázquez JM, Martínez EA (2005) Survival and in vitro fertility of boar spermatozoa frozen in the presence of superoxide dismutase or catalase or both. J Androl 26: 15-24.

Roca J, Hernández M, Carvajal G, Vazquez JM, Martinez EA (2006a) Factors influencing boar sperm cryosurvival. J Anim Sci 84: 2692-2699.

Roca J, H Rodriguez-Martinez, JM Vazquez, A Bolarín, M Hernandez, F Saravia, M Wallgren, Martínez EA (2006b) Strategies to improve the fertility of frozen-thawed boar semen for artificial insemination. In: Control of Pig Reproduction VII, Ashworth CJ, Kraeling RR (eds), Nottingham Univ. Press, Manor Farm, Thrumpton, UK: 261-275 (Soc Reprod Fert Suppl 62).

Roca J, Parrilla I, Rodriguez-Martinez H, Gil MA, Cuello C, Vazquez JM, Martinez EA (2011) Approaches towards efficient use of boar semen in the pig industry. Reprod Domest Anim 46 S2: 79-83.

Rodriguez-Martinez H (2007a) State of the art in farm animal sperm evaluation. Reprod Fert Dev 19: 91-101.

Rodriguez-Martinez H (2007b) Reproductive biotechnology in pigs: what will remain? In: "Paradigms in pig science". Wiseman J, Varley MA, McOrist S, Kemp B (eds), Nottingham University Press, Nottingham, UK, Chapter 15: 263-302 (ISBN 978-1-904761-56-3).

Rodriguez-Martinez H, Barth AD (2007) In vitro evaluation of sperm quality related to in vivo function and fertility. In: Reproduction in Domestic Ruminants VI. Juengel JI, Murray JF, Smith MF (eds), Nottingham University Press, Nottingham, UK: 39-54 (Soc Reprod Fert 64: 39-54, 2007).

Rodríguez-Martinez H, Saravia F, Wallgren M, Roca J, Peña FJ (2008) Influence of seminal plasma on the kinematics of boar spermatozoa during freezing. Theriogenology 70: 1242-1250.

Rodriguez-Martinez H, Kvist U, Saravia F, Wallgren M, Johannisson A, Sanz L, Peña FJ, Martinez EA, Roca J, Vazquez JM, Calvete JJ (2009) The physiological roles of the boar ejaculate. In: "Control of Pig Reproduction VIII" Rodriguez-Martinez H, Vallet JL, Ziecik AJ (eds), Nottingham University Press, UK: 1-21

Rodriguez-Martinez H, Wallgren M (2011a) Advances in boar semen cryopreservation. Vet Med Int 2011, Article ID 396181, 5pp (doi:104061/2011/396181), open access.

Rodriguez-Martinez H, Kvist U, Ernerudh J, Sanz L, Calvete JJ (2011b) Seminal plasma proteins: what role do they play? Am J Reprod Immunol (AJRI) 66 S1: 11-22.

Sanchez R, Risopatron J, Schulz M, Villegas J, Isachenko V, Kreienberg R, Isachenko E (2011) Canine sperm vitrification with sucrose: effct on sperm function. Andrologia 43: 233-241.

Sanchez-Osorio J, Cuello C, Gil MA, Parrilla I, Maside C, Almiñana C, Lucas X, Roca J, Vazquez JM, Martinez EA (2010) Vitrification and warming of in vivo-derived porcine embryos in a chemically defined medium. Theriogenology 73: 300-308.

Sanchez-Osorio J, Cuello C, Gil MA, Parrilla I, Almiñana C, Caballero I, Roca J, Vazquez JM, Rodriguez-Martinez H, Martinez EA (2010) In vitro post-warming viability of vitrified porcine embryos: Effect of cryostorage length. Theriogenology 74: 486–490.

Saravia F, Wallgren M, Nagy S, Johannisson A, Rodríguez-Martínez H (2005) Deep freezing of concentrated boar semen for intra-uterine insemination: effects on sperm viability. Theriogenology 63: 1320-1333.

Saravia F, Hernández M, Wallgren MK, Johannisson A & Rodríguez-Martínez H (2007) Cooling during semen cryopreservation does not induce capacitation of boar spermatozoa. Int J Androl 30: 485-499.

Saravia F, Wallgren M, Johannisson A, Calvete JJ, Sanz L, Peña FJ, Roca J, Rodríguez-Martínez H (2009) Exposure to the seminal plasma of different portions of the boar ejaculate modulates the survival of spermatozoa cryopreserved in MiniFlatPacks. Theriogenology 71: 662–675.

Saravia F, Wallgren M, Rodríguez-Martínez H (2010) Freezing of boar semen can be simplified by handling a specific portion of the ejaculate with a shorter procedure and MiniFlatPack packaging. Anim Reprod Sci 117: 279-287.

Saragusty J, Arav A (2011) Current progress in oocyte and embryo cryopreservation by slow freezing and vitrification. Reproduction 141: 1-19.

Saragusty J, Gacitua H, Pettit MT, Arav A (2007) Directional freezing of equine semen in large volumes. Reprod Domest Anim 42: 610-615.

Saragusty J, Gacitua H, Rozenboim I, Arav A (2009) Do physical forces contribute to cryodamage? Biotechnol Bioeng 104: 719-728.

Shaw JM, Trounson AO (2002) Ovarian tissue transplantation and cryopreservation: its application to the maintenance and recovery of transgenic mouse. In: Methods in Molecular biology, vol 180 Ch 11: 229-251 "Transgenesis techniques: Principles and Protocols" 2nd Ed, Clark A (ed) Humana press Totowa.

Shi LY, Jin HF, Kim JG, Mohana Kumar B, Balasubramanian S, Choe SY, Rho GJ (2007) Ultra-structural changes and developmental potential of porcine oocytes following vitrification. Anim Reprod Sci 100: 128-140.

Somfai T, Kashiwazaki N, Ozawa M, Nakai M, Maedomari N, Noguchi J, Kaneko H, Nagai T, Kikuchi K (2008) Effect of centrifugation treatment before vitrification on the viability of porcine mature oocytes and zygotes produced in vitro. J Reprod Dev 54: 149-155.

Somfai T, Noguchi J, Kaneko H, Nakai M, Ozawa M, Kashiwazaki N, Egerszegi I, Rátky J, Nagai T, Kikuchi K (2010) Production of good-quality porcine blastocysts by in vitro fertilization of follicular oocytes vitrified at the germinal vesicle stage. Theriogenology 73: 147-156.

Thurston LM, Holt WV, Watson PF (2003) Post-thaw functional staus of boar spermatozoa cryopreserved using three controlled rate freezers: a comparison. Theriogenology 60 101-113.

Vajta G, Holm P, Greve T, Callesen H (1997) Vitrification of porcine embryos using the Open Pulled Straw (OPS) method. Acta Vet Scand 38: 349-352.

Waterhouse KE, Hofmo PO, Tverdal A, Miller RRJr (2006) Within and between breed differences in freezing tolerance and plasma membrane fatty acid composition of boar sperm. Reproduction 131: 887–894.

Watson PF, Fuller BJ (2001) Principles of cryopreservation of gametes and embryos. In: Watson PF, Holt WV (Eds) Cryobanking the Genetic Resource. New York, NY: Taylor & Francis: 21–46.

Weitze KF, Rath D, Baron G (1987) New aspects of preservation of boar sperm by deep freezing in plastic tubes. Dtsch Tierärztl Wschr 82: 261-267.

Westendorf P, Richter L, Treu H (1975) Deep freezing of boar sperm. Laboratory and insemination results using the Hülsenberger paillete method. Dtsch Tierärztl Wochenschr 82: 261-267.

Willadsen S, Polge C, Rowson LEA (1978) The viability of deep-frozen embryos. J Reprod Fertil 52: 391-393.

Wilmut I (1972) The low temperature preservation of mammalian embryos. J Reprod Fertil 31: 513-514.

Woelders H, Chaveiro A (2004) Theoretical prediction of "optimal" freezing programmes. Cryobiology 49: 258-271.

Woelders H, Matthjis A, Zuidberg CA, Chaveiro AE (2005) Cryopreservation of boar semen: equilibrium freezing in the cryoscope and in straws. Theriogenology 63: 383-395.

Wongtawan T, Saravia F, Wallgren M, Caballero I, Rodríguez-Martínez H (2006) Fertility after deep intra-uterine artificial insemination of concentrated low-volume boar semen doses. Theriogenology 65: 773-787.

Wowk B (2010) Thermodynamic aspects of vitrification. Cryobiology 60: 11-22.

Yavin S, Aroyo A, Roth Z, Arav A (2009) Embryo cryopreservation in the presence of low concentration of vitrification solution with sealed pulled straws in liquid nitrogen slush. Human Reprod 24: 797-804.

Yi YJ, Cheon YM, Park CS (2002a) Effect of N-acetyl-D-glucosamine, and glycerol concentration and equilibrium time on acrosome morphology and motility of frozen-thawed boar sperm. Anim Reprod Sci 69: 91-97.

Yi YJ, Im GS, Park CS (2002b) Lactose-egg yolk diluent supplemented with N-acetyl-D-glucosamine affect acrosome morphology and motility of frozen-thawed boar sperm. Anim Reprod Sci 74: 187-194.

Yoneda A Suzuki K, Mori T, Ueda J, Watanabe T (2004) Effects of delipidation and oxygen concentration on in vitro development of porcine embryos. J Reprod Dev 50: 287-295.

Zeng WX, Terada T (2001) Effects of methyl-beta-cyclodextrin on cryosurvival of boar spermatozoa. J Androl 22: 111-118.

Zhou GB, Li N (2009) Cryopreservation of porcine oocytes: recent advances. Mol Hum
        Reprod 15: 279-285.
Zidovetzki R, Levitan I (2007) Use of cyclodextrins to manipulate plasma membrane
        cholesterol content: evidence, misconceptions and control strategies.   Biochim
        Biophys Acta 1768: 1311-1324.

Permissions

The contributors of this book come from diverse backgrounds, making this book a truly international effort. This book will bring forth new frontiers with its revolutionizing research information and detailed analysis of the nascent developments around the world.

We would like to thank Igor I. Katkov, for lending his expertise to make the book truly unique. He has played a crucial role in the development of this book. Without his invaluable contribution this book wouldn't have been possible. He has made vital efforts to compile up to date information on the varied aspects of this subject to make this book a valuable addition to the collection of many professionals and students.

This book was conceptualized with the vision of imparting up-to-date information and advanced data in this field. To ensure the same, a matchless editorial board was set up. Every individual on the board went through rigorous rounds of assessment to prove their worth. After which they invested a large part of their time researching and compiling the most relevant data for our readers. Conferences and sessions were held from time to time between the editorial board and the contributing authors to present the data in the most comprehensible form. The editorial team has worked tirelessly to provide valuable and valid information to help people across the globe.

Every chapter published in this book has been scrutinized by our experts. Their significance has been extensively debated. The topics covered herein carry significant findings which will fuel the growth of the discipline. They may even be implemented as practical applications or may be referred to as a beginning point for another development. Chapters in this book were first published by InTech; hereby published with permission under the Creative Commons Attribution License or equivalent.

The editorial board has been involved in producing this book since its inception. They have spent rigorous hours researching and exploring the diverse topics which have resulted in the successful publishing of this book. They have passed on their knowledge of decades through this book. To expedite this challenging task, the publisher supported the team at every step. A small team of assistant editors was also appointed to further simplify the editing procedure and attain best results for the readers.

Our editorial team has been hand-picked from every corner of the world. Their multi-ethnicity adds dynamic inputs to the discussions which result in innovative outcomes. These outcomes are then further discussed with the researchers and contributors who give their valuable feedback and opinion regarding the same. The feedback is then collaborated with the researches and they are edited in a comprehensive manner to aid the understanding of the subject.

Apart from the editorial board, the designing team has also invested a significant amount of their time in understanding the subject and creating the most relevant covers. They scrutinized every image to scout for the most suitable representation of the subject and create an appropriate cover for the book.

The publishing team has been involved in this book since its early stages. They were actively engaged in every process, be it collecting the data, connecting with the contributors or procuring relevant information. The team has been an ardent support to the editorial, designing and production team. Their endless efforts to recruit the best for this project, has resulted in the accomplishment of this book. They are a veteran in the field of academics and their pool of knowledge is as vast as their experience in printing. Their expertise and guidance has proved useful at every step. Their uncompromising quality standards have made this book an exceptional effort. Their encouragement from time to time has been an inspiration for everyone.

The publisher and the editorial board hope that this book will prove to be a valuable piece of knowledge for researchers, students, practitioners and scholars across the globe.

List of Contributors

E. Isachenko, P. Mallmann, G. Rahimi, J. Risopatròn, M. Schulz, V. Isachenko and R. Sànchez
Woman Hospital, University of Cologne, Germany

I.I. Katkov
CELLTRONIX and Sanford-Burnham Institute for Medical Research, San Diego, California, USA

I. Yakhnenko
CELLTRONIX and Sanford-Burnham Institute for Medical Research, San Diego, California, USA

V.F. Bolyukh and Y.I. Sokol
Kharkov National Technical University "KhPI", Kharkov, Ukraine

O.A. Chernetsov and A.B. Sushko
Animal Reproduction Center, Kulinichi, Kharkov region, Ukraine

P.I. Dudin
Raptor Restoration and Reintroduction Program, National Reserve "Galichya Gora", Voronezh region, Russia

A.Y. Grigoriev
Kharkov Zoo, Kharkov, Ukraine

V. Isachenko and E. Isachenko
Dept. Obstetrics and Gynecology, Ulm University, Germany

A.G.-M. Lulat and S.I. Moskovtsev
CReATe Fertility Center, Toronto, Ontario, Canada

S.I. Moskovtsev
Dep. Obstetrics and Gynecology, Toronto University, Toronto, Ontario, Canada

M.P. Petrushko and V.I. Pinyaev
ART Clinic, Kharkov, Ukraine

K.M. Sokol
Kharkov National Medical University, Kharkov, Ukraine

S.I. Moskovtsev, A.G-M. Lulat and C.L. Librach
CReATe Fertility Centre, Canada

S.I. Moskovtsev, A.G-M. Lulat and C.L. Librach
Department of Obstetrics & Gynaecology, University of Toronto, Canada

C.L. Librach
Division of Reproductive Endocrinology and Infertility, Department of Obstetrics & Gynaecology, Sunnybrook Health Sciences Centre and Women's College Hospital, Toronto, Ontario, Canada

Dayong Gao
University of Washington, Seattle, WA, USA

Xiaoming Zhou
University of Electronic Science and Technology of China, Chengdu, China

Raquel Martín-Ibáñez and Josep M. Canals
Departament de Biologia Cellular, Immunologia i Neurociències, Programa de Terapia Cellular, Facultat de Medicina, Universitat de Barcelona, Barcelona, Spain
Institut de Investigacions Biomèdiques August Pi i Sunyer (IDIBAPS), Barcelona, Spain
Centro de Investigación Biomédica en Red Sobre Enfermedades Neurodegenerativas (CIBERNED), Spain

Outi Hovatta
Division of Obstetrics and Gynecology, Department of Clinical Science, Intervention and Technology, Karolinska Institutet, K57, Karolinska University Hospital, Huddinge, Stockholm, Sweden

Ali Honaramooz
University of Saskatchewan, Canada

Juergen Liebermann
Director of Laboratory, Fertility Centers of Illinois, Chicago, IL, USA

Catherine Bigelow and Alan B. Copperman
Mount Sinai School of Medicine, New York, NY, USA

Wai Hung Tsang and King L. Chow
Division of Life Science, The Hong Kong University of Science and Technology, Hong Kong

Heriberto Rodriguez-Martinez
Department of Clinical and Experimental Medicine, Faculty of Health Sciences, Linköping University, Linköping, Sweden

www.ingramcontent.com/pod-product-compliance
Lightning Source LLC
Chambersburg PA
CBHW070737190326
41458CB00004B/1209